# The New Astronomy

# The New Astronomy

## Nigel Henbest
*Astronomy Consultant, New Scientist*

## Michael Marten
*Director, Science Photo Library*

## CAMBRIDGE UNIVERSITY PRESS

*Cambridge / London / New York / New Rochelle*
*Melbourne / Sydney*

Published by the Press Syndicate of the University of Cambridge
The Pitt Building, Trumpington Street, Cambridge CB2 1RP
32 East 57th Street, New York, NY 10022, USA
296 Beaconsfield Parade, Middle Park, Melbourne 3206, Australia

First published 1983

Printed in Great Britain

Library of Congress catalogue card number: 83–7716

*British Library Cataloguing in Publication Data*

Henbest, Nigel
  The new astronomy
  1. Astronomy
  I. Title    II. Marten, Michael
  520    QB43.2

ISBN 0 521 25683 6


Text: **Nigel Henbest**
Picture editor: **Michael Marten**
Design: **Richard Adams/ADCO**
Diagrams: **David Parker**

# Acknowledgements

Special thanks to: Heather Couper (expert
advice on galaxies), Mandy Caplin (typing),
Nigel Coke (picture sizing), Rosemary Taylor
and Salim Patel (background support).

This book would not have been possible without
the many individuals who located, supplied or
processed imagery especially for us. Particular
thanks for their time and trouble to:
  David Allen, R.J. Allen, Phil Appleton, John F.
Arens, Paul Atherton, Rainer Beck, Sidney van
den Bergh, Ralph Bohlin, Peter Clegg, Robin
Conway, Bryn Cooke, Karl Esch, Fred Espenak,
Giovanni Fazio, Eric Feigelson, John C. Geary,
Leon Golub, John C. Good, Steve Gull, Ted
Gull, Brian Hadley, Ken Hartley, Paul & Tony
Jorden, Daniel A. Klinglesmith, William Hsin-
Min Ku, Martha Liller, David Malin, Tom
Muxlow, Malcolm Niedner, P. Nisenson, Agnes
Paulsen, John Pye, Matthew Schnepps, Fred
Seward, Tom Stephenson, Larry D. Travis,
Patrick Wallace, John Walsh, Margaret B.
Weems, Richard Willingale.

# Contents

# 1 The New Astronomy

The 'new astronomy' is a phenomenon of the last two or three decades, and it has completely revolutionised our concept of the Universe. While traditional astronomy was concerned with studying the light – optical radiation – from objects in space, the new astronomy encompasses all the radiations emitted by celestial objects: gamma rays, X-rays, ultraviolet, optical, infrared and radio waves.

The range of light is in fact surprisingly limited. It includes only radiation with wavelengths 30 per cent shorter to 30 per cent longer than the wavelength to which our eyes are most sensitive. The new astronomy covers radiation from extremes which have wavelengths less than one thousand-millionth as long, in the case of the shortest gamma rays, to over a hundred million times the wavelength of light for the longest radio waves (Fig. 1.2). To make an analogy with sound, traditional astronomy was an effort to understand the symphony of the Universe with ears which could hear only middle C and the two notes immediately adjacent.

The rapid growth of the new astronomy is due partly to the accidental discovery in the 1930s of radio waves from beyond the Earth, which showed that there are non-optical radiations from space. But there have been two major barriers, overcome only in recent decades. First, there are technological problems. We must build new types of telescopes to gather other kinds of radiation, and focus them into an image. We must also develop new detectors to record the image and show it to us in a way we can comprehend. The other is a natural barrier. Earth's atmosphere absorbs most of the radiations from space before they reach the ground, and the new detectors for many wavelengths must be flown well above the atmosphere (Fig. 1.2). These branches of the new astronomy could not be pursued until telescopes could be launched by the rockets, and carried on the satellites, of the 'space age'.

As astronomers have broken the new ground of another wavelength region, they have generally started by surveying the sky for sources of this radiation, measuring their brightness and spectra, and making crude maps, often shown as contour diagrams. Only later on are the telescopes and detectors able to produce detailed images of the objects they are observing. As a result, images at wavelengths other than light have only become available in

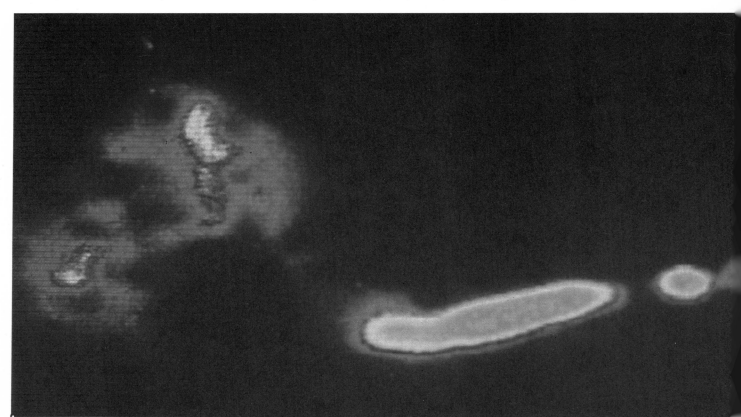

**Fig. 1.1** *This radio view of the galaxy 3C 449 shows jets extending more than a million light years from its core (central green oval). The radio brightness is coded by colour, the background sky being dark blue, and successively brighter regions of emission pale blue, pink, yellow and green;*

the past few years. In addition, computer techniques have been developed to process these images. Previous astronomy books had to rely on optical photographs for illustration, even when they were devoted to aspects of the new astronomy. This book brings together for the first time a collection of pictures which shows the balance of the new astronomy. Optical photographs are not ignored, but take their natural place as just one of the many kinds of view we can now obtain. All the pictures are orientated with north at the top, except where indicated otherwise.

In many cases, the optical picture is the least interesting. Dark clouds of dust in space completely hide the places where stars are born – regions whose details are readily seen at infrared and radio wavelengths. The gas in space is transparent, unseen by optical telescopes, but emits radio waves and gamma rays which makes it brilliant at these wavelengths. Distant clusters of galaxies trap pools of very hot gas, at a temperature of millions of degrees, and these can only be detected by their X-ray emission. Explosions at the centres of powerful galaxies eject beams of electrons which pump up enormous bags of magnetic field – the largest structures in the Universe, but invisible except to

astronomers using radio telescopes (Fig. 1.1).

The radiations producing the images in this book go under a variety of names – X-*rays* or radio *waves*, for example – but they are all basically similar in nature, 'waves' consisting of rippling electric and magnetic fields spreading out from a source – be it star, pulsar or quasar. The difference comes in the *wavelength* of the undulation, the distance from 'crest' to 'crest' of the electric wave, visualised very like the succession of waves at sea.

*Visible light* is radiation of an 'intermediate' wavelength – about 500 nanometres (one nanometre is a millionth of a millimetre). In everyday terms, the waves are certainly short: over a hundred wavecrests would be needed to span the thickness of this page. The human eye perceives light of different wavelengths as the various colours of the rainbow: red light has the longest waves, around 700 nanometres from crest to crest; and blue-violet is the shortest, with wavelengths of about 400 nanometres.

Radiation with shorter wavelengths is invisible to the eye. The range of radiations with wavelengths from 390 down to 10 nanometres is the *ultraviolet*. These are the rays in sunlight which tan our skins; for the astronomer,

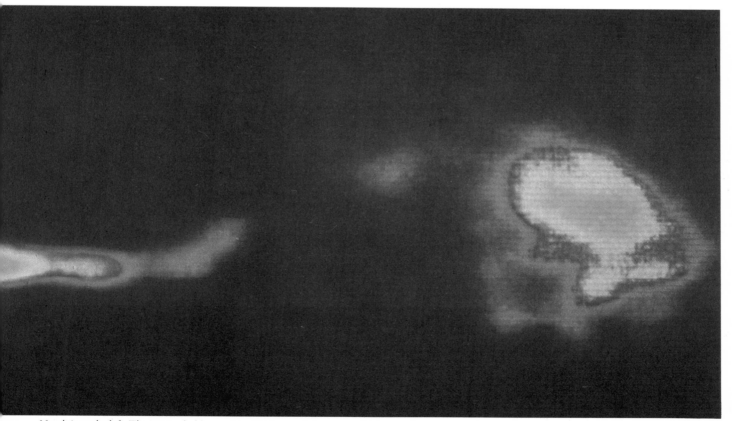

*North is to the left. The jets probably consist of high-speed electrons, which spread out to create large diffuse lobes of radio emission at their outer ends. The jets and lobes are completely invisible to optical telescopes.*

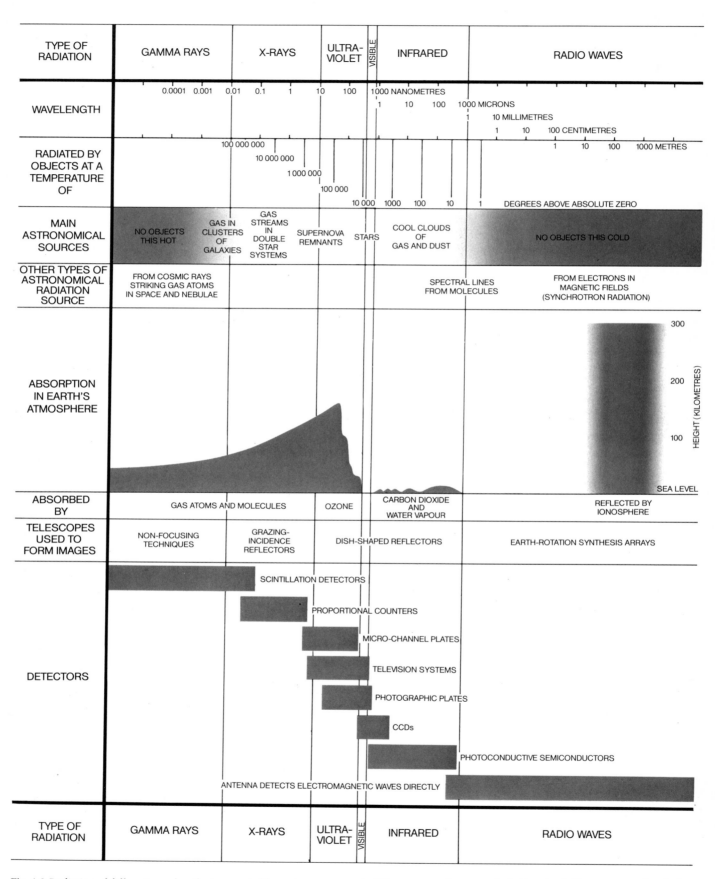

**Fig. 1.2** *Radiation of different wavelengths is generated by various processes in different types of astronomical object (top). The waves are absorbed to different extents in Earth's atmosphere (middle), and require different kinds of telescopes and detectors (bottom).*

**Fig. 1.3** *This X-ray view of the Hydra A cluster of galaxies does not show the galaxies themselves, but reveals a huge pool of gas between them. The gas is at a temperature of 100 million degrees, and is invisible except to X-ray telescopes.*

they are the 'light' from the hottest stars. During its passage through space ultraviolet radiation is imprinted with invaluable information about the tenuous gases between the stars.

At shorter wavelengths than the ultraviolet are the *X-rays*, whose crest-to-crest distance ranges from 10 down to only 0.01 nanometres – the latter is about one-tenth the size of an atom. X-rays from space are the hallmark of superheated gases, at a temperature of over a million degrees. X-ray sources can be the hot gases thrown out by exploding stars; or rings of gas falling down on to a pulsar or a black hole; they can be the hot gases of a quasar explosion, or the huge pools of gas which fill whole clusters of galaxies (Fig. 1.3).

Even shorter are the *gamma rays* – a name which encompasses all radiation whose wavelength is less than 0.01 nanometres. These come from the most action-packed of astronomical objects: from the compact pulsars, and the superexplosions of distant quasars. And gamma rays can spring from nuclear reactions, in regions of space where tremendously fast electrons and protons cannon into atoms in space and provoke nuclear reactions similar to those in the artificial particle accelerators which physicists use to probe the ultimate constituents of matter.

Moving the other way in wavelength, to radiations longer than those of light, we come into the realm of the *infrared*. These rays have wavelengths between that of red light – 700 nanometres – and around 1 millimetre. We think of infrared in everyday life as being heat radiation, the rays from an electric fire, for example. For astronomers, infrared radiation is the signature of the *cooler* objects in the Universe. An electric fire, at a temperature of a few hundred degrees Centigrade, is rather cool on the cosmic scale, where an average star has a temperature of several thousand degrees and some gas clouds have multi-million degree temperatures. Objects at room temperature produce infrared radiation too. We are constantly surrounded by this radiation on Earth, so we do

not notice it. But an infrared astronomer can pick up the radiation from a planet which is at everyday temperatures, and even below.

In fact, all objects produce radiation of some kind, and the lower the temperature the longer the wavelength of the resulting radiation. The hot filament of a light bulb naturally glows with visible light; an electric fire with shorter wavelength infrared; and our bodies with longer wavelength infrared. Objects which are cooled down until they almost reach the absolute zero of temperature $(-273.15°C)$ emit infrared radiation so long that it technically falls into the region of radio waves. Our view of the sky at different wavelengths is in many ways a portrait of different temperatures; and to the astronomer it makes more sense to measure temperatures upwards from absolute zero – instead of the rather arbitrary Centigrade system which is based on the melting point of ice and the boiling point of water at sea-level on our own planet. *Absolute temperatures* thus start at absolute zero: 0 K – where the symbol K stands for degrees Kelvin, named after the physicist Lord Kelvin who first realised the advantages of using absolute temperatures in science. The melting point of ice (0°C) is about 273 K; room temperature roughly 300 K; and the boiling point of water about 373 K – to convert Centigrade temperatures into absolute, add 273.15. The absolute scale is actually easier to visualise in one way, because it has no negative temperatures – nothing can be colder than absolute zero.

Infrared astronomers can 'see' cool clouds of dust in space, which are invisible at other wavelengths. These hidden dust clouds are the spawning ground for new stars, and infrared astronomers are privileged to see the first signs of star-birth.

Beyond the region of the infrared lies the last type of radiation. *Radio waves* cover a huge range of wavelengths: technically, they are radiation with a wavelength greater than 1 millimetre. Radio astronomers regularly observe the sky at wavelengths of a few millimetres or a few centimetres, but a few radio telescopes are designed to pick up waves as long as several metres – or even several kilometres. Natural radio broadcasters in the Universe are usually places of unbridled violence, where high-speed electrons – negatively charged subatomic particles – are whirled about in intense magnetic fields.

The telescopes used to observe and detect these radiations are described in the odd-numbered chapters. The results themselves are shown in the even-numbered chapters, with the images obtained at different wavelengths displayed as pictures. These images are routinely stored in a computer, and processed in a wide variety of ways, as discussed in more detail in Chapters 3 and 5.

Since the computers generally have colour TV monitors, it is possible to use colour coding to bring out the mass of

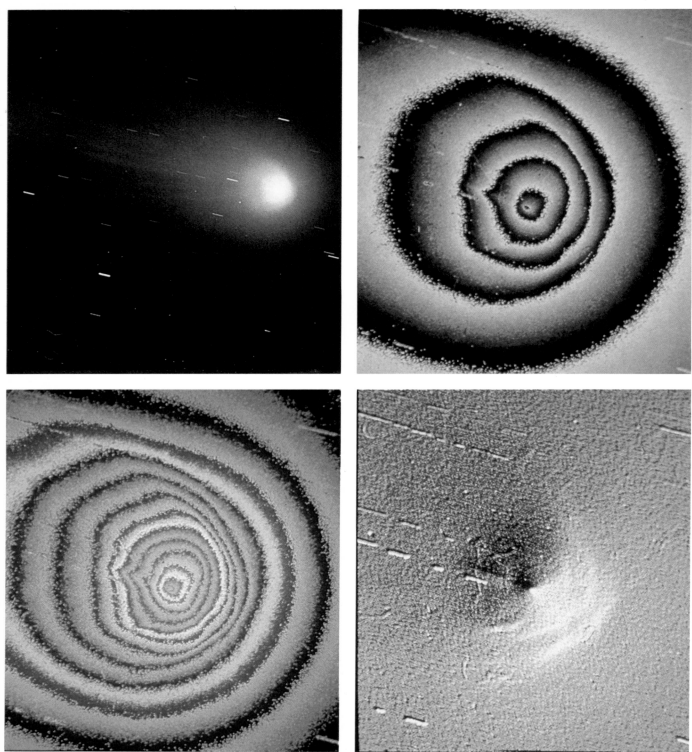

**Figs. 1.4–1.7** *'New astronomy' techniques of computer processing have been applied here to an old photograph of Halley's comet, to bring out hidden details. The original photograph, **Fig. 1.4** (top left), was taken in Egypt on 25 May 1910, during the comet's last return to the Sun. It shows details in the tail, but relatively little of the structure of the bright head. Four photographs taken that night have been scanned by a small light spot, and their electronic images added by computer to produce the remaining displays (each of which also shows four sections of several star images, trailed as the camera followed the moving comet). In **Fig. 1.5** (top right), brighter levels of the image have been coded by paler shades of blue, up to a level (white) where the coding jumps back to black, and then works up to white again. A succession of such jumps produces the dark and light fringes, like contour lines, surrounding a peak at the brightest, central part of the comet's head. This technique shows clearly a small jet, extending behind (to the left of) the head. **Fig. 1.6** (lower left) shows the same data, but with many more contour levels, and coded in several colours. The extra contours help to indicate the limited extent of the jet. The apparently three-dimensional image, **Fig. 1.7**, has been made by shifting the electronically stored image slightly, then subtracting the shifted image from the original. The comet's head appears as a peak, lit from the lower right, with apparent height indicating brightness. The technique reduces large scale contrasts, and emphasises small scale details. As well as the jet, Fig. 1.7 reveals bright arcs of gas in front of the comet's head (lower right), which are not easily seen in the other representations of the photograph.*

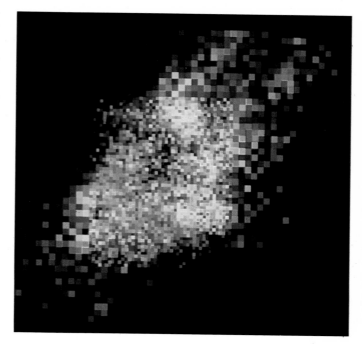

Fig. 1.8 *This radio view of a dense interstellar cloud is colour coded for velocity. The map of the Heiles 2 cloud was made by observing radio emission from carbon monoxide molecules at a wavelength of 2.6 millimetres. The cloud is moving away at an average speed of 6 kilometres per second, and regions actually moving at this speed appear yellow in this coding. The reddish regions at either end of the cloud are receding at 7 kilometres per second, while the blue central parts are moving away at only 5 kilometres per second.*

information in each image. Colour in reality comprises light of different wavelengths. When images are obtained at a wavelength in, for example, the X-ray or radio range, colour has no real meaning, and we can use colour coding in a variety of new and exciting ways (Figs. 1.4–1.7). Colour is employed in three main ways in *The New Astronomy*.

Observations at different wavelengths, combined into a single image, can be colour coded according to the wavelength. This technique is often used in infrared astronomy, where astronomers make observations of the same object at different wavelengths to reveal differences in temperature. Blue generally represents the shortest-wavelength view, green an intermediate wavelength and red the image at the longest wavelength.

Alternatively, colour coding can show velocity. The speed of a gas cloud is revealed in observations of a single spectral line from atoms or molecules of gas, usually obtained at radio wavelengths. By convention, gas coming toward us is shown in blue, and the colours shade through to red for gas moving away (Fig. 1.8).

But colour coding is most commonly used to show intensity. This technique is widely used at all wavelengths – including optical, when ordinary photographs are processed by computer. The astronomer assigns different colours, chosen at will, to the various levels of brightness in the picture. The results are not only picturesque. They overcome one of the problems of photographic representation, that it is impossible to show details in both the faintest part of a galaxy or nebula, and the brightest – which may be over a thousand times more brilliant. If the former are shown in a photograph, the bright regions are 'burnt out'; while a short exposure to reveal details in the bright regions would not show the faint parts at all. Intensity colour coding shows details of both bright and faint regions simultaneously, in different colours.

The new images often resemble works of art, with the Universe's natural artistry aided by the imagination of the astronomer at the computer. But the astronomer's main task is not to capture the unseen beauty of space, but to use this new information to help understand the structure and scale of the Universe, and how it is changing as time goes by.

In relation to the modern view of the Universe, our planet Earth is a mere speck along with the other eight planets circling the Sun. The planets range in size from Mercury, whose diameter is one-third the size of Earth's, to Jupiter with a diameter of eleven Earths. Saturn is almost as large as Jupiter, and these 'giants' of our Solar System are orbited by 23 and 16 moons, respectively.

Despite the size of the giant planets, they lie so far away that astronomers find it difficult to see details on them. The Earth orbits the Sun at an average distance of 150 million kilometres, but Jupiter's orbit is over five times

larger, and Saturn's nearly twice as big again. So these worlds appear very small in our sky: although our unaided eyes can see them shining brightly in reflected sunlight, they seem to be no more than points of light. It needs a telescope to show their globes, and Saturn's encircling girdle of rings.

Astronomers measure the apparent size of objects in the sky in terms of *degrees of angle* (°), and their subdivisions *arcminutes* and *arcseconds*: there are 60 arcminutes to one degree, and 60 arcseconds to one arcminute. This traditional system is undoubtedly cumbersome, but it becomes easier to understand if we take some examples. The entire sky is 180° across, from one horizon up to the zenith and down to the opposite horizon. Most of the traditional constellation patterns are around 20° across: the figure of the 'hunter' Orion (Fig. 4.2), for example, stands 15° tall. The Moon is surprisingly small. Because it shines so bright, and appears large when it is near the horizon, the Moon looks as though it should cover quite an area of sky; but in fact it is only half a degree across (30 arcminutes). It would take over three hundred Moons, put side-by-side, to stretch across the sky.

Our eyes cannot see details any finer than one or two arcminutes (about 1/20 the size of the Moon), and the planets all appear smaller than this. Even when the Earth is closest to Jupiter, the giant planet appears just under an arcminute across – 50 arcseconds – and Saturn only 20 arcseconds in size. Even the largest optical telescopes on Earth can see detail only as fine as an arcsecond or so, whatever magnification is used, because the images are blurred by Earth's atmosphere. So, ironically enough, astronomers have problems in examining even our nearest neighbours in space. These worlds are close enough, though, to be reached by unmanned spaceprobes, and the last two decades have seen their details revealed by dozens of these craft – from Mariner 10 which visited scorched Mercury, closest to the Sun, to the two Voyagers which have photographed in amazing detail the frozen realm of Saturn, its myriad rings, and its icy moons.

**Fig. 1.9** *Our Sun (purple spot) lies 30 000 light years from the centre of our Milky Way Galaxy, in a spur off one of its major spiral arms. Orange spots mark regions of star birth and of young stars, featured in Chapter 4, while blue spots are the planetary nebulae and supernova remnants of Chapter 6. Like the stars and gas, they concentrate in the spiral arms. Chapter 8 covers the galactic centre (red spot) and the Milky Way itself.*

On the everyday scale, Saturn is a very distant world. Lying some 1400 million kilometres away, it is so far off that the radio signals from the Voyager probes – travelling at the speed of light – took an hour and a half to reach the Earth. Saturn's orbit is, however, only a quarter of the way out to the farthest planet, the tiny frozen world Pluto. And even Pluto's distance shrinks into insignificance once we look out to the stars. The nearest star, Proxima Centauri (part of the Alpha Centauri star triplet), lies a quarter of a million times farther away than Pluto – over 40 million million kilometres. Such distances are just too large to comprehend, and the figures themselves become unwieldy in size.

Astronomers cope with star distances by discarding the kilometre as their standard length. A more convenient unit is the distance that a beam of light (or any other radiation) travels in one year. This standard length, the *light year*, works out to just under ten million million kilometres. So Proxima Centauri is about 4¼ light years away. With this new unit, star distances become more comprehensible. We can compare star distances, and construct a scale model in our minds, even if the distances themselves are too large for the human mind to encompass.

The bright star Sirius lies about twice as far away as Proxima Centauri, at 8.6 light years. It too is nearby on the cosmic scale; its neighbours as seen in the sky, Betelgeuse and Rigel in Orion, are far more distant, lying at 650 and 800 light years. They only appear bright in our skies because they are truly brilliant stars, each shining as brightly as thousands of Suns.

Optical astronomers have a rather archaic system for describing the apparent brightness of stars and other celestial objects as seen from the Earth. In this *magnitude* system, the faintest stars visible to the naked eye are of magnitude 6; while brighter stars have *smaller* magnitude numbers. A star of magnitude 1 – like Betelgeuse – is a hundred times brighter than a magnitude 6 star; the brightest star, Sirius, is ten times brighter still, and it has a negative magnitude: −1.4. Going brighter still, the Sun has a magnitude of −26.7!

Our Sun is, in fact, a typical star – middle-weight, middle-aged – born some 4600 million years ago. It appears exceptionally bright simply because it is near to us, and it is special to us on Earth because it supplies us with bountiful light and heat. The Sun is special to astronomers too, because its proximity means that we can study an average star in close-up detail. Conversely, investigations of other stars which are younger and older can tell us of the Sun's past, and of its likely future.

Star birth occurs because space between the stars is not entirely empty. There is tenuous *interstellar gas*, composed mainly of hydrogen, with tiny flecks of dark solid dust mixed in. In places, this gas and dust is compressed into sombre dark clouds. Within these, the interstellar matter is compressed by its own gravity into gradually-shrinking spheres, which heat up until they burst into radiance as new stars. The brilliant young stars light up the surrounding tatters of gas as a glowing *nebula*, like the famous Orion Nebula (Figs. 4.11–4.18), a mass of seething fluorescent gases lying some 1600 light years from us.

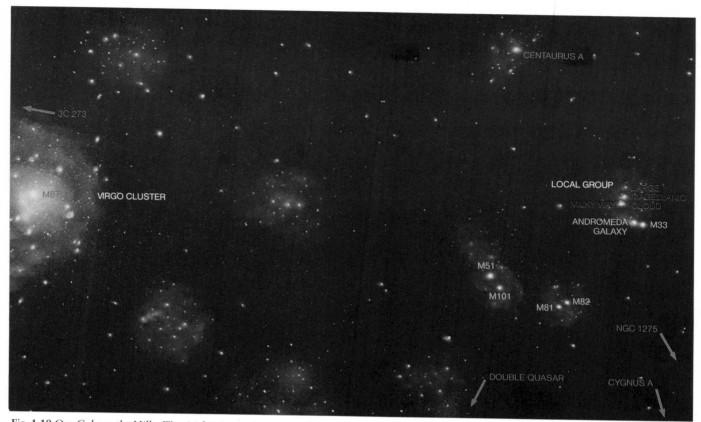

**Fig. 1.10** *Our Galaxy, the Milky Way (right), is a major galaxy of the Local Group, which is in turn part of the Local Supercluster of galaxies, centred on the Virgo Cluster (left). The Milky Way and its satellite the Large Magellanic Cloud (red lettering) are the subject of Chapter 8. Nearby normal galaxies (yellow lettering) feature in Chapter 10, while the active galaxies (green lettering and arrows) are covered in Chapter 12.*

Eventually, all stars die. They lose their outer gases, either in gentle cosmic 'smoke rings' or in violent supernova explosions; while their cores collapse to form tiny, compact – and very strange – objects. These cores weigh roughly as much as the Sun, but they can shrink down to the size of a planet to form a *white dwarf* star; even smaller as a *neutron star* or pulsar; or even collapse completely to form a *black hole*, whose gravitational field is so powerful that no radiation can escape from it.

Stars are grouped into huge star-islands, or *galaxies*. Our Sun (and the other stars and nebulae mentioned so far) is a member of the *Milky Way Galaxy* (Fig. 1.9), a collection of around 200 000 million stars – along with nebulae and collapsed star-corpses. The stars are arranged into a huge disc, some 100 000 light years across, all orbiting around the Milky Way's centre where the disc of stars is thicker. Our Sun, with its family of planets, lies about two-thirds of the way to the edge of the galaxy.

The Milky Way's nearest neighbours are two smaller galaxies, the Large Magellanic Cloud and the Small Magellanic Cloud. They can only be seen from the Earth's southern hemisphere, and the two Clouds were first described by the Portuguese navigator Ferdinand Magellan as he circumnavigated the Earth in 1521. Farther away lies the great Andromeda Galaxy, dimly visible from the Earth's northern hemisphere in autumn and winter months. At 2¼ million light years distance, it is the farthest object we can see with the naked eye.

But telescopes can reveal many other, much more remote galaxies (Fig. 1.10), their intrinsic brilliance

dimmed by their enormous distances. While the Andromeda Galaxy has a magnitude of five, rather brighter than the naked eye limit, modern telescopes, coupled with sensitive electronic detectors, can 'see' galaxies almost a hundred million times fainter, around magnitude 24. Such galaxies lie almost 10 000 million light years away from us. Within this vast region of space, the latest telescopes can pick out thousands of millions of galaxies. Many of them are clumped together into huge clusters, each cluster stretching over millions of light years.

Amongst this multitude of galaxies, a few display an intensely bright core, the site of a stupendous explosion – an explosion which can be a thousand times brighter than the galaxy itself. If the galaxy is extremely remote we cannot see anything but this central explosion – and astronomers have called such objects *quasars*. These exploding galaxy cores are so bright that astronomers can see them farther away than any other object in the Universe. The most distant quasars lie some 12 000 million light years away from us. The quasar explosions generate huge quantities of radio waves and X-rays too, and some of the many radio sources and X-ray sources discovered recently are probably quasars even more remote.

Centuries of observations with the eye and optical telescope have revealed the framework of the Universe, but their information on the planets, stars and galaxies has turned out to be only superficial. The story of the life and death of stars, of galaxies and of the Universe itself has only become apparent in recent years, with the advent of the new astronomy.

**2.1** *Ultraviolet, 28–31 nm, Skylab ATM extreme ultraviolet spectroheliograph*

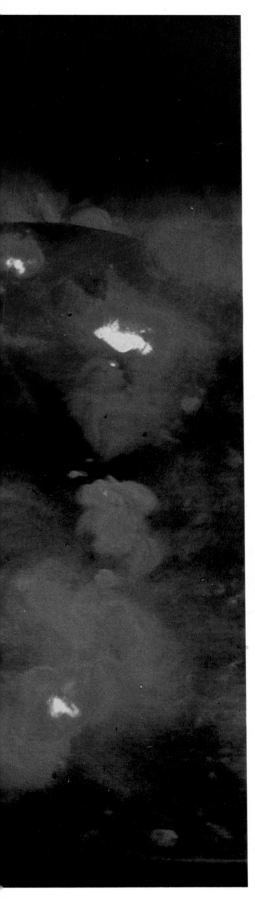

# 2 Solar System

The Solar System is dominated by the Sun – and by sunlight. The Sun contains 99.9 per cent of the mass of the system, so its gravitation controls the motions of the planets and the minor members of the system, like the asteroids and comets. The Sun is also the only important source of energy in the Solar System. Its central nuclear reactions are over a hundred million times more powerful than the interior heat sources of the planets combined. Most of this power is emitted from its surface as visible light.

Sunlight reflected from the planets makes them prominent in our night skies. Venus can shine ten times more brilliantly than the brightest star, while Mercury, Mars, Jupiter and Saturn all rival or exceed the half-dozen brightest stars. The appearance of the planets in visible light has become familiar from telescopic views and, more recently, from spaceprobe pictures.

But in solar system astronomy – as in the study of the Universe beyond – other wavelengths of radiation can reveal new aspects of familar objects. The Sun itself is no exception. Its powerful output of visible light produces a uniform glare from the whole disc which makes it difficult to study details of its surface or atmosphere.

Structure in the surface is best seen by isolating light of particular wavelengths, those emitted and absorbed by common atoms like hydrogen. Ultraviolet views show details of the lower atmosphere and X-rays the upper atmosphere – both virtually invisible to optical astronomers except during rare solar eclipses. These wavelengths, and radio waves, reveal that the Sun's surface and atmosphere teem with activity, and occasionally suffer powerful outbursts – unseen by optical astronomers – which can effect the Earth.

The long-term study of ultraviolet radiation and X-rays has required satellite observations above the Earth's absorbing atmosphere, and the manned space station Skylab was equipped with a specialised solar observatory. **Fig. 2.1** was taken at wavelengths around the middle of Skylab's range, in extreme ultraviolet radiation. The Sun's surface is not visible here: the picture instead contains several overlapping images of its lower and upper atmosphere. In Fig. 2.1, the Skylab instrument has spread the Sun's radiation out into a spectrum, with the wavelength increasing from right to left. This radiation comes from atoms in the atmosphere which radiate only at particular wavelengths – forming a complete image of the Sun at each wavelength of emission.

Within each image, dim regions are coded blue, and brighter parts red and yellow. (North is to the right in each image.) The full disc at the left is seen at a wavelength of 30.4 nanometres, in the radiation from helium atoms which have lost one electron. These occur in the lower atmosphere, and so the mottled disc shows the clumpy distribution of gas just above the visible surface – along with a huge prominence of gas projecting half a million kilometres – almost 50 Earth-diameters. The right-hand image, at 28.5 nanometres, comes from iron atoms missing 14 of their 26 electrons, because they lie in very hot gas clouds. The image reveals these isolated clouds in the outer atmosphere above the Sun's active regions, but not the cooler gas of the Sun's surface, and of the lower atmosphere and its prominence.

The planets have not been as intensely studied at non-optical wavelengths as the Sun, and the results have been overshadowed by detailed optical photographs from spacecraft. Ultraviolet and X-ray observations – from spaceprobes or from observatories in Earth orbit – can, however, reveal activity in planetary atmospheres which is invisible in their reflected sunlight. Astronomers have gleaned even more information from the longer wavelengths – partly because they can be observed from ground-based observatories. All the planets emit infrared 'heat radiation' simply because they are warmed by the Sun. Infrared astronomy also opens our view to the depths of the atmospheres of the large gaseous planets Jupiter and Saturn, which are warmer inside than at the visible cloud tops.

Radio astronomy plays two roles. Planets emit radio waves from their warm surfaces, or by *synchrotron* emission from electrons trapped in their magnetic fields. The latter reveals the extent, shape and strength of the field. In planetary *radar*, a beam of artificially generated radio waves is sent towards a planet, and the faint returning 'echo' is detected. This technique provides the most accurate distances to the planets, and can show details on the surfaces of solid planets. Since radio waves can penetrate clouds, radar astronomy has provided the first global views of the nearest planet, cloud-shrouded Venus.

# The Sun

'A star on our doorstep' is the astronomer's view of the Sun. Our local star is quite an average specimen in terms of its mass, size, temperature, and position in our Milky Way Galaxy (Fig. 1.9); it is halfway through its life span of some 10 000 million years. It has the advantage of being the one star we can study in great detail.

But in everyday terms, the Sun is colossal and awe-inspiring. It is a ball of hot gas, 1.4 million kilometres in diameter and containing the mass of 330 000 Earths. It consists mainly of hydrogen and helium, with a trace of the other elements, and at its core the temperature is so high – 14 million K – that nuclear fusion reactions convert hydrogen into helium. The reactions are similar to those in a hydrogen bomb, but self-controlled to produce a steady output of energy – at the rate of a million million hydrogen bombs every second. Since energy is equivalent to mass, every second the Sun loses four million tonnes of its matter as radiation. This energy starts at the Sun's core in the form of X-rays and gamma rays. The Sun's interior becomes cooler away from its centre, and the radiation's wavelength increases accordingly as it percolates outwards, until it emerges from the 5800 K surface in the form of 'light and heat'. Just over half the Sun's output is visible light, and almost all the rest is infrared with wavelengths less than two microns.

As seen optically (**Fig. 2.2**), the Sun is a uniformly shining sphere, with radiation emerging equally in all directions. The surface or *photosphere* seen in photographs like Fig. 2.2 is not a true boundary like the surface of a planet. It is the outer edge of the region of opaque gas, which in fact merges into the transparent gas beyond (the Sun's 'atmosphere'). The density of the Sun's gas decreases very abruptly in the photosphere, so the Sun appears to have a surprisingly sharp edge. The photosphere is about 500 kilometres thick, and near the centre of the Sun's disc we see down deeper into its hotter and brighter lower regions than we can towards the Sun's edge, where the radiation has to traverse more gas. Hence the Sun's disc appears brighter at the centre than at the edges.

The obvious features of Fig. 2.2 are the black blemishes called sunspots, the largest in Fig. 2.2 being as big as the Earth. The number of sunspots varies considerably, and Fig. 2.2 was taken when the Sun displayed an unusually large number. Sunspots are one manifestation of solar *active regions*, regions where strong magnetic fields break through the photosphere into the atmosphere. At optical wavelengths, the active regions appear in a negative role, as their magnetism blocks the Sun's general output of radiation to produce dark sunspots –

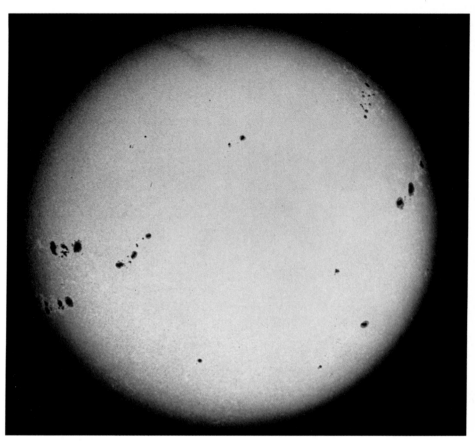

**2.2** *Optical (15 September, 1957), 46 m tower telescope, Mount Wilson Observatory*

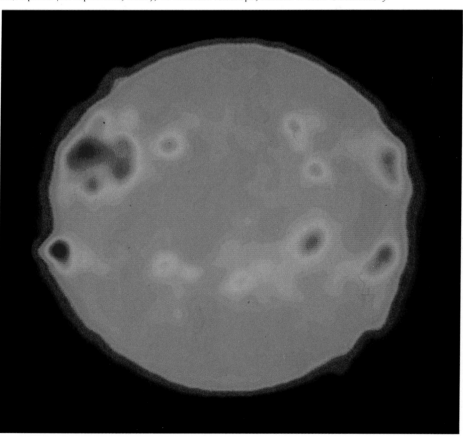

**2.3** *Radio (13 March, 1979), 2.8 cm, 100 m Effelsberg Telescope*

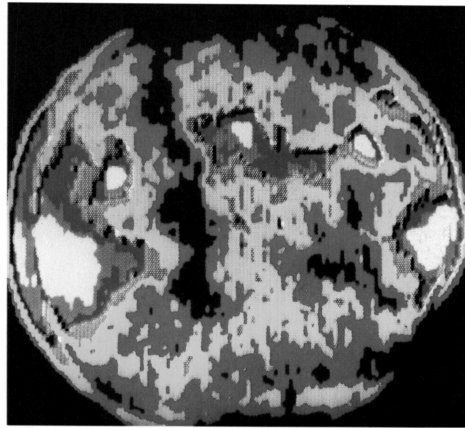

2.4 *Ultraviolet (30 May, 1973), 17–55 nm, Skylab ATM extreme ultraviolet monitor*

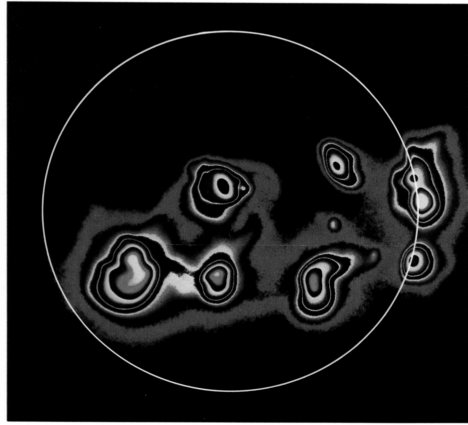

2.5 *X-ray (6 September, 1973), 0.6–3.3 nm, Skylab X-ray telescope*

although there are also slightly brighter areas, *plages*, surrounding the spots. But at most other wavelengths, the active regions are the brightest parts of the Sun.

The radio view (**Fig. 2.3**) shows emission both from the Sun's disc and from active regions. At this short wavelength (2.8 centimetres), the emission comes from hot gas just above the photosphere and the disc appears roughly the same size as in optical photographs (at longer wavelengths, however, the radio Sun appears several times larger). Faint regions are here colour-coded blue, with brighter regions green, yellow and red. The red regions are clouds of hot gas trapped in the magnetic field above active regions; a simultaneous optical photograph would reveal sunspots below each. Fig. 2.3 shows very clearly that active regions occur at the same latitude on the Sun, north and south of its equator.

Radiation of very short wavelengths comes from extremely hot gas, and the photosphere is so 'cool' that it appears black in ultraviolet and X-ray pictures. The photosphere and the lowest layer of the atmosphere, the *chromosphere*, are at about 5800 K; but at the top edge of the chromosphere the temperature rises abruptly through a thin *transition layer* to reach 1 million K or more in the outer atmosphere, the *corona*. The gases of the transition layer shine in the ultraviolet and the shorter extreme ultraviolet (EUV) radiation, while the corona is brilliant in X-rays.

**Fig. 2.4** shows our local star in the radiation from gas in the transition layer at a temperature of around 100 000 K. Less intense regions are coded red, and brighter parts yellow, green, blue, purple and white. The transition region does not cover the dark photosphere uniformly. Running down from the North Pole, there is a long dark rift, the lowest layer of a coronal hole – a low-density region of the Sun's atmosphere (Fig. 2.15). The Sun's poles (top and bottom) also have no overlying transition layer. The magnetic fields which generally confine the Sun's atmosphere are here directed straight out into space – like the lines of force from the ends of a bar magnet – and the gas escapes freely. The gas is most confined by the strong fields over active regions, which therefore appear brightest in Fig. 2.4

Farther out, the fields of the active regions trap the hotter, X-ray emitting gas of the corona (**Fig. 2.5**). In this contoured map, regions of successively greater brightness are enclosed in red, yellow, green, blue, pink and white (with black between the colours); the edge of the photosphere is marked by the thin white circle. This gas is generally at 2 to 3 million K, but the intense white regions are at 5 million K, almost a thousand times hotter than the photosphere below.

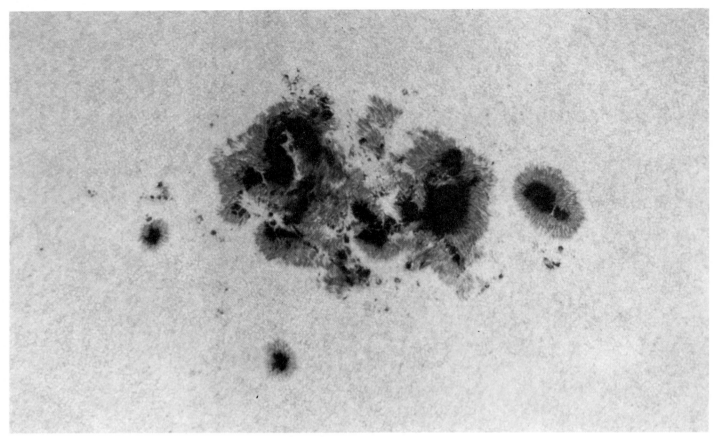

2.6 *Optical (17 May, 1951), 46 m tower telescope, Mount Wilson Observatory*

Sunspots are best studied in visible light, at the wavelengths at which the photosphere and details within it are most easily discerned. The number of sunspots comes and goes in a cycle of 11 years. New spots first appear at high latitudes, and later spots in the cycle nearer to the Sun's equator. A small spot will last a few weeks, while the largest may take several months to grow and then shrink away again.

Spots generally come in pairs, or in complex groups of spots. The straightforward optical photograph (**Fig. 2.6**) shows a typical large complex – several times bigger than the Earth. Each spot has a dark core, the *umbra*, and a paler outer *penumbra* composed of alternating bright and dark filaments of gas. (The small-scale mottling outside the spot group is the pattern caused by energy 'bubbling up' through the photospheric gas.) The blackness of a sunspot's umbra is only a contrast effect. It is about 2000 K cooler than the photosphere, and so emits much less radiation – but a large sunspot is in fact as bright as the full Moon.

**Fig. 2.7** reveals details inside a sunspot's umbra by the technique of speckle interferometry, which reduces blurring by the Earth's atmosphere. In each image, the brightest regions are coded pale pink, and darker regions red, yellow, blue and black. The top frame shows a very small spot, about 4 arcseconds in apparent size (3000

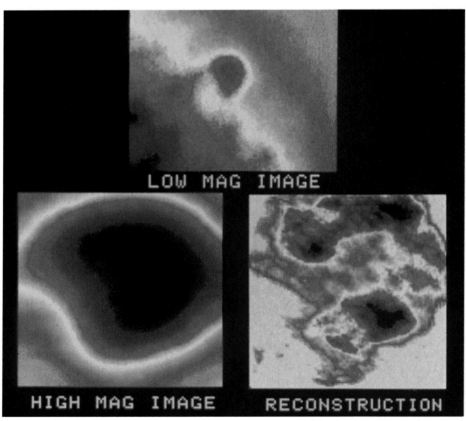

2.7 *Optical (17 July, 1981), 405–415 nm, speckle-processed, CID camera, McMath Solar Telescope, Kitt Peak National Observatory*

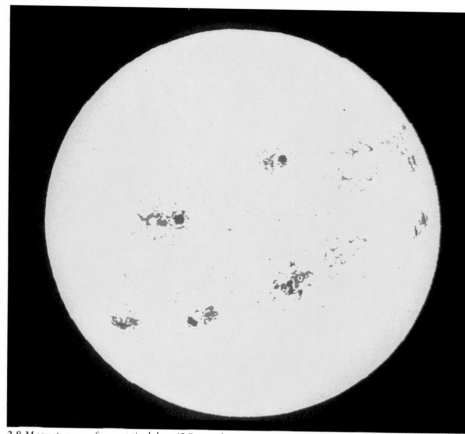

2.8 *Magnetogram, from optical data (5 September, 1973), Kitt Peak National Observatory*

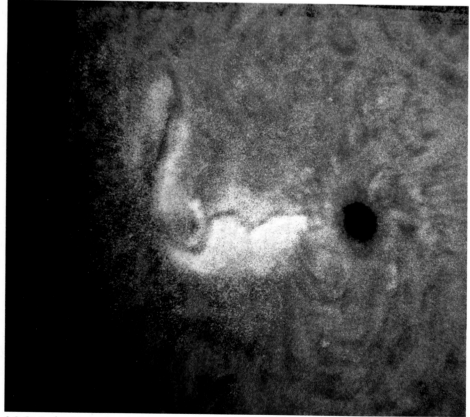

2.9 *Optical (5 November, 1978), 656 nm hydrogen line: 656.32 nm (red), 656.28 nm (green), 656.24 nm (blue), 0.1 m telescope, Culgoora Observatory*

kilometres in diameter) comparable to the very smallest spots visible in Fig. 2.6. An electronic detector, which responds to a wide range of brightness from the 'black' umbra to the photosphere, took a series of 46 very short exposures to 'freeze' the atmosphere's blurring. One of these appears as the magnified image at lower left in Fig. 2.7. The exposures were added to produce the final image of the spot seen at lower right. Cearly, the interior of a sunspot is not uniformly dark: this particular spot is darker around the edges, and slightly less dark in the centre. There is also smaller structure, down to a size of 0.3 arcseconds (200 kilometres), including striations stretching to the top left.

Sunspots occur where powerful magnetic fields break through the photosphere, and prevent interior light and heat from escaping. The Sun's general magnetic field is similar in strength to the Earth's, but the field in sunspots is over a thousand times stronger. The field is concentrated in many smaller ropes, seen as they spread out from the umbra as the filaments in a spot's penumbra. The fine details in Fig. 2.7 may indicate that the individual ropes are narrower and contain a more concentrated field than had previously been thought.

The magnetic field in sunspots affects the spectral lines from gas in them. Optical spectral studies not only show the field strength, but also the magnetic polarity — whether it is a 'north pole' or a 'south pole'. The magnetogram (**Fig. 2.8**) shows the magnetism of all the spots on the Sun, colour-coded red for 'north pole' and blue for 'south pole', and superimposed on a yellow disc indicating the photosphere. Each pair of sunspots clearly includes one spot of each polarity. The magnetic fields must run between them like the lines of force of a horseshoe magnet, stretching up into the atmosphere in an arch.

In a group of sunspots, the field patterns are more complex. **Fig. 2.9**, taken in the light from hydrogen atoms, shows a large spot (right) and a smaller spot (left). The large bright white region is a plage of denser gas in the chromosphere just above the spots, trapped by the magnetic field. Running along its centre, and lying above it, is a dark line of cooler gas, a prominence which is supported along the neutral line where magnetic fields from the different spots are in balance. The wavy line of the prominence shows that the field does not follow a simple pattern in this spot group. The picture is colour coded so that the exact wavelength emitted by hydrogen would appear green, slightly shorter wavelengths blue and slightly longer radiation red. If the gas were moving fast, the Doppler shift would cause it to appear blue or red. The lack of strong colours shows that this group of spots, plage and prominence is relatively stable (compare with the eruptive prominence, Fig. 2.11).

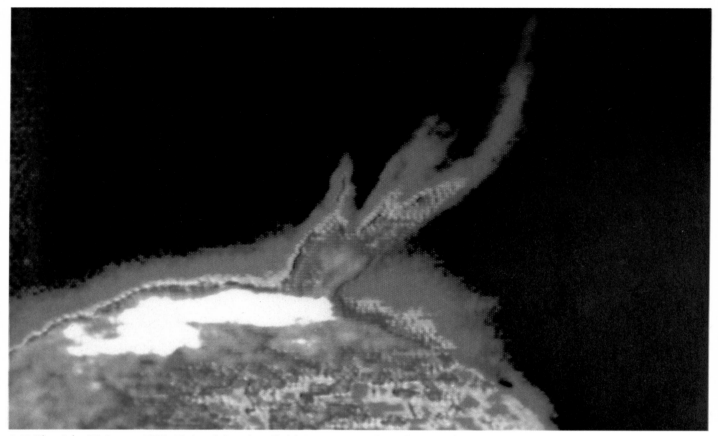

**2.10** *Ultraviolet (21 August, 1973), 30.4 nm helium line, Skylab ATM extreme ultraviolet spectroheliograph*

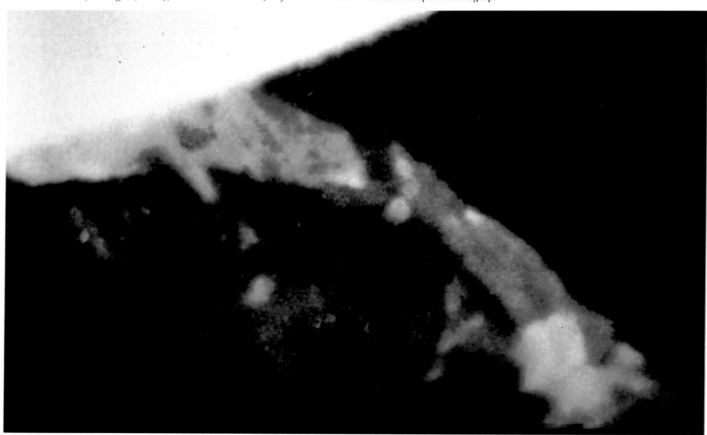

**2.11** *Optical (27 May, 1978), 656 nm hydrogen line: 656.33 nm (red), 656.28 nm (green), 656.23 nm (blue), 0.1 m telescope, Culgoora Observatory*

'Active regions' on the Sun not only experience the gradual build-up of magnetic field which gives rise to sunspots, prominences and loops in the corona; they also produce powerful outbursts and explosions. These great eruptions – virtually invisible at optical wavelengths – send gas and fast electrons streaming through the Solar System. When they reach Earth, the particles produce aurorae in the atmosphere and radio blackouts as they disrupt the ionosphere; and they present a radiation danger to astronauts above the protection of Earth's atmosphere.

Eruptive prominences are one type of solar outburst. They begin as normal prominences, dense curtains of gas at a relatively cool temperature (about 10 000 K) hanging high up in the much hotter corona. Prominences are supported by the magnetic fields that stretch from one sunspot or sunspot group to another. Occasionally, this magnetic field will change its shape and structure quite markedly in just a few minutes. As it twists and tears itself away from the Sun at one end, the magnetic field catapults the prominence's gas out into space. The quiescent prominence becomes an eruptive prominence.

**Fig. 2.10** is a view of an eruptive prominence, taken in the light from helium which reveals the chromosphere (seen also in Fig. 2.1). The faintest regions here are coded red, and brighter regions yellow, blue, lilac and white (North is to the right). The white area is an active region associated with a sunspot group, and the magnetic fields have flung a prominence out into space as the multi-stranded jet of matter stretching out half a million kilometres.

The erupting prominence in **Fig. 2.11** is only half as long, but even so it extends twenty Earth-diameters from the Sun's surface. This picture was taken at optical wavelengths, with a ground-based telescope tuned into the red light from hydrogen atoms, and it is colour coded to show velocities. The radiation from parts of the prominence which are moving towards or away from us have their wavelengths shifted by the Doppler effect, and Fig. 2.11 has been constructed from three simultaneous observations at closely adjacent wavelengths to reveal these motions. Red colour-coding shows gas which is moving towards us at a speed of 30 kilometres per second, while blue indicates regions of the filament moving away at this speed. Green shows gas which is moving neither towards or away from us, but it is erupting away from the Sun perpendicular to our line of sight.

Fig. 2.11 reveals the little-understood complexities of motion within an erupting prominence. Although the curve of gas looks like a continuous ejection, the colours show clearly that while the lower

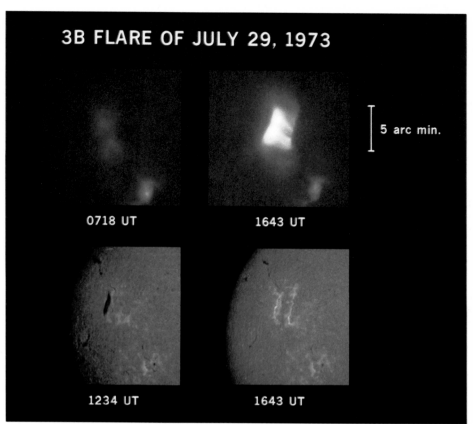

2.12 *X-ray (29, July 1973), 0.2–1.7 nm (blue), Skylab X-ray telescope; optical (29, July 1973), 656 nm hydrogen line, ground-based solar telescope*

part of the prominence and the extreme tip are moving towards us, the middle region is moving away. They may be separate sprays of gas ejected as a fan, which we are seeing from the edge. A 'time-lapse' series of photographs like this would reveal the sideways movement of the ejected gas streamers too, and enable astronomers to study the motion of eruptive prominences in three dimensions.

More powerful even than eruptive prominences are solar *flares*. These outbursts generate a huge amount of radiation, and accelerate electrons up to high speeds, but they usually eject relatively little gas into space. A flare starts as a small very intense source of radiation and spreads along the lines of magnetic field in a matter of minutes, like the flame from a match tossed into a pool of petrol, and sets a region larger than the Earth shining with the brilliance of a million hydrogen bombs. Despite this output of radiation, a flare cannot match the photosphere's output of light, and flares are usually visible at optical wavelengths only by isolating the light from a single type of atom which is set glowing in the flare. But at other wavelengths, a flare totally dominates the Sun.

**Fig. 2.12** compares views taken in X-rays (blue) with optical pictures in the spectral line emitted by hydrogen (red). The time of each picture is shown in Universal Time (UT, formerly Greenwich Mean Time).

Before the flare (left-hand images), the small active region at the centre is quiescent. The optical image shows some plage regions, and a prominence in silhouette, but this region has little in the way of hot gas above it to produce X-rays. Suddenly, and unpredictably, a flare begins. In a few minutes, it has spread and lit up the plage regions, and the prominence. Although this brightening is visible in the optical (red) photograph, it is not very obvious. The flare has, however, heated up the gas in the prominence until it shines brilliantly at X-ray wavelengths (blue). In a flare like this, the coronal gas can reach a temperature of 20 million K – even hotter than the Sun's core.

The flare's power comes from the energy pent up in the magnetic fields of sunspots regions. When magnetic field loops of opposite polarity approach one another too closely, they can annihilate. This magnetic field reconnection is very similar to an electrical short-circuit, and like a short-circuit it produces a sudden flash of energy – the solar flare. Once a reconnection is made, it can spread rapidly along the divide between two magnetic field regions – often marked by a prominence – liberating energy all along the way. Magnetograms show that an active region's magnetic field structure is simplified after the trauma of a flare has short-circuited the kinks in its complex magnetic pattern.

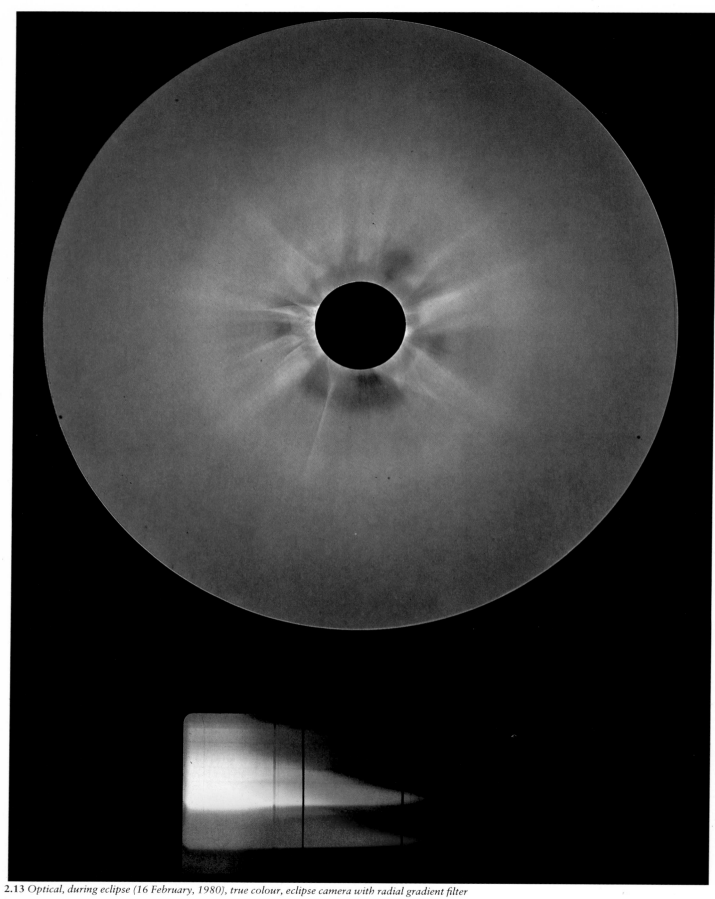

**2.13** *Optical, during eclipse (16 February, 1980), true colour, eclipse camera with radial gradient filter*

The Sun's huge outer atmosphere of hot gases, the *corona*, is a million times fainter than the brighter photosphere. It is normally invisible, because it is only one-hundredth as bright as the daytime sky. But an eclipse shows the corona in its true glory (**Fig. 2.13**), when the Moon blocks out the photosphere's light, and with it the blue sky of scattered sunlight. The corona's brightness falls off sharply with distance from the Sun, and this photograph compensates with a filter which is dark at the centre and progressively more transparent to the edges. As a result, details are visible from the inner atmosphere, the chromosphere, to a distance of two Sun-diameters – 3 million kilometres. The chromosphere displays reddish prominences at top left; above, in the lower corona, are loops and 'helmets' of gas held in the magnetic fields of active regions, extending far into space. The 'rainbow' (bottom) is a crude spectrum of the corona, obtained simultaneously. Note the strong green light from iron atoms, and red light from hydrogen.

Eclipses occur only rarely and last but a few minutes. Continuous observation of the corona's light requires a coronagraph, a telescope which blocks the photosphere's light. **Fig. 2.14** is from the coronagraph on the Solar Maximum Mission satellite. The photosphere is hidden behind the circular disc at lower left, and the brightness of the corona is shown as bands of colour (isophotes) enclosing regions of successively higher intensity. The faintest parts are delineated by blue isophotes, with green for brighter regions (the central, brightest regions of these being shown by a single blue isophote). Fig. 2.14 is taken in the green light from iron atoms which have been stripped of thirteen electrons, in the hot (1 to 2 million K) gas. Such pictures have shown that flares and eruptive prominences can eject the overlying coronal gas into space at speeds up to 1000 kilometres per second. These clouds may be larger than the Sun, but only contain a million-millionth of an Earth-mass of gas.

An X-ray view of the Sun (**Fig. 2.15**) reveals the hot coronal gas – which naturally emits these wavelengths – without the need to hide the photosphere, which is dark in X-rays. Hence we can see the corona across the Sun's face. The corona is extremely patchy, the hot gas being only noticeably present in the active regions, where it is concentrated into the loops of magnetic field joining sunspots.

The long dark gap is a *coronal hole*. Here the magnetism of the Sun's surface does not form closed loops, but extends straight out. Hence hot gas is not trapped near the Sun, but constantly streams away from the coronal holes. This solar wind of particles blows outwards through the whole Solar System, so all the planets effectively lie within the Sun's atmosphere.

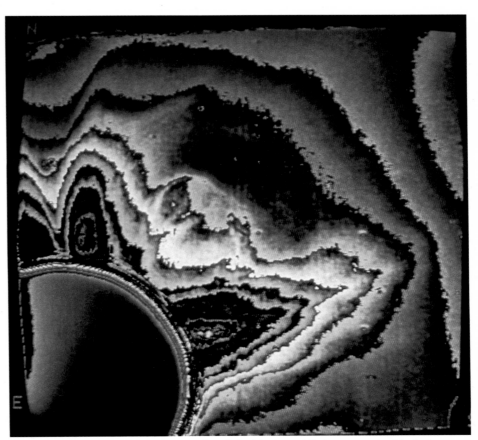

2.14 *Optical (6 May, 1980), 530 nm iron line, Solar Maximum Mission coronagraph/polarimeter*

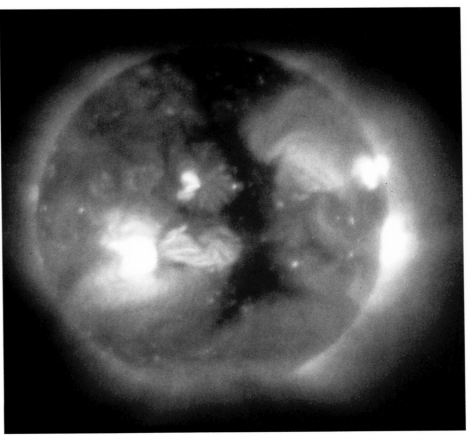

2.15 *X-ray (1 June, 1973), 0.2–6 nm, Skylab ATM X-ray telescope*

**2.16** *Ultraviolet, 365 nm, Pioneer Venus Orbiter cloud photopolarimeter*

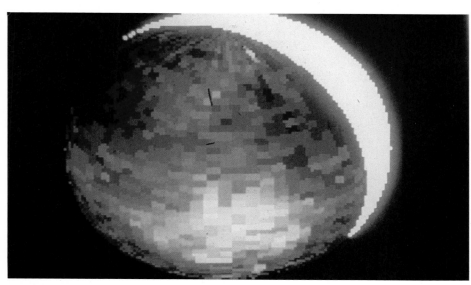

2.17 *Ultraviolet, 198 nm, Pioneer Venus Orbiter ultraviolet spectrometer*

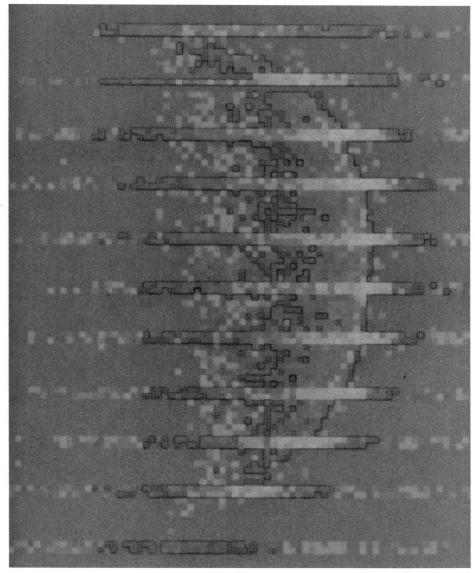

2.18 *Ultraviolet, 135 nm oxygen line (main image) and 122 nm hydrogen line (interspersed strips), Pioneer Venus Orbiter ultraviolet spectrometer*

# Venus

With a diameter only 5 per cent less than that of the Earth, Venus has long been regarded as our twin. It is also the planet which can come closest to Earth (within 40 million kilometres) and as a dazzling 'Morning Star' or 'Evening Star' it is by far the brightest and most beautiful. For these reasons, the romantic image of Venus as a moist, Carboniferous-era version of our own world was slow to die. But probes and Earth-based radio and radar investigations have now revealed Venus to be the hottest, driest, roughest and most hostile planet.

Venus shines brightly because clouds in its dense atmosphere are particularly good at reflecting sunlight, and hide its surface from the view of optical astronomers. Many Soviet and American probes have investigated the planet and its atmosphere, both by direct sampling and by images at ultraviolet, infrared and radio wavelengths. **Fig. 2.16** is a photograph of Venus's cloud-tops, taken from a distance of 65 000 kilometres by the Pioneer Venus Orbiter craft in January 1979. This ultraviolet photograph emphasises faint cloud details. The streaks and bands are evidence of rapid circulation. Venus's atmosphere rotates in only four Earth-days (retrograde), as compared to the surface's 'day' (also retrograde) of 243 Earth-days. Since Venus completes an orbit about the Sun in 225 Earth-days, its day is longer than its year.

The rapid circulation of the upper atmosphere, which is still not fully understood, gives rise to strong, high-altitude winds. One result of these winds is shown in **Fig. 2.17**. At top right is the planet's sunlit crescent. The glow on the night-time side, coded for intensity ranging from yellow (bright) to black (faint) is caused by atoms of nitrogen and oxygen combining together fluorescently to produce nitric oxide. The atoms have been carried round by high winds from the daytime hemisphere, where they were originally produced by the action of sunlight on molecules of carbon dioxide and nitrogen. The ultraviolet glow, whose distribution gives clues as to the patterns of the high-altitude winds, may be related to the 'ashen light' reported by visual observers on the Earth.

**Fig. 2.18** reveals more of Venus's outer atmosphere. It is another false-colour view through the Pioneer's ultraviolet spectrometer, in which the crescent shows the distribution of oxygen atoms in the upper atmosphere with amounts increasing from blue to yellow. The horizontal bars are strips of a simultaneous view revealing atoms of hydrogen, surrounding the planet in a huge slowly-escaping cloud. However, these are minor constituents of Venus' colossal atmosphere, which is 100 times more massive than that of the Earth, and consists almost entirely of carbon dioxide.

When compared to fast-spinning planets like the Earth and Jupiter, Venus's atmosphere has a relatively simple circulation pattern. The upper clouds spiral slowly towards the poles from the equator, where they descend in two eddies (the polar vortices). However, beneath this there are at least four other levels of circulation, together with other large-scale motions. **Fig. 2.19** shows some of these patterns in a colour-coded temperature map made by the Pioneer Orbiter infrared radiometer. At the centre is Venus's north pole, while the surrounding hemisphere, like a clock, shows the time of day. Noon is at the bottom, midnight at the top, morning on the right and evening on the left. The colours represent temperature, ranging from dark blue (cold) to dark red (hot), at a level of the atmosphere whose pressure is one-tenth of the Earth's at sea-level. Not surprisingly, the atmosphere is hottest directly under the noonday Sun. But it is also warmer at midnight than in the morning or afternoon: a sign that winds or large-scale circulation patterns have carried warm air round to the dark side.

Two of the hottest patches are at the poles themselves, where the polar vortices create 'holes' in the atmosphere, and we see through to the deeper, hotter layers. Venus's surface temperature of 750 K is higher than that of any other planet. The reason lies in the atmosphere's enormous

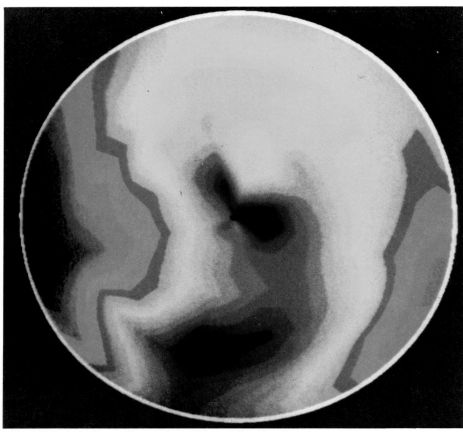

**2.19** *Infrared, 11.5 μm, Pioneer Venus Orbiter infrared radiometer*

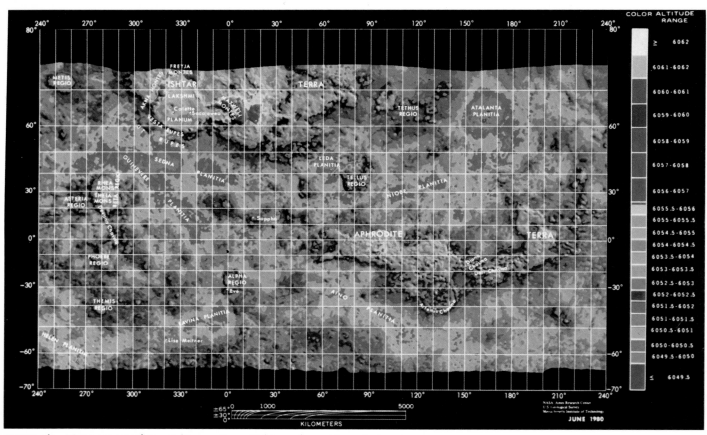

**2.20** *Radar, 17 cm, topographic map, Pioneer Venus Orbiter radar mapper-experiment*

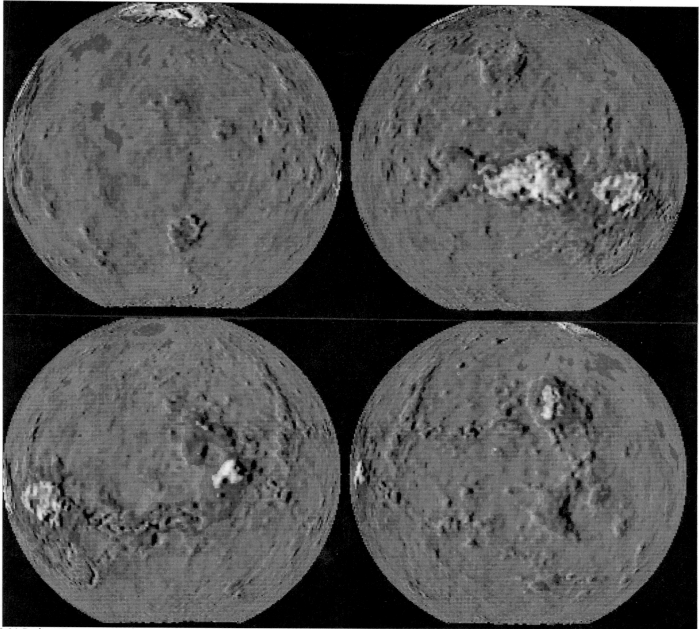

2.21 *Radar, 17 cm, topographic map, Pioneer Venus Orbiter radar mapper experiment*

carbon dioxide content, which inhibits the absorbed sunlight from being re-radiated as infrared radiation. This 'greenhouse effect' has led to the immense heat build-up.

It is not known if Venus ever had seas. Instead of ocean basins and mountains, 60 per cent of its surface is covered by a rolling plain. This is coded blue in **Fig. 2.20**, a view of the surface obtained by bouncing radio waves off the planet, through the sulphuric acid clouds which are opaque to other radiation. Heights are colour-coded according to the scale on the right, where the figures represent the excess or shortfall of height compared with Venus's mean radius of 6051 kilometres. Although there are a few low regions (dark blue or purple), they are much smaller than ocean basins on

the Earth and are probably ancient craters. Venus has two main 'continents': Ishtar Terra in the North, and Aphrodite Terra around the equator. Both are probably younger areas, and the reflected radar waves show the surface in these regions to be very rough. Ishtar Terra, a high plateau equivalent to Australia in size, contains the highest point on the planet, Maxwell Montes. These mountains reach to 11 kilometres above the mean surface altitude, and enclose what may be an ancient volcanic crater.

The four colour-coded views in **Fig. 2.21** were also obtained from radar mapping by the Pioneer Orbiter, but are shown in an orthographic projection which represents more clearly the true proportion of

'continent' to 'plain'. Ishtar Terra (top left) is seen to be far smaller than Aphrodite (top right and lower left), and the overall smoothness due to the extensive rolling plains is more obvious. There is little evidence for moving crustal plates – 'continental drift' – as we find on the Earth, but this may be a result partly of the coarse resolution of the radar mapping. However, Venus is not completely dead. Beta Regio, the high region in the lower right-hand image, appears to be a pair of volcanoes each 4000 metres high, separated by a huge rift valley. These may still be active. Lightning strokes which have been observed in this region are believed to come from electrically-charged clouds of dust ejected by the eruptions.

**2.22** *Optical, true colour, Voyager 1*

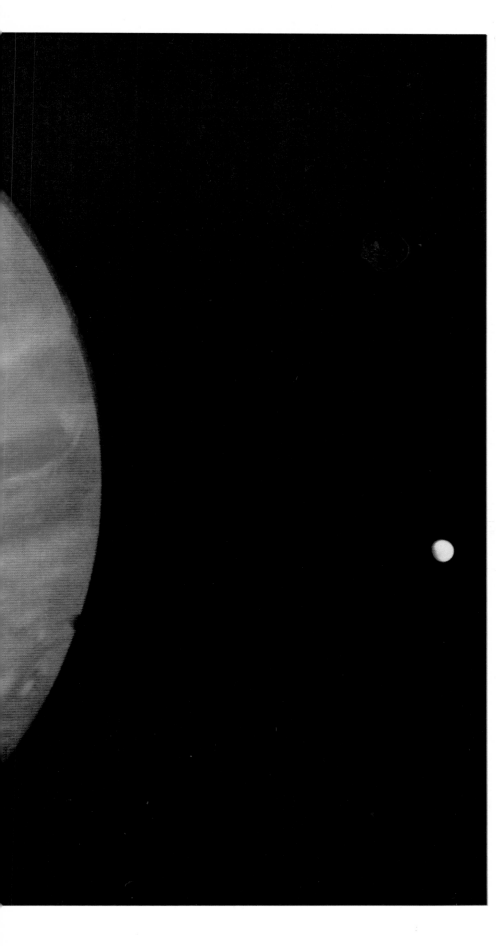

# Jupiter

The giant of the planets, Jupiter, has a diameter of 11 Earth-diameters, and a mass 318 times that of our planet. Jupiter may have a rocky core, similar in size and composition to the Earth, but most of its bulk is composed of lighter substances, mainly hydrogen and helium, but with some methane, ammonia and water. The planet's gravity compresses its own matter so that it behaves like a liquid rather than a gas, apart from a thin outer gaseous atmosphere containing layers of coloured cloud.

Jupiter is large enough that its general shape and features are easily seen with Earth-based optical telescopes. The planet appears noticeably flattened, because its equatorial regions bulge outwards under the effect of the 'centrifugal force' due to Jupiter's fast rotation. Turning once in only 9 hours 55 minutes, Jupiter rotates faster than any other planet. The main features seen on Jupiter are bands of different coloured clouds running parallel to the equator. Within these bands are finer streaks and ovals. The largest oval, the Great Red Spot, is very long-lived for an atmospheric feature, having been observed as early as the 1660s. Earth-based telescopes have also revealed a dozen satellites; four are comparable in size with the planets Pluto and Mercury.

But our knowledge of Jupiter has recently been revolutionised by spaceprobes studying the planet and its satellites at close quarters. **Fig. 2.22** is one of the 35 000 photographs returned by the two Voyager probes in 1979. At the extreme right is the white satellite Europa, a smooth ice-coloured world; while in front of Jupiter is the orange moon Io, coated in sulphur ejected by its active volcanoes – one of the major discoveries of Voyager 1.

Sequences of Voyager photographs reveal that the light and dark bands are travelling around the planet at different speeds, up to 400 kilometres per hour, with bands at successively higher latitudes moving round in opposite directions. The ovals lie between these oppositely directed wind currents, and rotate with the relative motion – like ball bearings between two moving surfaces. The Great Red Spot (lower left in Fig. 2.22) seems to be just a particularly large oval.

Fig. 2.22 shows that the clouds also contain turbulent swirls, or eddies. These eddies are transporting energy upwards within the atmosphere, and meteorological studies have indicated that it is the energy of the eddies which is eventually harnessed to drive the regular wind currents at different latitudes. The Earth's atmosphere also has jet streams fed by the energy of turbulent eddies, and studies of Jupiter's atmosphere are providing clues to the Earth's weather and climate.

2.23 *Infrared, 1.6 (blue), 2.2 (green) and 4.8 (red) μm, 3.9 m Anglo-Australian Telescope*

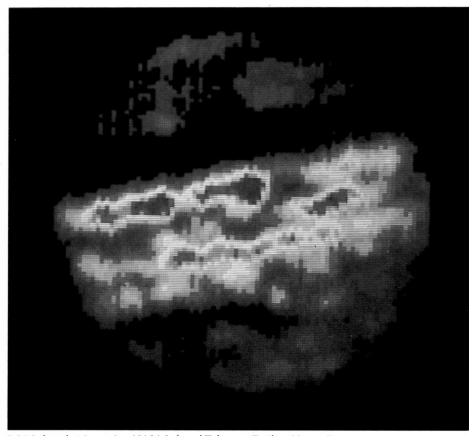

2.24 *Infrared, 4.8 μm, 3 m NASA Infrared Telescope Facility, Mauna Kea*

Other wavelengths can reveal features of Jupiter both above and below the layers of clouds visible in optical telescopes and which showed up in stunning detail when viewed by the Voyagers' TV cameras.

**Fig. 2.23** combines images of Jupiter at three successively longer wavelengths, colour-coded and superimposed. The blue image, made at 1.6 microns, shows the planet's cloud tops much as we see them in optical pictures, brighter in the equatorial and temperate regions than at the poles. The green, 2.2 micron, image is at a wavelength absorbed by methane, and Jupiter reflects very little radiation of this wavelength except from its methane-poor poles. The red image, however, is the most interesting. This 4.8 micron radiation does not consist of reflected sunlight, but is generated by the warmth of Jupiter's interior, and we detect its emission from regions of the planet where the deeper layers are not hidden by higher clouds.

This radiation is seen more clearly in **Fig. 2.24**, a view at 4.8 microns which is coloured so that the faintest regions are dark blue, and brighter regions pale blue, yellow, orange and red. The planet's own radiation clearly escapes mainly from two belts parallel to the equator, corresponding to the dark brown belts seen in optical pictures like Fig. 2.22. The comparison of optical and infrared pictures reveals the height and temperature of the cloud layers, and hence gives clues to their composition. The brown clouds are relatively low, and the infrared measurements show that the planet's internal heat warms them to a temperature of 230 K (no colder than an arctic winter); for comparison, the average temperature of Jupiter's cloud tops is 125 K. Given their temperature, the appropriate pressure and Jupiter's composition, it is most likely that the brown clouds are composed of ammonium hydrosulphide, a white compound which is probably tinged brown by small amounts of other sulphur compounds. Small holes in this cloud deck show us deeper, hotter regions of Jupiter — as warm as 'room temperature', 300 K. These appear as brilliant spots on the infrared maps, and the Voyager cameras saw them as hazy blue spots, because light from below was being scattered by the denser gas, much as Earth's air scatters sunlight to cause a blue sky.

Some 50 kilometres above the brown clouds lie the white zones of cloud covering much of Jupiter's disc and blocking off most of the 4.8 micron radiation from below. Infrared measurements show that these clouds are at a temperature of 130 K, and they probably consist of ammonia crystals. One of the coldest regions of Jupiter, and hence one of the highest, is the Great Red Spot — or to be accurate, its top, because its longevity indicates that the Great Red Spot also penetrates far down into the atmosphere. Its colour may result from phosphine gas being brought up from the depths to its top, where the gas is decomposed by sunlight into red phosphorus.

To keep at a steady temperature, a planet should radiate away as much energy as it absorbs from sunlight. Jupiter's infrared emission, at wavelengths of 4.8 microns (Fig. 2.24) and longer, however, amounts to 70 per cent more than this. The planet must therefore have its own interior heat source, probably because it is still contracting slightly from its formation out of a gas cloud, and the consequent compression heats up its interior.

Shorter wavelengths reveal other details of Jupiter's atmosphere. High-speed electrons trapped in Jupiter's magnetic field (Figs. 2.27–2.29) are funnelled down at the planet's north and south magnetic poles, where they hit molecules in the atmosphere. The collisions produce aurorae – like the Earth's Northern and Southern Lights – which were seen by the Voyagers. The aurorae also shine in the ultraviolet, and surprisingly strongly at X-ray wavelengths. **Fig. 2.25** is the Einstein Observatory's X-ray view of Jupiter. The planet's disc has been marked in as the white oval, and the plane of its equator by the straight lines. The X-ray emission has been coded dark blue for the faintest regions, and pale blue and white for the more intense parts. The aurorae clearly occur in the thin outer atmosphere above Jupiter's magnetic poles, which are 10° away from the poles of rotation.

The aurorae, and thunderstorms in Jupiter's clouds, provide concentrated energy sources which can weld the simple molecules of Jupiter's atmosphere into more complex varieties, including organic molecules. Some dozen molecules have now been identified here, generally from their spectral lines at infrared wavelengths. Organic molecules also have characteristic lines in the ultraviolet, and some show up in **Fig. 2.26**, an ultraviolet spectrum of Jupiter's clouds. Each diagonal band (from upper left to lower right) is a segment of a highly magnified spectrum; the complete spectrum would be obtained by running them consecutively. The intensity within each runs from grey and white for the faintest regions, through blue, to green and red for the brightest. (The white spot at the upper right is a flaw.) The spectrum of reflected sunlight would comprise red or green bands, crossed by broad dark absorption lines. But the spectrum in Fig. 2.26 also contains many narrow lines due to absorption by molecules in Jupiter's atmosphere. Ultraviolet studies can also reveal the emission spectrum of gases around Jupiter heated by sunlight (like Venus's hydrogen envelope, Fig. 2.18), and the glow of gases in Jupiter's aurorae.

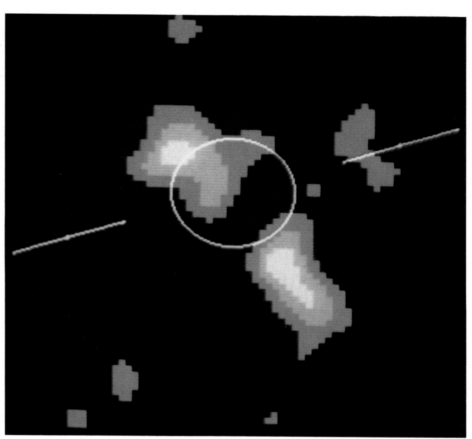

2.25 *X-ray, 0.4–8 nm, Einstein Observatory HRI*

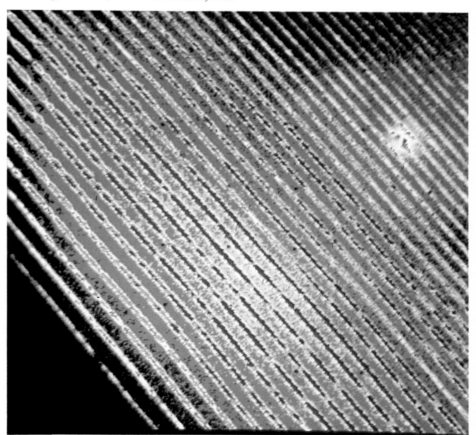

2.26 *Ultraviolet, spectrum, 180–320 nm, International Ultraviolet Explorer satellite*

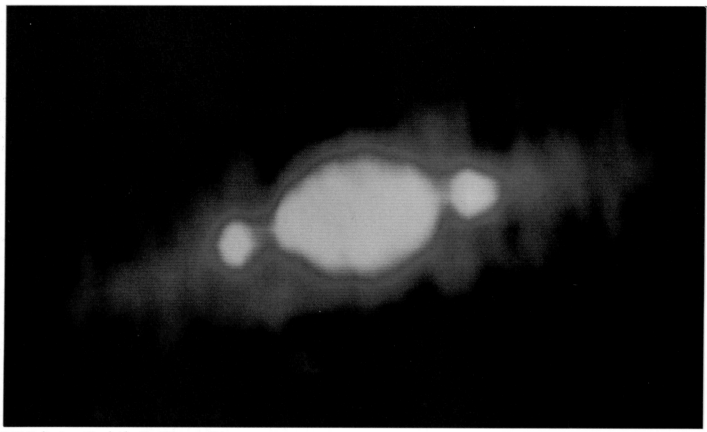

**2.27** *Radio, 6 cm, Cambridge Five Kilometre Telescope*

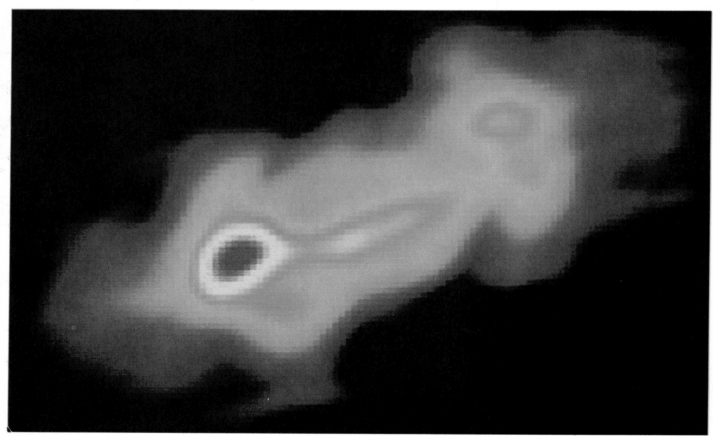

**2.28** *Radio, 20 cm, Very Large Array*

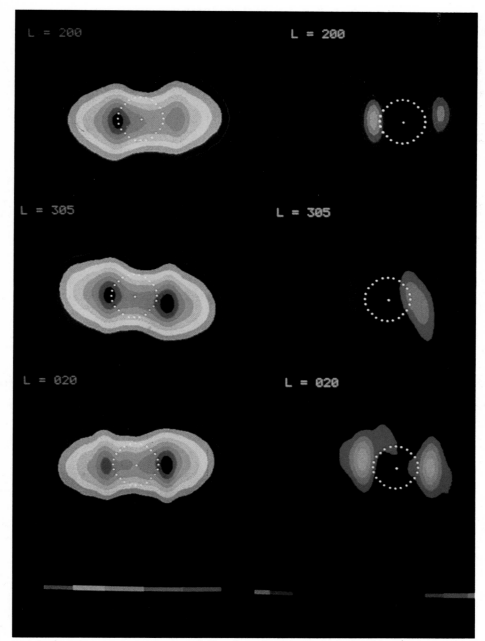

L = 200     L = 200

L = 305     L = 305

L = 020     L = 020

*2.29 Radio, 21 cm continuum, intensity (left) and circular polarisation (right), Westerbork Synthesis Radio Telescope*

Jupiter is the brightest of the planets when observed at longer radio wavelengths, its lightning storms and aurorae (Fig. 2.25) producing intense bursts of radio emission at wavelengths of 10 metres and more. But Jupiter also produces strong steadier emission at shorter radio wavelengths, as seen in Figs. 2.27–2.29.

Fig. 2.27, at 6 centimetres, is colour-coded so that the faintest regions are blue, and brighter parts are red, yellow and white. The bright oval in the centre is Jupiter itself, emitting radiation at these wavelengths just because of its innate warmth of 125 K. The wings of emission to either side are entirely different. They are caused by Jupiter's intense magnetic field.

This field, about 20 000 times stronger than the Earth's magnetic field, traps fast electrons in a belt around Jupiter which is similar to the Earth's Van Allen belts. As the electrons whirl through the magnetic field they produce radio emission, by the synchrotron process.

Fig. 2.28 is a radio view at a longer wavelength (20 centimetres), coded so that faint regions are dark blue, and brighter portions of the image pale blue, green, yellow and red. Jupiter itself is fainter, appearing as the central blue and green disc, while the comparatively strong radiation from the belt reveals its structure clearly. Regions of different brightness indicate irregularities either in the

concentration of electrons or in the magnetic field. The regions of the belt to either side of Jupiter probably contain similar concentrations of electrons, but the magnetic field at left is 'beaming' the strongest radiation at the Earth, while it concentrates the radiation from the right-hand part into a beam which misses our planet, and hence the latter region appears fainter.

The three images in the left-hand column of Fig. 2.29 show Jupiter's changing radio appearance as its rotates, with colour-coding running from green for the faintest parts, through yellow, orange, red to black for the brightest regions. In each picture, Jupiter's equator is horizontal and the planet's globe indicated by the dotted lines. The second image was after the planet had rotated by 105° (just over a quarter of a turn), and the third after 180° (half a turn).

The sequence reveals that Jupiter's radiation belt (marking the *magnetic equator*) does not girdle the planet's actual equator; it is angled by about 10°, so that the regions of radio emission tilt up and down during Jupiter's rotation. In the top and bottom images of Fig. 2.29, the north magnetic pole is angled respectively towards and away from us, and the belts look symmetrical with respect to Jupiter's equator. In the middle image, Jupiter has turned so that the magnetic poles lie at 'one o'clock and seven o'clock', and the belts are seen at their maximum tilt.

The magnetic field runs in loops between the magnetic poles, and the electrons spiral around the field lines on paths like the coils of a spring. As seen from one end of a 'line' of magnetic field, the electrons are moving anticlockwise, and from the other end, clockwise. The synchrotron radiation from these electrons is an electromagnetic wave whose direction of maximum intensity (electric vector) circles around in pace with the electrons, to produce radiation which is *circularly polarised*; either left-handed or right-handed.

In the right-hand images of Fig. 2.29, radiation with left-handed polarisation is coded blue, with successively brighter regions in paler shades while right-handed radiation is red, shading to orange for more intense regions. When the magnetic north pole is inclined towards us (top image), the radiation (from both sides of Jupiter) should have left-handed circular polarisation; in the lowest image, the circular polarisation should be right-handed. Both these predictions are borne out by observation. In the intermediate position (middle image), there should be no circular polarisation; but Fig. 2.29 reveals definite polarisation, of opposite kinds to either side of Jupiter. The planet's magnetic field is clearly rather more complex than was at first believed: the magnetic equator is apparently not a flat plane, but is warped like a potato crisp

2.30 *Optical, true colour, Voyager 1*

# Saturn

Second only to Jupiter in size and mass, Saturn is a giant planet almost ten times as wide as the Earth and 95 times as massive. Its fame, however, rests not on its dimensions, but on its rings. Although Jupiter and Uranus have faint ring systems, Saturn's wide bright rings are unique, making it the most beautiful planet as seen by telescope or passing spaceprobe.

**Figs. 2.30** and **2.31** were taken by the Voyager 1 and Voyager 2 probes respectively, four days after each had passed closest to the planet. The Voyagers were able to see Saturn as we can never view it from the Earth, with the planet only half-illuminated by the Sun. Saturn's shadow is thrown across the rings behind, while the rings in front of the planet cast shadows onto Saturn's sunlit hemisphere. The great difference between the rings' appearance in Figs. 2.30 and 2.31 occurs because Voyager 1 looked back at the sunlit side of the rings, and Voyager 2 saw their other side, the dark face tilted away from the Sun.

The rings' structure is seen in Fig. 2.30 much as it appears from Earth – except in far finer detail. A dark band running round the planet divides the system into two main rings. Outside this dark Cassini Division lies the A-ring; while inside is the brighter B-ring. Within the B-ring, Fig. 2.30 shows the faint C-ring – or crepe ring – best seen just to the right of the centre of the planet's lit-up hemisphere.

Earth-bound observations had indicated that the rings are not solid objects girdling Saturn, but must consist of huge numbers of small ice blocks each orbiting the planet as a small moon. Even the Voyagers' close-up pictures could not show the individual blocks, but they did reveal – to the amazement of the astronomers – that this ring material is concentrated in about 100 000 individual ringlets. Fig. 2.30 shows some of the more prominent ringlet structure in the B-ring – although each 'ringlet' seen here is in fact composed of hundreds of narrower ringlets. The picture also demonstrates dramatically that the rings are indeed not solid: at the bottom, Saturn's globe is visible through even the densest part of the B-ring.

The Voyagers could determine the size of the blocks comprising the rings in several ways. Their effect on the Voyagers' radio transmissions as they passed behind the rings indicate that the main blocks are about 1 metre across in the C-ring, and are progressively larger farther out from the planet, to reach 10 metres in the A-ring. They are separated by a distance roughly ten times the block's diameter.

But Fig. 2.31 indicates the presence of much smaller ice crystals too. The dark side of the larger ice blocks must be completely black, and the eerie glow from the rings in Fig. 2.31 comes from particles so small that they scatter light forward, appearing dark in direct sunlight but bright when observed into the Sun – like cigarette smoke. The rings in Fig. 2.31 are almost the negative of Fig. 2.30; from the outside inwards, the bright bands are the outer edge of the A-ring, the Cassini Division, and a group of ringlets at the outer edge of the C-ring. The B-ring, brightest in direct sunlight, is the darkest region of the rings in Fig. 2.31.

The ice blocks appear yellowish-white in true colour pictures like Fig. 2.30, but when the colour differences are extremely enhanced by computer (**Fig. 2.32**), the rings are seen to differ slightly in colour. The C-ring (left) and the particles in the Cassini Division (towards top right) reflect short wavelengths better, and appear dark blue in Fig. 2.32. The B-ring changes colour smoothly from the outside (blue-green) to the inside (orange), which reflects long wavelengths most efficiently. There is a surprisingly sharp change in colour between the B- and C-rings, much more pronounced than the change in brightness (compare with Figs. 2.30 and 2.31). The colour differences are still unexplained, but the ice is probably coloured either by impurities or by damage to its crystalline structure from high-speed electrically-charged particles trapped in Saturn's magnetic field.

**2.31** *Optical, true colour, Voyager 2*

**2.32** *Optical and ultraviolet, enhanced colour, Voyager 2*

In ordinary optical photographs, Saturn's atmosphere and clouds display none of the complex whirls, spots, and strong colours of Jupiter's cloud layers. But computer-enhancement can emphasise its subtle features. **Fig. 2.33** shows enhanced photographs of Saturn from Voyager 1 (bottom) and Voyager 2 (top). The original pictures were taken at wavelengths of green and violet light, and in the ultraviolet. They have been coded in colours corresponding to the wavelength sequence, but at longer wavelengths: green becomes orange; violet, green; and ultraviolet, blue. The enhanced false colours still, however, approximate the true colours – with regions reflecting short wavelengths appearing blue, and those reflecting longer wavelengths orange.

The most striking difference between the pictures is the appearance of the rings, because during its approach Voyager 2 (top) saw the rings tilted up more towards the Sun, and hence more brightly lit than Voyager 1's approaching view (bottom). The Voyager 2 picture (top) shows clearly some of the dark 'spokes' crossing the rings (on the left), probably fine particles of ice raised above the main rings by electrical forces. But the images show in particular the structures of Saturn's atmosphere, which have changed very little in the nine months between the two photographs.

The edge of the planet and most of the southern hemisphere appear blue, for the same reason that the Earth's sky is blue. Molecules in the atmosphere above the Saturn's clouds scatter short-wavelength (blue and ultraviolet) radiation more efficiently than longer wavelengths, and towards the edge of the planet our line of sight passes through more of this overlying atmosphere. At the centre of the planet's disc, it is easier to look through to the ammonia clouds below. They are tinged orange in Fig. 2.33 – yellowish-orange in true colour – possibly by impurities containing sulphur.

These clouds are probably composed of ammonia, like the white upper clouds of Jupiter (Fig. 2.22). On the colder Saturn, however, the ammonia clouds do not form thin narrow bands at particular latitudes, but deck the entire planet. This may well be the main reason for Saturn's bland appearance. The Voyager 1 (lower) view in Fig. 2.33 reveals a dark blue band in the northern hemisphere, where we are probably seeing through a greater depth of atmosphere down a rift in the ammonia clouds – a narrower version of the wide gaps in Jupiter's white clouds which reveal the orange and brown clouds below. But such rifts on Saturn seem to be temporary, as the Voyager 2 picture (top) shows white clouds encroaching on this region.

Towards the pole from this bluish band, the picture shows some faint ovals and whirls. As in the case of Jupiter, the gas swirls or eddies are short-lived, but the

*2.33 Optical and ultraviolet, false colours: green light (orange), violet light (green), ultraviolet (blue); Voyager 2 photo (above) and Voyager 1 (below)*

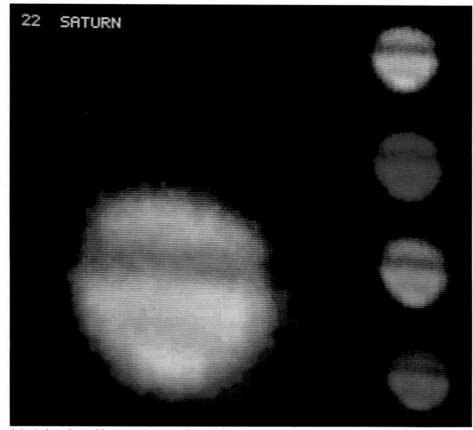

*2.34 Infrared, 10 (blue), 12 (green) and 20 (red) μm, 2.34 m Wyoming Infrared Telescope*

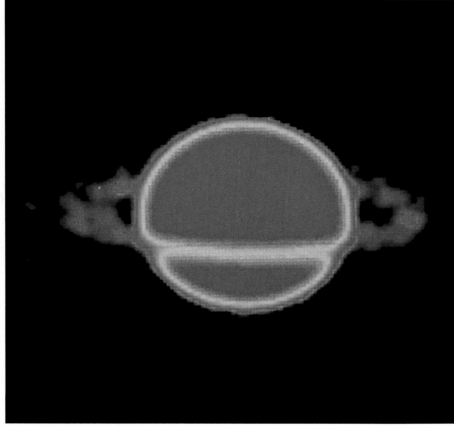

*2.35 Radio, 2 cm, Very Large Array*

ovals can survive for at least one year – the interval between the first and last photograph from the two Voyagers. The swirls seem to play the same role as Jupiter's eddies (Fig. 2.22) in driving strong winds around the planet. Despite Saturn's smaller size, it has winds far stronger than Jupiter: at the equator, they are four times faster, with a speed of 1700 kilometres per hour. The wind speed falls off to almost zero towards higher latitudes, but the bright (yellow) band near the poles is a jet stream with a speed of 400 kilometres per hour.

Long-wavelength views of Saturn show the planet's output of heat energy. Part of this is due to the Sun's heating, but, like Jupiter, Saturn has an internal source of heat. This produces as much heat as Saturn receives from the Sun. The infrared images (**Fig. 2.34**) are at three different wavelengths, as shown by the coloured images on the right: 10 microns (blue), 12 microns (green) and 20 microns (red). The combined image is seen at the top right, and to a larger scale as the main image.

The rings, too faint to show in these observations, were then tilted up so they would cross Saturn to the north of the equator, and the dark band crossing Saturn in these images is the shadow of the rings falling on the cloud tops. In this shaded region, the clouds are cooler than the average cloud-top temperature of 95 K. They emit only weakly at the shorter infrared wavelengths (blue and green), but still appear reasonably bright at the longer wavelength of 20 microns (red), and as a result the shadowed region is tinged red in the composite. The colour is otherwise fairly uniform over Saturn, showing that the planet's cloud tops have roughly the same temperature from equator to pole – unlike Venus (Fig. 2.19) for example. The slight changes in colour and intensity may reveal details of Saturn's interior heat distribution, or regions of differing height and thus temperature.

Short radio wavelengths also reveal Saturn's heat energy. The 2 centimetre view (**Fig. 2.35**) is colour-coded so that the brightest regions are red, and fainter parts yellow, green and blue. Around Saturn's bright globe, appearing red, Fig. 2.35 shows a fringe of weaker emission (yellow, green and blue) from its cooler outer atmosphere. But what is most striking is the radio appearance of the rings. They are blue and green to either side of the planet, where we detect them by their emission of heat energy. But the rings are cooler than Saturn itself. Where they run across in front of the planet, they absorb radiation coming from Saturn to appear as a yellow band, fainter than the red of Saturn's globe.

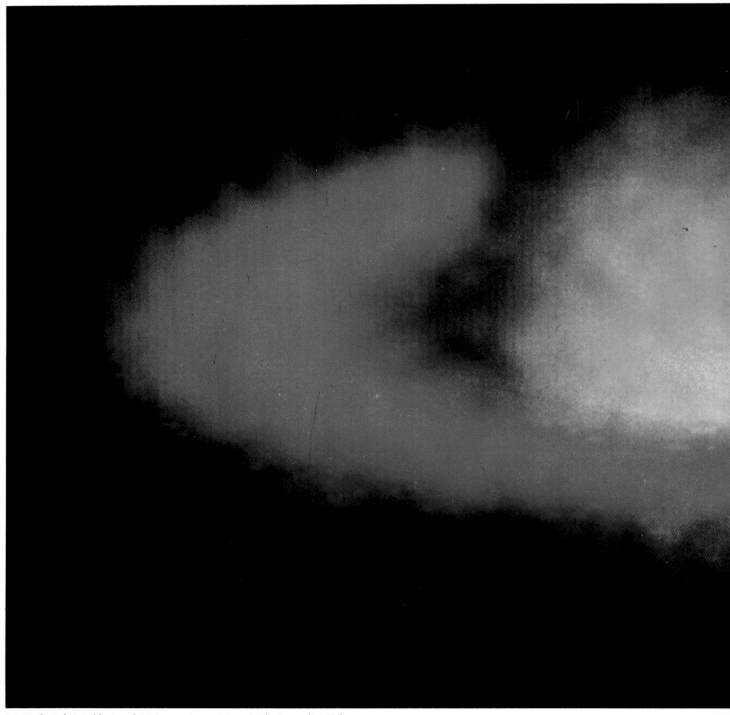

2.36 *Infrared, 3.8 (blue) and 4.8 (orange) μm, 3.9 m Anglo-Australian Telescope*

The striking composite picture (**Fig. 2.36**) reveals Saturn at near infrared wavelengths, with the rings reflecting sunlight and the planet glowing with its own heat.

The shorter wavelength image was made at a wavelength of 3.8 microns, and is coded blue. The ice blocks in the rings reflect 'sunlight' of this wavelength, and hence the ring system appears bright. But the methane gas in Saturn's atmosphere absorbs radiation with a wavelength of 3.8 microns, and the planet itself thus reflects no radiation of this wavelength. The methane also absorbs any heat radiation of this wavelength coming from within the planet, so Saturn's globe appears dark at 3.8 microns.

The 4.8 micron image, coded orange, is entirely different. The Sun emits relatively little radiation at this longer wavelength, so the icy rings appear dark. But Saturn's interior heat makes it shine brightly at 4.8 microns, and the planet thus appears prominently.

The supply of heat which makes Saturn glow in Fig. 2.36 is probably a result of rearrangements deep in its interior. Like Jupiter (Fig. 2.22), Saturn is made up largely of hydrogen and helium – substances with such a low density, even when compressed into a planet, that Saturn would float in water, given an ocean large enough! While Jupiter's interior heat comes from slow shinkage, originating from the time when the planets first formed some

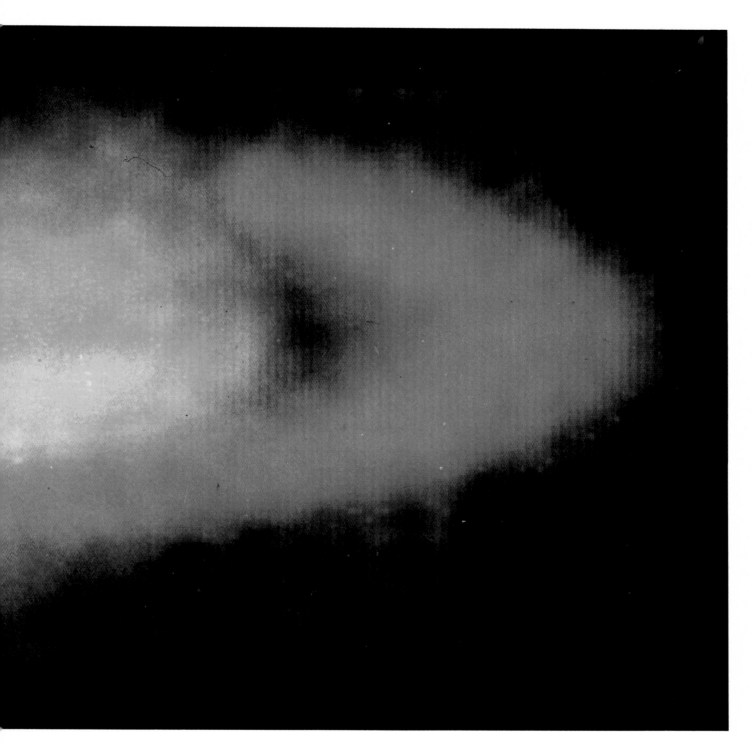

4600 million years ago, calculations show that the smaller Saturn should by now have finished its contraction and should have cooled down throughout.

Saturn's heat must come from a different process. At its formation, the hydrogen and helium atoms would have been well mixed together throughout the planet. But Saturn's core has now cooled to the point where the atoms would naturally separate, like oil and vinegar in a salad dressing. The less-abundant helium atoms should be gathering together as helium droplets within the surrounding hydrogen. Because helium is denser than hydrogen, these droplets fall towards the Saturn's centre, to make a helium core within the predominantly hydrogen planet. In their fall, the droplets lose gravitational energy, which is converted to heat.

This theory would mean that Saturn's outer layers should by now have been drained of about half their original helium. The Voyager spaceprobes have indeed confirmed this prediction. While Jupiter's atmosphere contains 11 atoms of helium to every 100 hydrogen atoms – like the Sun and gas clouds in space – Saturn has only 6 helium atoms to 100 atoms of hydrogen. Hence it seems that Saturn's infrared glow is indeed the result of helium rain in the planet's interior, dripping into a hidden central helium ocean larger than planet Earth.

# 3 | Optical Astronomy

Optical astronomers have built their biggest and best new telescopes far from civilisation and high above sea-level. The view of the skies from mountain peaks is unrivalled. Here above the turbulence, mists and clouds of the lower atmosphere, they can look up to a clear, black sky. The stars are steady, hard points of light, free from the incessant twinkling familiar at sea-level as starlight is capriciously deflected by air currents in the lower atmosphere.

Look through a telescope at sea-level and the 'twinkling' distorts the images of stars and planets, much like our view of an object seen through an undulating water surface. From a mountain peak, the same telescope shows us crisp, sharp, steady images. Photograph the sky from sea-level

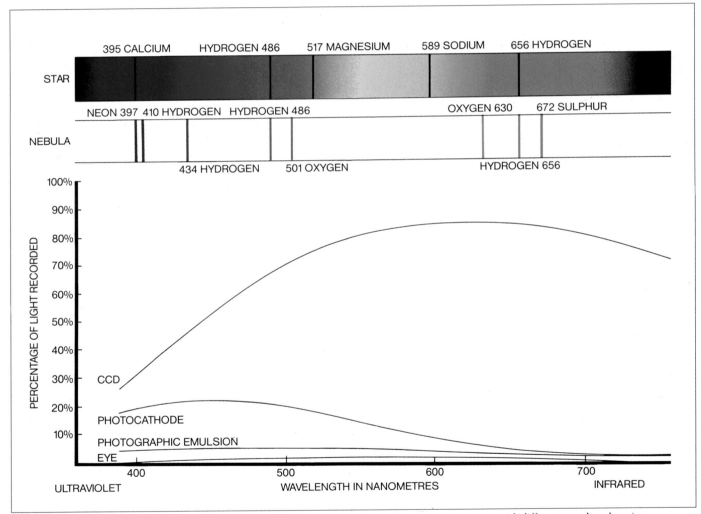

**Fig. 3.1** *Visible light from astronomical objects encompasses a range roughly from 400 to 700 nanometres, with different wavelength regions appearing to the eye as the various colours of the rainbow. Dense hot gas, as in a star, produces a continuous spectrum. More diffuse gas shows radiation of only a few particular wavelengths – spectral lines – characteristic of the common elements in the gas. A star's outer gases absorb the continuous spectrum, to give dark absorption lines. The gases of a nebula produce bright emission lines. The prominence of the different lines depends on the gas's temperature and density, as well as its composition. Astronomers have a choice of detectors to record light. The eye's performance is even poorer than it appears here, because it cannot build up a long-exposure image. Photographic emulsions are now being overtaken by photocathodes, the light sensitive electrodes of detectors like the Image Photon Counting System; and by the extremely sensitive CCDs (charge-coupled devices), which are light-recording silicon chips. Photocathodes are most efficient at the shorter wavelengths, and are also used in ultraviolet detectors like microchannel plates (Fig. 9.5). Unlike most light detectors, CCDs are particularly sensitive to longer wavelengths. They are not only providing detailed optical observations of faint objects, but also pioneering the study of the 'photographic infrared' wavelengths, up to 1100 nanometres (Fig. 5.1). The latter wavelengths have so far been studied only with relatively inefficient photographic plates, but they are an important extension of the visible spectrum because such long wavelength radiation is dimmed less by the dust in space.*

**Fig. 3.2** *The Canada–France–Hawaii Telescope is one of several instruments perched atop the 4200 metre peak of Mauna Kea in Hawaii. With a mirror measuring 3.6 metres in diameter, it is the seventh largest optical telescope in the world.*

and the 'twinkling' blurs all the stars into fuzzy blobs. Moreover, the lower atmosphere is contaminated with background light – radiation from artificial lights scattered by minute dust particles in the air. It makes the city nights so bright that the stars are swamped, and even in the countryside there is a slight background of scattered light. It may be too faint to see, but when astronomers take long-exposure photographs to record the faintest possible stars and galaxies, the background light eventually builds up to fog the photograph and obscure the faintest images.

So modern observatories are no longer built conveniently close to home. Instead, astronomers seek the best sites for astronomy. In the past, the world's major

observatories were generally built where they were easily accessible. For example, King Charles II of England founded the Royal Greenwich Observatory to assist navigators, and he placed it on a knoll overlooking the busy shipyards of Greenwich and Deptford on the River Thames, only a few miles from the royal court in London. The world's greatest lens telescope, the Yerkes refractor, was sited on the shores of a lake near Chicago in 1897. Its founder, G.E. Hale, was an astronomy professor at the University of Chicago and the telescope was financed by the local millionaire Charles Tyson Yerkes who had made his money on the Chicago trolley-car system.

These days, however, astronomers are prepared to fly

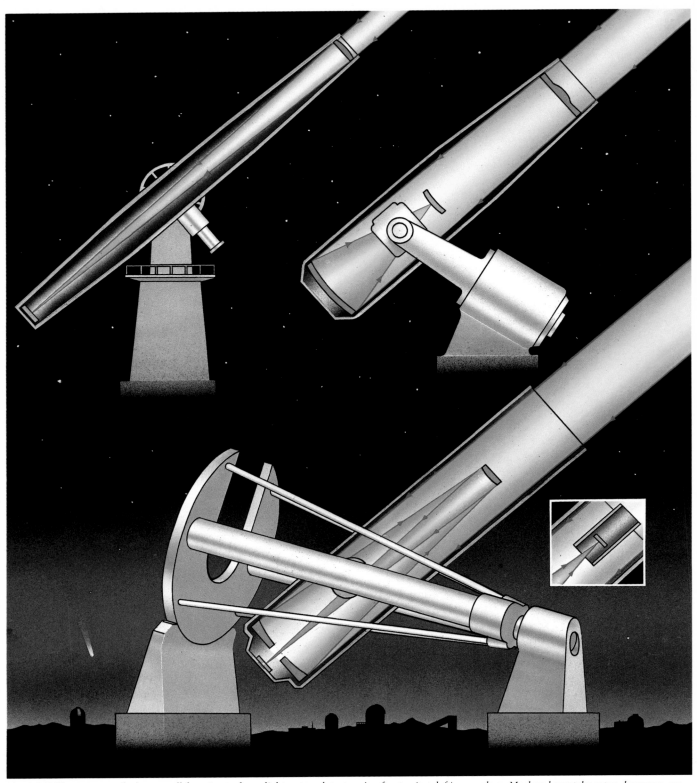

**Fig. 3.3** *Different types of telescope all function to focus light onto a detector. A refractor (top left) uses a lens. Modern large telescopes, however, are reflectors (lower left), focusing light with a curved mirror. The light can be detected at the prime focus, in a cage supported within the top of the telescope's framework tube. Alternatively, a secondary mirror placed just below this focus can reflect the light back through a hole in the centre of the main mirror, to the Cassegrain focus. A Schmidt telescope can 'see' a much wider region of sky. Its main mirror is part of a sphere (rather than a paraboloid) and a thin lens at the top corrects the distortions such a mirror would normally produce.*

Fig. 3.4 *The world's largest refracting telescope, the 1 metre instrument at the Yerkes Observatory, Wisconsin, completed in 1897.*

halfway round the world, and take a jeep up remote barren mountains to reach their telescopes. The wild heights of the Andes, for example, are superb sites for studying the southern regions of the sky. In 1976, astronomers from the United States opened what was then the world's third-largest telescope on the remote Chilean mountain of Cerro Tololo; and in the same year a consortium of European countries unveiled another monster telescope only a hundred kilometres away at Cerro La Silla. The United Kingdom is constructing its latest telescopes at a new international observatory on the peak of La Palma, in the Canary Islands.

One of the best observatory sites in the world is the huge extinct volcano Mauna Kea, which towers over the island of Hawaii. At its flat summit a huddle of telescope domes is growing amid a bleak lunar-type landscape of fractured lava, with no living plant or animal in sight. The University of Hawaii blazed the trail to this mountain peak; it now shares one of the world's largest telescopes here with Canada and France (Fig. 3.2). The American space agency NASA has a large telescope here and so do the British – both of them designed to study infrared 'light' (see Chapter 5).

The peak of Mauna Kea stands 4200 metres (14 000 feet) above sea-level. To take advantage of its unsurpassed view of the sky, astronomers have to face the problems of working at an altitude almost half the height of Mount Everest. The beaches of Hawaii may be a tropical paradise, but the peak of Mauna Kea is always cold. In winter the observatory is buffeted by blizzards with winds of over 100 kilometres per hour; the observatory operates Hawaii's only snowplough! The air is so dry that skin and lips are constantly chapped – and the extreme dryness can play such havoc with electronic equipment that astronomers must sometimes boil kettles beneath the equipment racks to keep the air moist. Worst of all, though, is altitude sickness. The first dome on Mauna Kea was provided with oxygen equipment, but it proved too cumbersome to use, and not actually essential. Most astronomers find they can work in the thin, oxygen-poor atmosphere – but only slowly. Thinking becomes a problem. It can take a quarter of an hour to correct an equipment fault whose solution might be obvious in a minute at sea-level.

However, the rewards of working on remote inhospitable mountain peaks far outweigh the disadvantages. There the huge sophisticated modern telescopes seek out the faint rays of light coming in from planets, stars and distant galaxies.

Telescopes can gather light in two different ways (Fig. 3.3). A *refracting telescope* is the type that springs to most people's minds. Like the seaman's telescope of Nelson's day, a refractor has a lens at the top end of a tube, which collects light and focuses it to form an image

at the lower end. The earliest telescopes were refractors. The most famous Renaissance scientist, Galileo Galilei, was the first to investigate the sky systematically with his little refractor – or 'optick tube' – which had a lens only 4 centimetres across. With this telescope he discovered the satellites of Jupiter and the phases of Venus, and resolved the Milky Way into stars.

Later refractors used larger and larger lenses, culminating in the huge Yerkes refractor (Fig. 3.4) near Chicago which has a lens slightly over a metre in diameter (40 inches). But there is a natural limit to the size to which you can build a refractor. The lens can be supported only at its edge, and a very large lens will sag out of shape under its own weight. The Yerkes lens weighs almost a quarter of a tonne, and is at just about the limit.

There is, however, another way to make a telescope, using a curved mirror (like a shaving mirror) to focus light to a focal point in front of it. This type of *reflecting telescope* has its mirror at the bottom of the tube. The tube – which may be just an open lattice work – supports some kind of device to intercept the focused light. In the very biggest telescopes, there is enough space for a small cabin at the focal point inside the telescope, where an astronomer can sit – making observations at the *prime focus*. More often, though, the tube is used to support a *secondary mirror* just below the focus. This mirror reflects the focused image to a point where it can be examined more easily. In the small 'backyard telescopes' used by many amateur astronomers, the secondary mirror is set at an angle so that the image is formed at the side of the tube,

*OPTICAL ASTRONOMY 43*

**Fig. 3.5** *Kitt Peak National Observatory, Arizona, is a national facility for US astronomers. It comprises 14 telescopes, some of which are operated by individual universities. The largest (centre) is a 3.8 metre reflector, the world's sixth biggest, opened in 1973. Kitt Peak also hosts the McMath Solar Telescope. This 1.5 metre telescope is the largest solar instrument in the world.*

and the astronomer looks into the side of the tube rather than through the end. This type is called a Newtonian reflector; it is the arrangement used by Sir Isaac Newton in the first reflecting telescope, which he produced in 1672.

The Newtonian arrangement is not very convenient for large telescopes, however, because it means that the observer still has to sit a long way up the tube. The most common type is the *Cassegrain reflector,* in which a curved secondary mirror reflects the image back down the telescope tube and through a hole in the centre of the main mirror to the observer. The astronomer thus sits behind the main mirror, looking up the telescope tube to the secondary mirror at a reflection of the sky reflected from the main mirror. (This design is now becoming popular with amateur astronomers too, and often there is a correcting lens at the top of the tube to make a *catadioptric telescope.*) Another alternative for large professional telescopes is to use a succession of mirrors to reflect the image down to spectrographs (see below) situated in rooms below the telescope dome.

Reflectors have one crucial advantage over refractors. A mirror can be supported over the whole of the back surface, and so it does not sag under its own weight. Sir William Herschel, the famous Prussian astronomer working in England, built a reflecting telescope as early as 1789 which was larger than the Yerkes refractor of a

century later. The first of the great modern reflectors was the 2.5 metre (100 inch) diameter telescope at Mount Wilson, California, completed in 1917. Although it was named the Hooker telescope after its financier, credit for the Mount Wilson instrument really belongs to the great American astronomer George Ellery Hale, who persuaded businessman John D. Hooker to fund it; it was Hale who had previously persuaded the Chicago businessman Yerkes to fund the world's largest refractor.

Hale was convinced that an even larger reflector could be built, and he applied his persuasive powers to the Rockefeller Foundation. In 1948, ten years after Hale's death, the project was complete. A 5 metre (200 inch) telescope named the Hale Telescope, was opened on Palomar Mountain in California.

Since then, the Hale reflector has remained the king of telescopes. In 1976, the Soviet Union completed a 6 metre telescope at the Zelenchukskaya Astrophysical Observatory in the Caucasus; but although it surpasses the Hale Telescope in sheer size the mirror is so inferior that the Russian giant is no match for the Californian telescope. Most telescopes built in the past thirty years have been about 4 metres in size, including a 3.8 metre at Kitt Peak in Arizona (Fig. 3.5) and the 3.9 metre Anglo-Australian Telescope in New South Wales (Figs. 3.7–3.9). This trend away from larger telescopes has come about because new ways of detecting the light that reaches telescopes have been developed.

Traditionally there have been two reasons for aiming towards the construction of larger telescopes. The first is the need to see more detail. If you look at the Moon through a backyard telescope, you can see a multitude of craters of all sizes. You can see finer detail by changing to a higher-power eyepiece, and it may seem that a succession of stronger eyepieces would result in higher and higher magnification, until you could see the footprints left behind by the Apollo astronauts. In fact, there is a limit to the sharpness of the image, the diffraction limit, which is set by the size of the telescope's lens or mirror. If you magnify too much, you do not see finer detail, but merely blow up the intrinsically fuzzy image to make it fuzzier – astronomers call this 'empty magnification'. The only way to see finer detail is to use a larger telescope – with a wider lens or mirror.

That is the theory, and it would certainly be entirely true for a telescope observing the sky from space. But from the Earth's surface we are looking up through the constantly shifting atmosphere, and the 'twinkling' from atmospheric turbulence produces a blurring of its own. Astronomers describe the resolving power of a telescope in terms of the size of the blurred image of a point source of light. This is called the *seeing disc*. The smaller the seeing disc, the finer the details the telescope can resolve. If it could be taken outside the Earth's atmosphere, the Hale Telescope would have a seeing disc, set by its 5 metre diameter, of only 0.02 arcseconds in diameter. Even in the best atmospheric conditions, however, the air above the telescope makes the image twinkle and wobble about. The human eye can follow some of this motion, and the seeing disc is perceived to be about 0.2 arcseconds, around ten times larger than the diffraction limit. It is far worse when the long-exposure photographic (or electronic) images are taken to study faint, distant stars, galaxies or quasars. The image's wobbling spreads the seeing disc to a couple of arcseconds or more. The best telescope sites in the world (like the observatories in Chile, La Palma and Hawaii) have seeing discs of about one arcsecond or a little smaller, which is the theoretical limit for a typical amateur astronomer's telescope with a 15 centimetre mirror!

The second reason for building large telescopes is simpler. Most of the interesting objects in the Universe lie so far away that they appear very faint in our skies. To study them, it is necessary to pick up and analyse as much as possible of their dim, scarce light, and by building a larger telescope, more light can be collected. The mirror of the 5 metre Hale Telescope, for example, has four times the area of its predecessor, the 2.5 metre at Mount Wilson, so its image of a star will be four times brighter. This cry for 'more light' has been the driving force behind the construction of bigger and bigger telescopes.

**Fig. 3.6** *An astronomer guides the 1.2 metre UK Schmidt Telescope at Siding Spring, Australia. This instrument has recently completed a thorough photographic survey of the southern sky.*

The traditional caricature of an astronomer has always been a white-haired, white-bearded old man peering myopically through his long telescope tube. For the past century, however, astronomers have only rarely looked through their telescopes. The human eye is simply not a very good light detector. It is sheer waste to put an astronomer's eye at the business end of an expensive, sophisticated telescope. The eye's main drawbacks are that it is not very sensitive to faint objects, and it cannot store the images it sees for analysis later; also the eye cannot measure brightnesses and positions of stars with any precision. For all serious astronomy, telescopes now pour their light onto a photographic plate – or, more recently, into an electronic detector. A modern telescope is less like Nelson's spyglass than a huge telephoto mirror system for a photographic camera or a sensitive television camera.

Photographic plates have long been the astronomer's mainstay. He can expose a plate one night, and then take as long as he likes to analyse the positions and brightnesses of stars, the shapes of galaxies, and so on. If he discovers an interesting star or quasar, he can search back through old 'archival' plates to discover how bright it has been in the past, when it may have been recorded accidentally on a plate taken for another purpose.

There is a special type of telescope designed specifically to photograph the sky. The Estonian optician Bernhardt Schmidt invented the type of wide-angle telescope now known as the *Schmidt telescope* (or Schmidt camera) in the 1930s. It is a strange cross between a reflector and a

Fig. 3.7 *The 3.9 metre Anglo-Australian Telescope (AAT) is the world's fifth largest reflector, with the most sophisticated and accurate computer control of any. The 'horseshoe' structure is part of the equatorial mounting which allows the telescope to track stars as the Earth rotates.*

Fig. 3.8 *The long-exposure photograph shows stars trailed into circles by the Earth's rotation, behind the AAT dome.*

refractor (Fig. 3.3): a mirror at the bottom reflects starlight onto a curved photographic plate halfway up the tube, as in an ordinary reflector. But normally the images of objects at the edge of such a wide-angle view would be severely distorted. Schmidt corrected this distortion by putting a thin, only slightly curved lens at the top of the telescope tube.

The first large Schmidt telescope – with a top lens 1.2 metres across – was installed at Palomar Mountain in 1948, to 'scout' for the Hale Telescope. Astronomers could search for interesting objects on the Schmidt plates, and then point the tunnel-visioned Hale giant to these particular objects. In the 1950s, the Palomar Schmidt was used to photograph systematically the whole of the sky visible from California. Each of its long-exposure plates recorded stars and galaxies as faint as magnitude 21 – a million times fainter than the dimmest star the eye can see. Each plate is 6° square – about twelve times the apparent width of the Moon. This Palomar *Sky Survey* gave for the first time a really detailed atlas of the sky to guide the new generation of giant telescopes.

Similar Schmidt telescopes stationed in the southern hemisphere have now surveyed the regions of the sky which cannot be seen from California. The work has been

Fig. 3.9 *David Malin, who produced many of this book's optical photographs, sits in the prime focus cage of the AAT (top right of Fig. 3.7). In front of him is the large photographic plate-holder.*

divided between the European Southern Observatory's 1 metre Schmidt telescope in Chile, and the slightly larger UK Schmidt (Fig. 3.6), a 1.2 metre telescope sited at the Anglo-Australian Observatory in New South Wales.

Schmidt telescopes have given us some of the most stunning photographs in modern astronomy. Their wide field of view can encompass huge vistas of glowing gas, revealing the intricate sinuous tendrils of bright gaseous nebulae crossed by the narrow bands and 'elephant trunks' of dark obscuring dust clouds. But there is a problem in reproducing astronomical photographs – whether from a Schmidt or from a conventional reflector. Modern plates can record outer faint tendrils of nebulosity which are ten thousand times dimmer than the nebula's bright central regions, and most of this brightness range is lost in printing. Either the faintest tendrils do not register, or the bright regions are overexposed.

One elegant solution is the technique of *unsharp masking*, used in many of the photographs of nebulae (and galaxies) reproduced in this book. The photographer uses the original plate – a negative – to produce a positive image on film, at the same scale as the original but without the same degree of contrast between light and dark. By putting this positive film mask over the original negative plate, the photographer effectively darkens the bright areas of the original negative image, and so decreases the range in contrast such that the faintest regions are only thirty times fainter than the centre of the nebula. This range in contrast will come out very clearly on ordinary photographic paper. In practice, astronomers like David Malin of the Anglo-Australian Observatory – a pioneer of the method – make the positive mask a little blurred ('unsharp'). This means that the mask and original plate do not have to be lined up quite as accurately, and it also brings out the finest details in the original plate in crisp relief (Fig. 3.10).

Faint details can also be revealed with the aid of computers. The brightness of images on a photographic plate are routinely measured by means of a microdensitometer – an instrument which scans a small spot of light back and forth over the image, and measures the amount of light transmitted, thus revealing the darkness of the plate at each point. The microdensitometer scans in a two-dimensional 'raster' pattern (like the electron beam in a television set); and if its output is connected to a simple computer the image can be reproduced by feeding the electrical message to the scanning beam of a television screen.

In its simplest form, such a system would reproduce the original black and white image. Since a TV screen has only a limited contrast range, this would be no more useful than taking a print from the original negative. But the computer can process the signals from the microdensitometer to show up faint details and minute changes in contrast, in

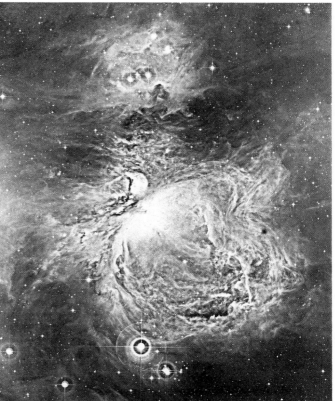

Fig. 3.10 *The Orion Nebula, photographed by David Malin using the UK Schmidt telescope, and his 'unsharp-masking' technique (described in the text).*

**Fig. 3.11** *The nearby spiral galaxy M83, photographed with the 3.8 metre Kitt Peak Telescope. The image has been slightly computer-enhanced to reveal details (such as the concentrations of hot young stars in the spiral arms), but the colours are natural.*

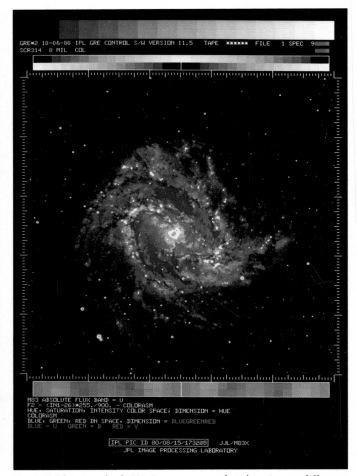

**Fig. 3.12** *Photograph of M83, image-processed so that views at different wavelengths appear in false colour. Red zones are regions of old stars and dust; green shows up ordinary stars; while blue areas are ultraviolet-emitting regions of very young stars.*

vivid eye-catching colour (Figs. 1.4–1.7).

The computer 'remembers' an image by storing a record of its brightness at each point, as measured by the scanning microdensitometer spot. Instead of reproducing this information as a shade of grey on the TV screen, colours can be assigned to each level of intensity: if the range of intensity were 100 to 1, the operator could assign red to the faint end of the range, say 1 to 25, yellow to the levels 26 to 50, green to 51 to 75 and blue to the brightest levels 76 to 100. The image on the screen is now multi-hued: all regions are the same brightness, but they differ in colour. The black of the background sky glows red; the most brilliant stars are blue – no brighter on the screen, but distinguished by a different hue – while regions of intermediate brightness form concentric green and yellow rings around the islands of blue.

This particular display would not tell the astronomer anything new. But in a search for very faint scraps of nebulosity, the four colours could be assigned differently, say to the levels 1, 2, 3 and 4, with the levels 5 to 100 black. Now all the bright images disappear from the screen as regions of uniform black: the background sky is still red, and the faintest wisps of nebulosity stand out in strident colour.

There are many variants on this theme, but the technique itself – *false-colour image-processing* – is

becoming more and more important to astronomers. It is part of an increasing interaction between the astronomer and the raw data, with a computer as 'marriage-broker'. Instead of straining their eyes peering at a photographic plate, modern astronomers can sit in front of a colour TV screen, and manipulate the image electronically (Figs. 3.11–3.12). If they want to know the brightness of a star, they can point to it on the screen and the computer will work it out. During coffee breaks, they can call up other programmes on the computer – space invaders or TV Rubik's cube!

False-colour images taken from black-and-white photographs can be beautiful as well as informative, as many of the pictures in this book show. But it is, of course, possible to take real colour photographs of the sky – and the Universe is full of colour. The hottest stars burn a fierce steely blue-white, while old cool stars glow orange or dull red. Nebulae come in a range of colours. Where their light comes mainly from hydrogen atoms, they shine a vivid crimson; where there is oxygen, the tint becomes apple green. Clouds of dust lit up by a star appear as sky-blue 'wraiths'.

Astronomers occasionally take colour photographs on ordinary colour film; but these emulsions are not designed for the very faint images and long exposures used in astronomy, and the colour balance can end up entirely

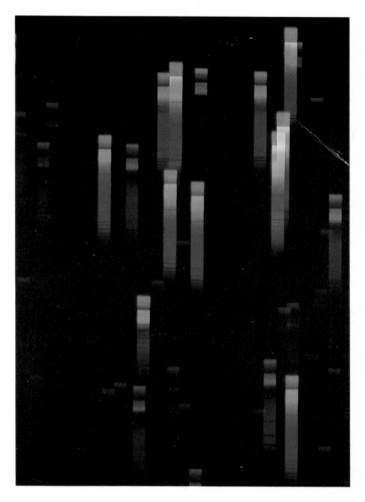

**Fig. 3.13** *Spectra of stars in the Hyades (Fig. 4.36), photographed through a prism which spreads out their light. Note the dark, absorption lines produced by different gases in the stars' atmospheres.*

wrong. Most of the spectacular true-colour photographs in this book are composites of three different black-and-white photographs, taken successively through filters which pass only red, green, and blue light. The colour photograph is built up at the printing stage. The three black-and-white negatives are printed in turn onto one piece of colour photographic paper. The first is illuminated with red light, the second with green and the third with blue light. The result is a true-colour photograph of the sky.

Astronomers generally use a colour filter anyway when photographing the sky for research, so they can be sure of exactly what colours (wavelengths) they are studying. How do you decide, for example, whether a particular bluish star is brighter or fainter than a neighbouring red star? Different black-and-white photographic emulsions would have different colour-responses from one another (and from the human eye), and so give conflicting results. The only answer is to observe through a filter which passes only one colour; and when comparing brightnesses, to stipulate which filter has been used.

The differences can be quite remarkable. The well-known constellation Orion (the hunter) boasts two of the dozen or so brightest stars in the sky, reddish Betelgeuse at the top left of the hunter's outline, and bluish-white Rigel at the lower right. But when seen (or photographed) through a blue filter (Fig. 4.2), Betelgeuse becomes one of the fainter stars in the constellation, its intense ruddy glow extinguished by the filter. Conversely, a red filter (Fig. 4.3) shows Betelgeuse dominating the constellation; it dims Rigel's blue fire until it is no rival. If our eyes had evolved to be more sensitive to red than blue (or vice versa), we would have a very different idea of the hierarchy of the brightest stars in the sky.

The colours we perceive are a crude representation of the wavelength of the light radiation (Fig. 3.1). Our eyes are sensitive to a range of wavelengths from 390 nanometres (blue-violet) to 700 nanometres (red), a total range of some 310 nanometres. The filters normally used in astronomy pass quite a wide range in wavelengths: the standard 'B' (blue) filter, for example, lets through all radiation with wavelengths between 385 and 485 nanometres. Such filters are very useful for studying stars, which do not produce radiation of a single wavelength but emit light over a wide range of wavelengths – the whole rainbow spectrum of colours. A gas produces a continuous spectrum of this kind when it is relatively dense, so that the atoms are comparatively close together. When atoms are relatively isolated from one another, however, as they are in a nebula, they emit light of just a few, very specific wavelengths. Each element produces its own particular wavelengths, which act as its spectral 'fingerprint'. Hydrogen emits intense red light at a wavelength of precisely 656.28 nanometres and also fainter blue-green

and blue light (at 486 and 434 nanometres). Oxygen produces intense green light at 496 and 501 nanometres wavelength.

An astronomer can take advantage of these specific wavelengths. If investigating faint nebulosities, a 'narrow band' filter can be put in front of the telescope to pass only light of a particular wavelength. A common ploy is to study the bright red hydrogen light with a filter that passes only light whose wavelength is within five nanometres of the theoretical wavelength of the hydrogen light. Such filters have revealed huge faint nebulosities filling the constellation of Orion, and hundreds of dim loops of gas in our neighbouring galaxy, the Large Magellanic Cloud (Fig. 8.18).

The techniques described so far all produce a two-dimensional *image* of some object in the sky. But this is only one way to investigate the Universe – and an image is of little use in the study of a star so distant that it appears as no more than a point of light in even the largest telescope. The second major spearhead is *spectroscopy*. Here the telescope is used to gather light from just one object – a star perhaps, or a small region of a nebula or galaxy. This light is passed into a *spectrograph*, which spreads it out as a spectrum of wavelengths – a band running from the blue-violet wavelengths at one end to red at the other, like a precise section through a rainbow. Within this band, the emissions from particular elements fall at their characteristic wavelengths, as *lines* cutting across the spectral colours (Fig. 3.1). The lines from the glowing gas of a nebula are always bright *emission lines*; but the lines in a star's spectrum are generally dark *absorption lines* – silhouettes against the star's bright continuous spectrum of colours (Fig. 3.13).

Spectroscopy thus can identify the elements that are present in a star. A detailed analysis of the intensities of the lines reveals the relative abundances of the different elements in the star, the star's temperature, and whether it has a strong magnetic field.

By studying the precise wavelengths of the lines in a spectrum, it is possible to determine the speed at which a star is travelling towards or away from us. If a star or galaxy is moving away from us, its radiation is slightly stretched out and all the wavelengths are very slightly longer than they should be; if it is coming towards us, the wavelengths are bunched together and shortened (Fig. 3.14). This distortion of wavelengths is known as the *Doppler effect*; it is exactly the same effect which raises the pitch of an ambulance siren as it comes towards us, and abruptly drops the pitch as the ambulance passes us and moves

away at speed. Most stars in our Galaxy travel at speeds of around 20 kilometres per second (70 000 kph) – very fast by everyday standards, but only enough to shift their spectral lines by less than a tenth of a nanometre. One very odd 'star' called SS 433 (Figs. 6.34–6.38), however, is ejecting streams of gas at 81 000 kilometres per second, so rapidly that the spectral lines appear in totally the wrong place in the spectrum – and they move up and down the spectrum as the gas streams change direction.

Beyond our Galaxy, the other galaxies show Doppler shifts in their spectra. All the distant galaxies are moving away from us, carried by the general expansion of the Universe so that the more distant ones are moving faster. Their motion away from us stretches their radiation to longer wavelengths, with the result that any particular spectral line is moved towards the red end of the spectrum, and generally the Doppler shift of a galaxy is called its *redshift*. Knowing how rapidly the Universe is expanding, astronomers can convert the measured redshift of a galaxy into its actual distance from us. For the most distant detectable objects of all, the exploding disrupted centres of

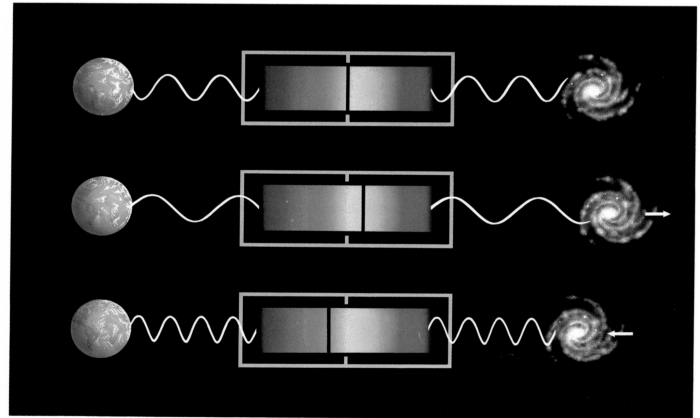

**Fig. 3.14** *Lines in the spectra of stars and galaxies reveal how fast they are moving towards or away from us. At the top, the galaxy is at rest. If a galaxy is receding (middle), its wavelengths are 'stretched' so that its spectral lines are shifted towards the red end. The approaching galaxy's wavelengths (bottom) are compressed, shifting its lines towards the blue.*

far-off galaxies which are called quasars, the redshift is the only guide we have to their distances.

Originally, spectra were recorded on photographic plates; and the wavelengths and intensities of spectral lines were measured with travelling microscopes and microdensitometers. But the past two decades have seen a revolution in astronomy. Electronic devices can now measure spectra directly, with high precision and sensitivity.

Researchers started using electronic light detectors soon after the Second World War. The first were *photomultipliers*, simple devices which could measure the brightness of a star very accurately, but could do little more. They have been succeeded by various types of electronic *image-tube*. Light falls onto a sensitive metal surface in an evacuated glass tube, and it ejects electrons from the surface at a rate which depends on the light intensity at each point. Electric fields speed up these electrons to a high energy, and magnetic fields focus them to form an image at the other end of the tube, in energetic electrons rather than weak photons. The electron image can be recorded either by a phosphor screen – like an ordinary TV screen – or by a photographic emulsion which is sensitive to electrons rather than light.

Early image-tubes were small, so they did not rival photographic plates when it came to taking direct pictures of the sky; but their accuracy and sensitivity to faint light levels made them ideal for recording the spread-out light in the spectrum of a dim star, galaxy or quasar.

The most spectacularly successful image-tube detector (Figs. 3.15–3.16) is the Image Photon Counting System (IPCS) developed by the British astronomer Alec Boksenberg. Foremost among its many achievements, the IPCS has shown fine structure in the spectra of distant quasars, disclosing details of the immense explosions there, and also revealing that the light has passed through many intervening galaxies on its way to us. After this epic journey, the light is captured by one of the large new reflecting telescopes, and focused into a spectroscope which marshals the radiation into a spectrum. This spectrum falls onto the first part of the IPCS, an ordinary image-tube which produces an image of the spectrum a million times brighter on a phosphor screen. The image on an image-tube screen, however, always suffers from unwanted bright specks – like 'snow' on a TV screen – and the IPCS is unique in being able to 'clean up' the image. The intensified image is viewed by an ordinary TV camera, backed by a computer which analyses every 'blip' on the screen and decides whether it is the right shape and intensity to be caused by incoming light, or is simply 'snow'. The computer rejects all the 'snow', and draws up a true spectrum from the rest of the image.

The IPCS can also record direct images of the sky – like a photograph – and Boksenberg's team used it to discover

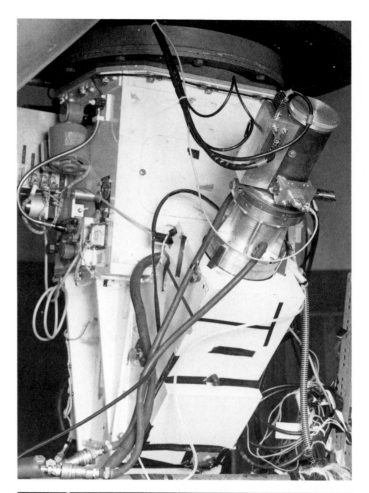

**Figs. 3.15–3.16** *The IPCS detector seen in position on a telescope* (**Fig. 3.15**). *The large white box bolted directly to the mirror (top) is a spectrograph, and the IPCS is attached at the front. Light is first intensified by the image tube (white section, lower right) before being scanned by a computer-interfaced TV camera (metallic casing, top). The camera 'sees' an output (**Fig. 3.16**) which is a mixture of real 'photon events' and electrical noise. Computer analysis removes the noise.*

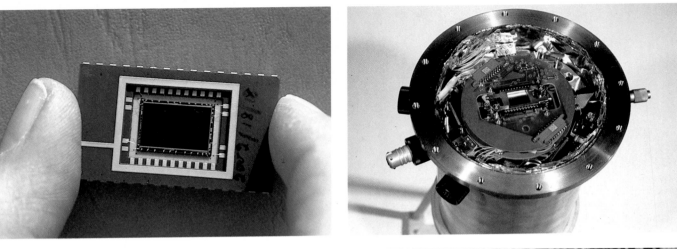

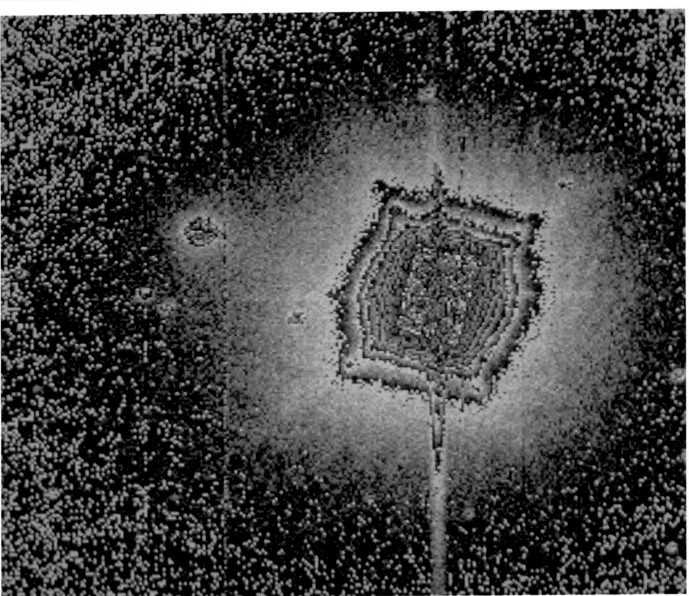

**Figs. 3.17–3.19** *The photograph of the 'Red Rectangle' (**Fig. 3.19**), in the wavelength of red light, was obtained with a CCD camera. The 'Rectangle' is a region of very faint nebulosity surrounding a bright star, and the camera's wide dynamic range reveals the faint extensions, colour coded here for intensity, without contamination by the star's light. At the camera's heart is a minute CCD 'chip' (**Fig. 3.17**). It has 221 760 pixels each 0.02 millimetres square, arranged in a matrix 385 down by 576 across.* **Fig. 3.18** *shows a similar chip in position in the specially cooled CCD camera jacket. This camera was attached to the Anglo-Australian Telescope for the photograph in Fig. 3.19, which required an exposure of only 3¹/₃ minutes.*

the faintest pulsar that produces light, the Vela pulsar (Fig. 6.32). But for direct imaging the IPCS is being challenged by a newer development, the *charge-coupled device* or CCD (Figs. 3.17–3.19). This is really just an oversized silicon chip, of the kind used these days in everything from pocket calculators to programmable washing machines. Instead of incorporating miniature transistors and resistors, the CCD is an array of small light-sensitive regions called *pixels*. The image from a large modern telescope is focused onto the silicon slab, which is only a couple of centimetres across, and builds up a charge in each light sensitive pixel proportional to the brightness of light falling on it. When the 'exposure' is complete, the charges can be read off into a computer memory.

The output from the IPCS or a CCD is stored ready in a computer memory, so astronomers can directly apply to it all the display tricks of false-colour imagery. The latest CCDs have 640 000 pixels, arranged in 800 columns each containing 800 pixels. They can register a view of the sky in finer detail than a standard British television set (with 625 lines). They are ten times more sensitive than astronomical photographic plates in the blue to yellow parts of the spectrum, and compare even better in the red and the 'photographic infrared', which is of slightly longer wavelength (Fig. 3.1).

The present generation of telescopes, somewhat smaller than the great 5 metre Hale Telescope, is a direct result of these amazing advances in light-detectors. When it was completed in 1948, the Hale Telescope was using photographic plates which only recorded one three-hundredth of the light falling on it. Photographic plates have been improved to detect about one-thirtieth of the available light; but the latest CCDs respond to three-quarters of the light falling upon them – two hundred times more sensitive than the photographic plates of the 1940s. Hale had been building larger telescopes to gather more light – and the available 'detectors' had been throwing away over 99 per cent of it. He designed the 5 metre telescope to be four times more sensitive than the 2.5 metre Mount Wilson telescope; yet if he had been presented with a modern CCD, he could have achieved the same result with a 0.4 metre telescope – the size of many amateur astronomers' instruments!

But in the 1980s we are beginning to reach the limit in detector sensitivity. Already CCDs can detect 75 per cent of the light falling on them; even if we could increase their sensitivity to the theoretical limits of 100 per cent it would be only a small improvement in light grasp. The great step from 'around one per cent' to 'around 100 per cent' has already been made. As a result, astronomers are now once again dreaming of larger telescopes.

The University of Texas is currently the front-runner, with plans for a 7.6 metre (300 inch) leviathan which would surpass both the Hale Telescope and the Soviet 6

metre. It is very difficult – and extremely expensive – to make a single mirror much larger than this, and telescope designers have other schemes in mind. The University of California is designing a 10 metre telescope, with a mirror made up of 36 smaller hexagonal segments, fitted together like bathroom tiles.

The alternative is a 'multiple aperture telescope': several ordinary large mirrors held in the same framework, with secondary mirrors which reflect the light from each to the same focal point. The United States already has such an instrument, the futuristic looking Multiple Mirror Telescope on Mount Hopkins in Arizona (Fig. 3.20). It has six mirrors, each 1.8 metres in diameter; together they have the same area as a single mirror 4.5 metres in diameter, so the MMT is effectively the world's third-largest telescope at present. The design team sees no problems in constructing an enlarged MMT: with eight mirrors, each 5 metres across, it would be effectively a 14 metre telescope.

Although these larger telescopes will all collect more light, they will eventually suffer from 'fogging' by stray light in the atmosphere during very long exposures – photographic or electronic. And they will not be able to see finer details, because the seeing disc will still be determined by our atmosphere's 'twinkling'. There is a way around both these problems: the revolutionary plan to launch the Space Telescope (Fig. 3.22).

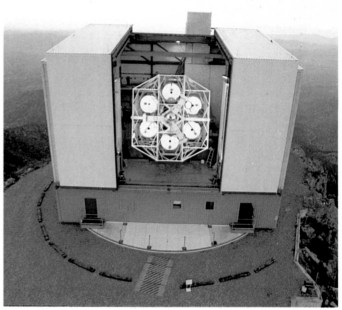

Fig. 3.20 *The Multiple Mirror Telescope on Mount Hopkins, Arizona, is the world's third largest. Its six computer controlled 1.8 metre diameter mirrors are equivalent to a single mirror 4.5 metres across.*

A telescope orbiting the Earth has the blackest skies of any observatory. Without a turbulent atmosphere above, it can focus details right down to the theoretical limit for the size of its mirror – and when looking at compact objects like stars or quasars, the extra concentration of light increases the contrast with the sky, and so allows the telescope to 'see' even fainter objects.

The Space Telescope – funded 85 per cent by the United States and 15 per cent by the European Space Agency – is well under way, and should be launched in 1986. It has a main mirror 2.4 metres across (Fig. 3.21), smaller than the fifteen or so largest ground-based telescopes, but its unique observing position above the atmosphere means that it will be able to detect stars, galaxies and quasars fifty times fainter than those visible from Earth. It is another breakthrough in the quest for 'more light'. In addition, the telescope is capable of seeing finer details. A 2.4 metre mirror – without atmospheric blurring – can see details as fine as 0.06 arcseconds, more than ten times smaller than the finest details visible from ground-based telescopes. To put it another way, the Space Telescope can look at what is a minute blur to a ground-based telescope, and separate it out into a hundred stars.

Optical astronomy may be the oldest branch of astronomical research – with its roots stretching back thousands of years into prehistory – but it has certainly

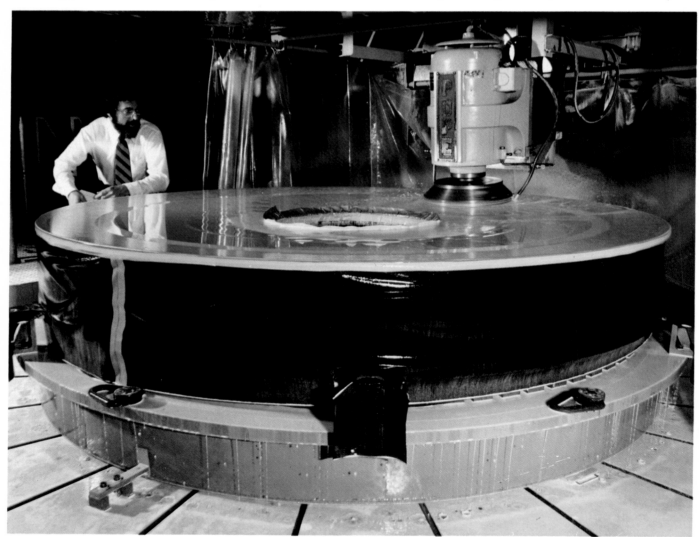

**Fig. 3.21** *Cutting operations commence on the 2.4 metre diameter glass disc which will become the mirror of the Space Telescope. The 0.5 metre cutting wheel (right) will remove over 130 kilograms of glass from the disc before it acquires its correct degree of curvature. The mirror will then be finely polished and coated before it is positioned in the telescope.*

moved with the times. The new telescopes need precision engineering and computer control; the new light-detectors are amongst the most complex new pieces of solid-state electronics; the analysis of observations again involves the latest in computer technology; and the Space Telescope is more precise than any previous instrument that has been blasted aloft by the brute power of rocket engines.

Optical astronomy is not the only 'astronomy', as it was until the 1930s; but it is not in danger of succumbing to competition from the others. All have their part to play. Optical astronomers have the advantage that their telescopes (apart from the Space Telescope) are sited on the Earth. Although it may be quite a journey to Chile, it raises less problems than controlling an X-ray or ultraviolet telescope in space. Also Earth-based telescopes are cheap. You can build a dozen world-class telescopes of around 3.5 metres in size for the cost of a single major X-ray observatory in space.

Many youngsters are spurred to a career in astronomy by the thrill of seeing a dark star-studded sky. And although, as professional astronomers, the naked eye or small telescope may be superseded by huge steel and glass reflectors with a silicon chip 'retina', there remains much knowledge to be gleaned from the optical radiation which reaches us from space.

**Fig. 3.22** *Artist's impression of the Space Telescope, due to be carried into orbit aboard the Shuttle (lower left) in the late 1980s. Instruments on board will be powered by solar panels (left and right of cylinder). The telescope will transmit data to Earth via a Tracking and Data Relay Satellite (top right), whose high orbit will always allow links with the ground station (lower right).*

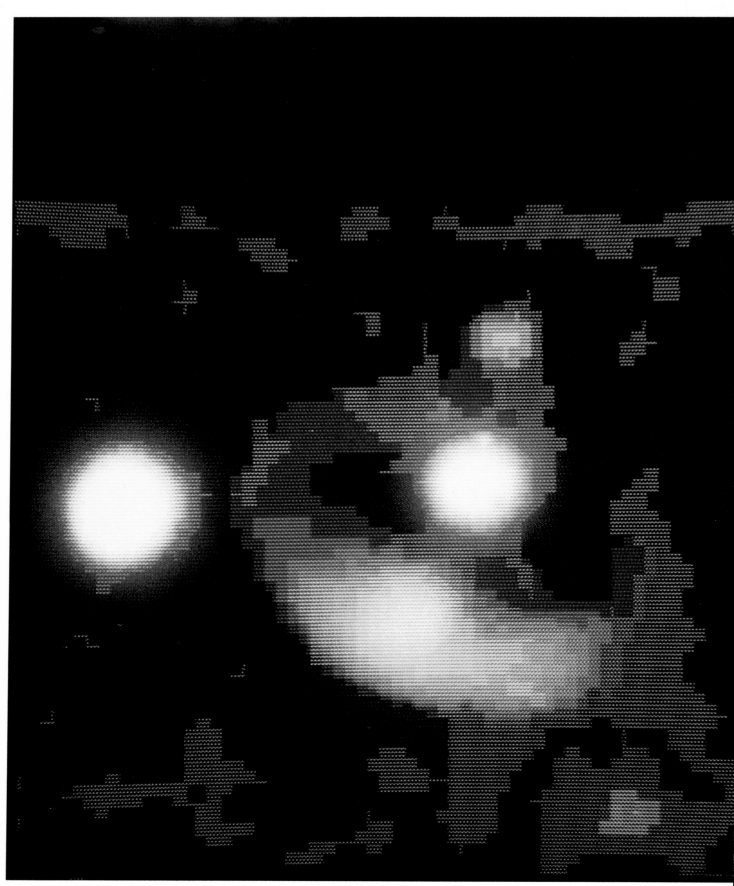

**4.1** *Monoceros R2, infrared, 10 (green) and 20 (red) μm, 2.34 m Wyoming Infrared Telescope*

# 4 Starbirth

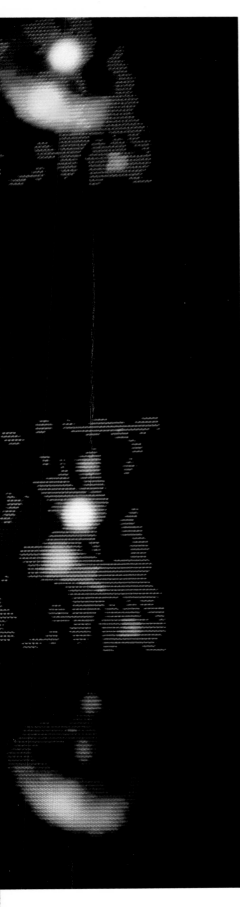

Our eyes can detect about 6 000 stars while large optical telescopes show us thousands of millions of fainter and more distant stars. These stars lie at different distances; they differ from one another in mass, size, brightness and temperature. But almost all have one thing in common: they are in the prime of life.

There must, however, be some stars forming and others dying (Chapter 6). At the present time this happens at a rate of about ten stars born and ten dying in our Galaxy each year. New stars are born when the very tenuous gas in between the existing stars is compressed, either by travelling through one of the Galaxy's spiral arms or by the shock wave from the explosion of a supernova. But it is very difficult to see star birth in progress. The interstellar gas contains small grains of dust, like cosmic soot, and where the gas is compressed the dust builds up until it almost totally blocks the light coming from the 'star nursery' within. The dust absorbs and scatters light very effectively because the grains are about the same size as the wavelength of visible light – a few hundred nanometres. The longer wavelengths of infrared and radio, however, can penetrate the dust quite easily, and reveal star formation within the clouds.

Radio waves show up various stages of the birth process, in 'lines' of particular wavelengths from molecules. These groups of atoms can only exist in dense, dusty clouds where they are protected from disruptive ultraviolet radiation, so surveys of the sky for radio emission from molecules reveal the dense clouds – *molecular clouds* – where stars are about to form. Stars which have already formed in such clouds disturb the gas around, and its motion shows up as doppler shifts in the molecules' spectral lines; while the infrared radiation from young stars can make molecules in the surrounding gas shine more brilliantly, as *masers* (the radio equivalent of lasers). The hot gas around new stars, at a temperature of some 10 000 K, produces radio emission of all wavelengths.

Infrared telescopes, however, probably bring us closest to seeing star birth as it occurs. In the collapsing cloud, the temperature is only 10 K, and the dust in it emits infrared of about 300 microns wavelength. As concentrations of dust shrink in size, they should become warmer.

When the central temperature reaches 10 000 000 K, nuclear reactions begin and the *protostar* becomes a star, shining with a surface temperature of a few thousand degrees. So a protostar should have an intermediate surface temperature, of a hundred to a few hundred degrees, and radiate at infrared wavelengths of 10 to 20 microns.

Early infrared astronomers found such sources inside molecular clouds, and thought they had discovered protostars. The sources brightest at 10 microns have, however, turned out to be stars which have already formed, but so enveloped in dust that even their powerful short infrared output cannot escape. And many of the '20 micron' sources are just dense patches of dust within the cloud, warmed by a neighbouring newly born star. Astronomers have yet to find a true protostar.

The infrared view of a star formation region (**Fig. 4.1**) shows the two kinds of source. Lying almost 3 000 light years away, Monoceros R2 appears in the sky next to the Orion region of star formation (which is rather closer at only 1 600 light years). The stars here have condensed from a dense molecular cloud, which appears in views of the region made at the radio wavelength emitted by carbon monoxide molecules (Figs. 4.4 and 4.5), and at gamma ray wavelengths (Fig. 4.6).

On the optical photograph, Fig. 4.3, it would be visible just within the lower-left corner, but the dust dims it entirely. The US Air Force's infrared survey of the sky, however, found it to be one of the twelve brightest nebulae when observed at 20 microns wavelength. Fig. 4.1 is a detailed view of the innermost 1 arcminute of Monoceros R2 at wavelengths of 10 microns (coded green) and 20 microns (red). The separate images are shown at bottom right, and combined at top right and in the enlarged main image. Two young stars appear as the round whitish-coloured images. The right-hand star emits more at the shorter wavelength and appears greener in the composite. Radiation from the young stars is heating a surrounding shell of gas and dust (orange) to a temperature of 150–200 K, making it shine brightly at a wavelength of around 20 microns.

# Orion

The stars of Orion (the hunter) make up the most brilliant of all the constellations, containing one-tenth of the seventy brightest stars in the sky. Bright red Betelgeuse forms one of the hunter's shoulders; below, a line of three stars makes up his Belt, with a nebulous Sword hanging from it; and two lower stars make the bottom of his tunic – the one on the right being Rigel, the seventh brightest star in the sky. The figure of Orion shows clearly on the blue-filtered photograph (**Fig. 4.2**), although the filter dims the red star Betelgeuse.

In the direction of Orion, we are looking along one of the Galaxy's spiral arms – or to be accurate, a spur off one of the major spiral arms (Fig 1.9). Spiral arms are the major sites for star formation, so all along our line of sight through Orion we see young stars, including the hot, luminous stars which have only a short lifetime. (Betelgeuse is an exception, being a red giant star nearing the end of its life.) These O and B type stars have temperatures of 20 000 K and above, and they shine bluish-white.

Brilliant Rigel, 800 light years away, is the nearest of Orion's B stars. Twice as far away, we come to a group which is extremely young. These stars of the Orion OB Association include the stars of the Sword, the Belt and the fainter stars surrounding them and extending to the upper right of the Belt. The stars have condensed from a single large gas cloud. Judging by the stars' ages, star formation started at the top right of the cloud some 12 million years ago, and has gradually spread along it. The stars of the Belt region formed 8 million years ago, and the youngest in the Sword are only 2 million years old.

As stars begin to shine, they heat up and disperse the residual gas around them. The gas reaches a temperature of 10 000 K, and the hydrogen atoms glow at particular wavelengths of the spectrum. The strongest line in the visible spectrum is hydrogen-alpha, which is red light at a wavelength of 656 nanometres. The photograph taken through a red filter (**Fig. 4.3**) thus not only emphasises the few red stars like Betelgeuse, but brings out the hot gas as clouds and rings of nebulosity. The great arc of glowing gas to the left of Orion was discovered by the American optical astronomer E. E. Barnard in 1894 and is known as Barnard's Loop; it is 300 light years across and has been expanding outwards for 2 million years.

**4.2** *Optical, blue light, 8 cm camera lens, Harvard College Observatory at Agassiz*

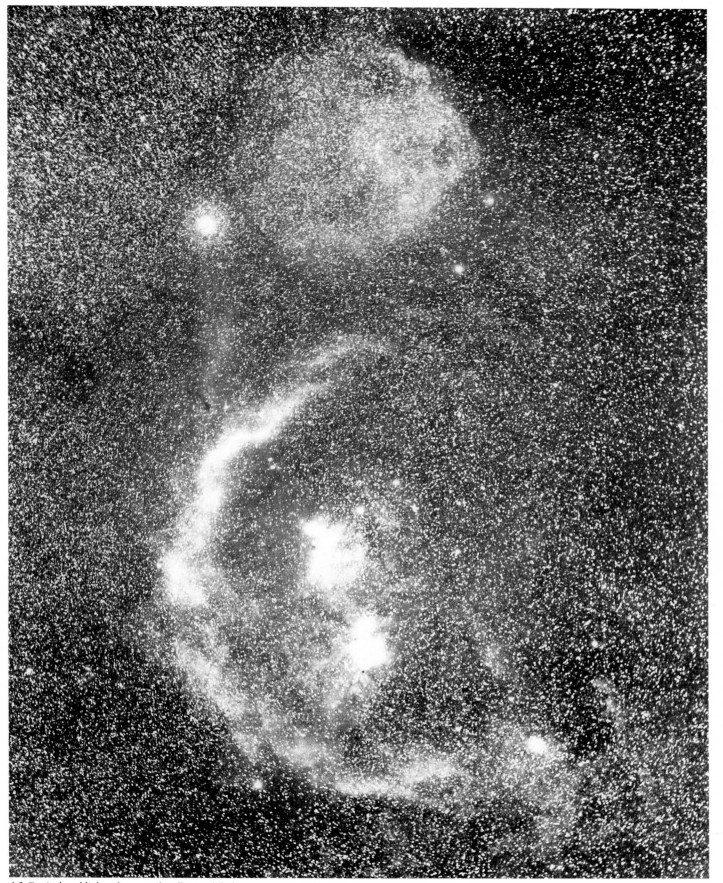

**4.3** *Optical, red light, photographically-amplified, camera mounted piggy back on 0.4 m telescope, Siding Spring*

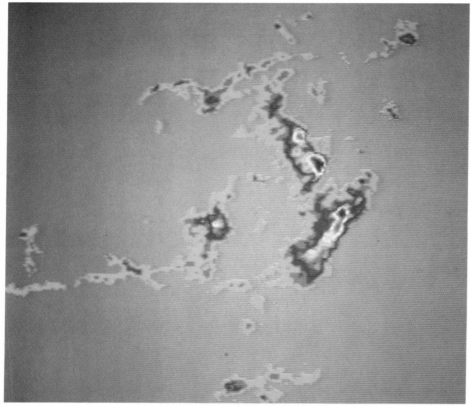

**4.4** *Radio, 2.6 mm carbon monoxide line, 1.22 m Sky Survey Telescope, Columbia University*

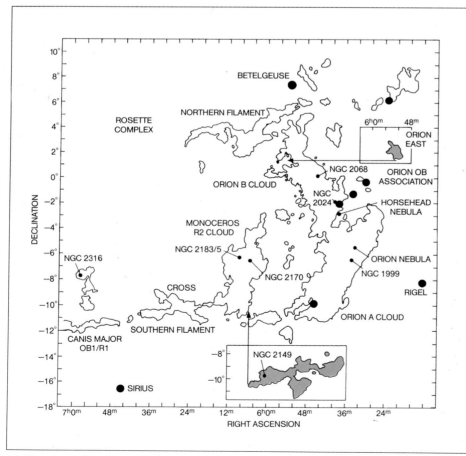

**4.5** *Diagram, carbon monoxide features in Orion (displaced features coloured green)*

Orion contains dense, cold gas clouds which have yet to condense into stars, and these are seen dramatically in **Fig. 4.4**, made with a radio telescope tuned to the wavelength emitted by carbon monoxide molecules. Here the background 'dark' sky is coded turquoise; feebly-emitting regions of the clouds are pink, and successively more intense regions blue, yellow and red. The large area of sky mapped is shown in **Fig. 4.5**, where the brightest stars have been superimposed. (The two regions shaded green in Fig. 4.5 have been moved from their indicated positions for clarity: their velocities show they lie at a different distance from the clouds they would coincide with here.)

Most of the mass in these clouds is in the form of hydrogen molecules, with the tell-tale carbon monoxide as only a minor constituent. The two most massive clouds (A and B in Fig. 4.5) are each some 150 light years long, and contain enough matter to form over 100 000 new stars. There are also many previously unknown clouds, including two peculiar straight filaments (lower left and across top) which probably lie slightly behind the Orion clouds and are 500 light years long but only 20 light years thick.

The gas atoms in space emit gamma rays when struck by high-speed cosmic ray particles, and so the intensity of gamma ray emission (**Fig. 4.6**) also indicates the density of the invisible clouds in Orion. Fig. 4.6 covers a similar area to Fig. 4.4; here the most intense regions are black and the background white. Despite the different emission process, and a wavelength a million million times shorter, the gamma ray map shows a distribution of gas clouds very similar to that seen in the carbon monoxide picture.

In many places, the gas is turning into stars. Some new star clusters are marked in Fig. 4.5, for example the visible cluster of O and B stars Canis Major OB1, and the hidden infrared cluster Monoceros R2 (shown in detail in Fig. 4.1). In other cases,

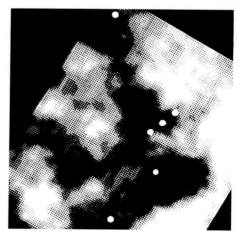

**4.6** *Gamma-ray, 0.000 000 2–0.000 01 nm, negative print, COS-B satellite*

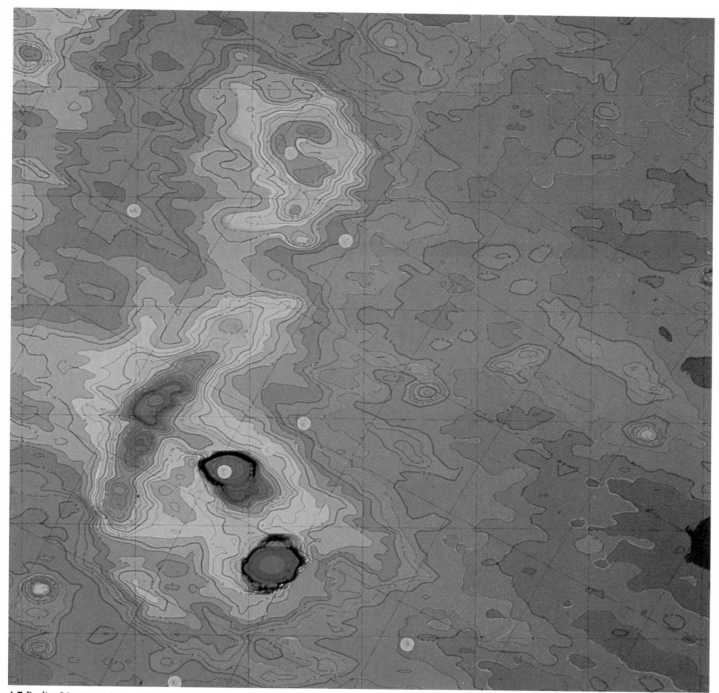

**4.7** *Radio, 21 cm continuum, 25 m Stockert Telescope, Bonn University*

the hot gas nebulae lit up by the stars are most conspicuous, and are shown by their NGC numbers.

The main region of star formation has been at the top right, producing the Orion OB association (Fig. 4.2), and is now continuing where the gas is densest at the right-hand end of the Orion A and B clouds – the region of the Orion Nebula, and of the Horsehead Nebula and its bright neighbour NGC 2024. The very massive, brilliant stars here are lighting up the less dense gas around, and dispersing it into space. The hot gas shows up in optical

photographs taken through a red filter (Fig. 4.3); and it also emits radio waves powerfully, over the whole range of wavelengths.

The radio map (**Fig. 4.7**) was made at a wavelength of 21 centimetres, but not looking specifically at the hydrogen line. The background sky is blue, and regions of successively brighter radio emission are coded yellow, orange and red. Orion's main stars are again superimposed. (Note the background brightening towards the Milky Way band at the top left.) Hot gases in the Orion Nebula and Horsehead region

are the two brilliant sources here; but fainter emission comes from the huge semicircle of Barnard's Loop (centre) and the ring of gases expelled from young stars centred on lambda Orionis (top) – the star marking Orion's 'head'. Fig. 4.7 is very similar to the red light Fig. 4.3, but the radio map has two advantages: it does not suffer from a confusing number of star images, and radio sources cannot be concealed from us by dust in space.

**4.8** *Optical, true colour, 1.2 m UK Schmidt Telescope*

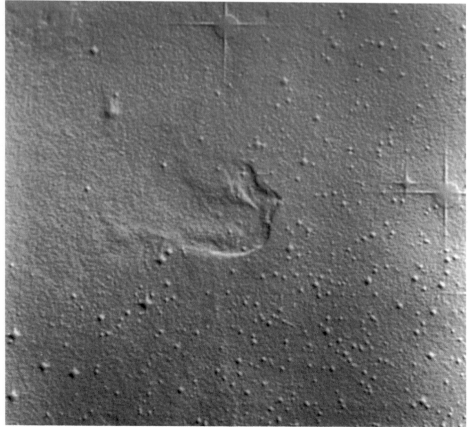

# Horsehead Nebula

Silhouetted against the bright nebulosities of Orion is one of the most famous dark nebulae: the Horsehead Nebula. Its environment is shown in the colour picture (**Fig. 4.8**). The brilliant blue star is Alnitak, the left-hand of Orion's three belt stars. It is an extremely hot, supergiant star, with a temperature of 30 000 K, shining as brightly as 15 000 Suns. The other bright blue star (lower right), sigma Orionis, lies at the same distance of 1 400 light years. It is at about the same temperature, but is only one-fifth as bright; it is a normal hydrogen-burning (main sequence) star rather than a supergiant. The photograph covers 2° of sky (four Moon-breadths) North to South — about one-half the separation of Alnitak from the Orion Nebula.

This region is the borderline between young stars (to the right) and the cool dark molecular cloud (to the left), as delineated by the carbon monoxide and gamma ray emission from the cloud (Figs. 4.4–4.6). The division shows clearly in the colour photograph. The dust in the dark cloud on the left is obscuring all the stars behind; the few stars on this side of the picture lie in front. On the right, we can see many more stars lying beyond.

Within the dark cloud, dense accumulations of gas are turning into stars, and lighting up the gas around. The bright nebula just to the left of Alnitak is NGC 2024: this cloud of hot gas — bisected in the optical picture (Fig. 4.8) by a foreground dust lane — generates most of the radio emission from this region, as seen in the Bonn radio map (Fig. 4.7). The large, faintly glowing nebula stretching down from Alnitak is catalogued as IC 434. It is 30 light years in total size, and radio measurements show that it contains as much glowing gas as 250 Suns. IC 434 is itself neither interesting nor spectacular. What draws our attention to it is the distinctive protuberance from the dark cloud to its left, appearing from our viewpoint as a horse's head in silhouette.

The Horsehead Nebula looks small in Fig. 4.8, but in reality it measures 3 light years from nose to mane — almost the distance from the Sun to the nearest star. Its intricate structure is revealed in the more-detailed black-and-white photograph (**Fig. 4.9**).

Astronomers at the Royal Greenwich Observatory have computer processed the central regions of Fig. 4.8, 'raising' bright regions and 'lowering' dark regions to create the strange three-dimensional bas-relief (**Fig. 4.10**). Small peaks (some with spikes caused by the telescope) are stars, while the dark Horsehead creates a sunken hollow.

**4.9** *Optical, red light, 3.6 m telescope, European Southern Observatory*

**4.10** *Optical, bas-relief representation, 1.2 m UK Schmidt Telescope*

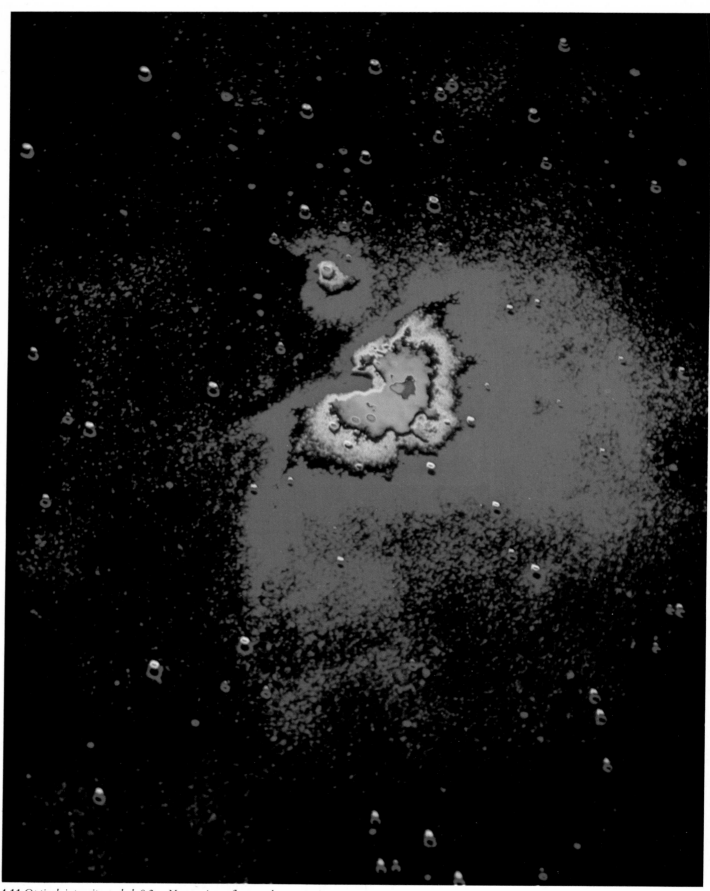

**4.11** *Optical, intensity-coded, 0.2 m Newtonian reflector telescope*

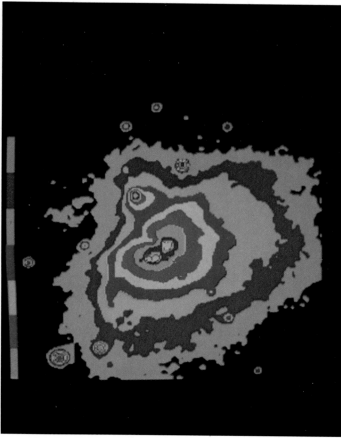

4.12 *Optical, true colour, 3 m telescope, Lick Observatory*

4.13 *Ultraviolet, 170–190 nm, 0.3 m rocket-borne telescope*

# Orion Nebula

The fan-shaped traceries of the Orion Nebula consist of hot gases spread out over 15 light years of space. Although the Nebula lies 1600 light years away, it is large enough to appear the size of the full Moon (30 arcminutes across) in our skies.

The Orion Nebula is also called M42, as the forty-second entry in Charles Messier's 1784 catalogue of nebulous objects. The small, round nebula just above is known as M43. Both nebulae are lit up by ultraviolet radiation from hot, newly-born stars at their centres. As a result, each nebula is very much brighter at the centre than at the edge, making it difficult to photograph the whole nebula in a single exposure. American astronomer Fred Espenak has taken an ordinary long exposure photograph, and coded the different levels of brightness in different colours to create an isophote picture (**Fig. 4.11**). The most brilliant regions, around the central stars, are shown blue, while the surrounding, slightly fainter gas is coded green. This is roughly the extent of the nebula as seen through a small telescope. Fainter levels are shown yellow and red.

David Malin, of the Anglo-Australian Observatory, has developed a very different technique which suppresses this range in brightness, so that fine details show up in

ordinary photographic form from centre to edge. The unsharp-masked photograph (**Fig. 4.14**) has been made by printing the negative through an unfocused positive copy of itself to reduce the brightness range without losing detail. Three separate exposures with different coloured filters have been unsharp-masked to show the nebula's true colours.

The weaker radiation reaching the outer regions of the nebula causes hydrogen to glow in its characteristic red light, at a wavelength of 656 nanometres. Towards the hot stars at the centre, more energetic short wavelength ultraviolet lights up many other atoms. The strongest emission comes from oxygen atoms which have lost two electrons, and shine at green wavelengths.

Malin's photograph shows up the stars responsible for lighting up the nebulae. In M43, it is the single star in the centre, NU Orionis, a star at a temperature of about 25 000 K and shining as brightly as several thousand Suns. The Orion Nebula itself, M42, is centred on a small group called the Trapezium. Its four main stars are seen even more clearly in the close-up of the nebula's central region (**Fig. 4.12**) taken by George Herbig at the Lick Observatory in California; this photograph mimics closely the nebula's appearance as seen through a small telescope.

Photographs of the nebula at visible

wavelengths show mainly the distribution of hot gas. But M42 also contains much of its original dust. This reflects shorter wavelengths better than longer, and Malin's photograph shows vividly the bluish-grey dust lanes sweeping forward around the top left edge of the blister burnt by the Trapezium in the dark dense cloud behind.

The ultraviolet photograph (**Fig. 4.13**) is dominated by radiation from dust. It was taken by a 31 centimetre telescope with a microchannel plate detector (page 154) which was lofted above the Earth's ozone layer by an Astrobee rocket. The image (originally recorded on photographic film) has been converted to an isophotal map, with the brightness levels running down the scale on the left: from green to purple, and then a second cycle of these colours, to correspond to levels from faintest to brightest.

In the ultraviolet, radiation reflected from dust reduces the contrasts in the nebulae: the optically dark dust lane between M42 and M43 actually glows (at the 'blue' level), while the 'bite' out of the left of M42 is hardly noticeable. Because dust is spread all around M42 and M43, the nebulae together appear more circular in the ultraviolet picture.

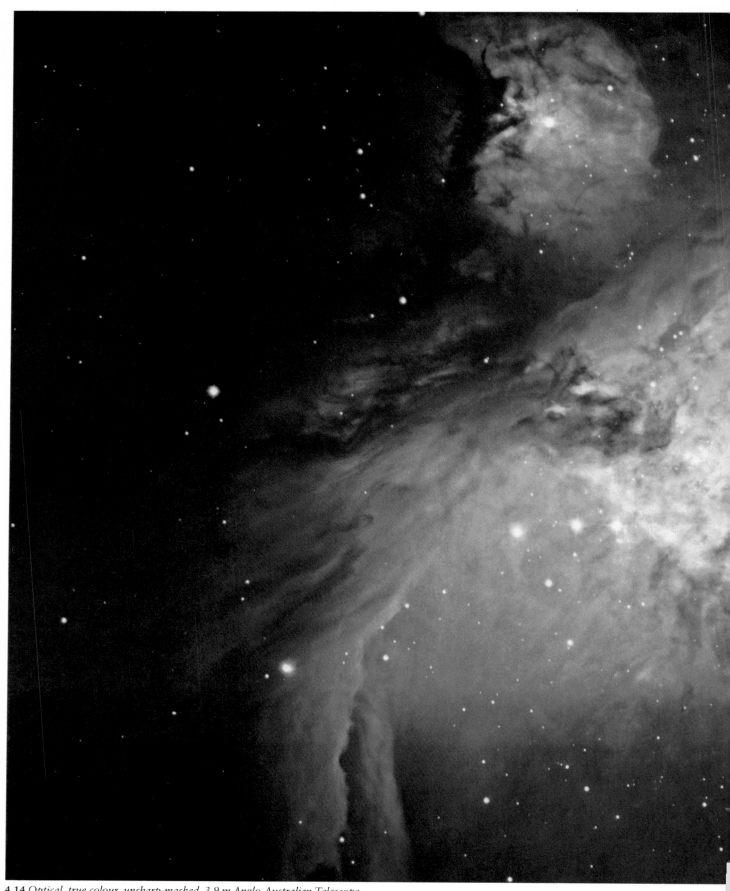

**4.14** *Optical, true colour, unsharp-masked, 3.9 m Anglo-Australian Telescope*

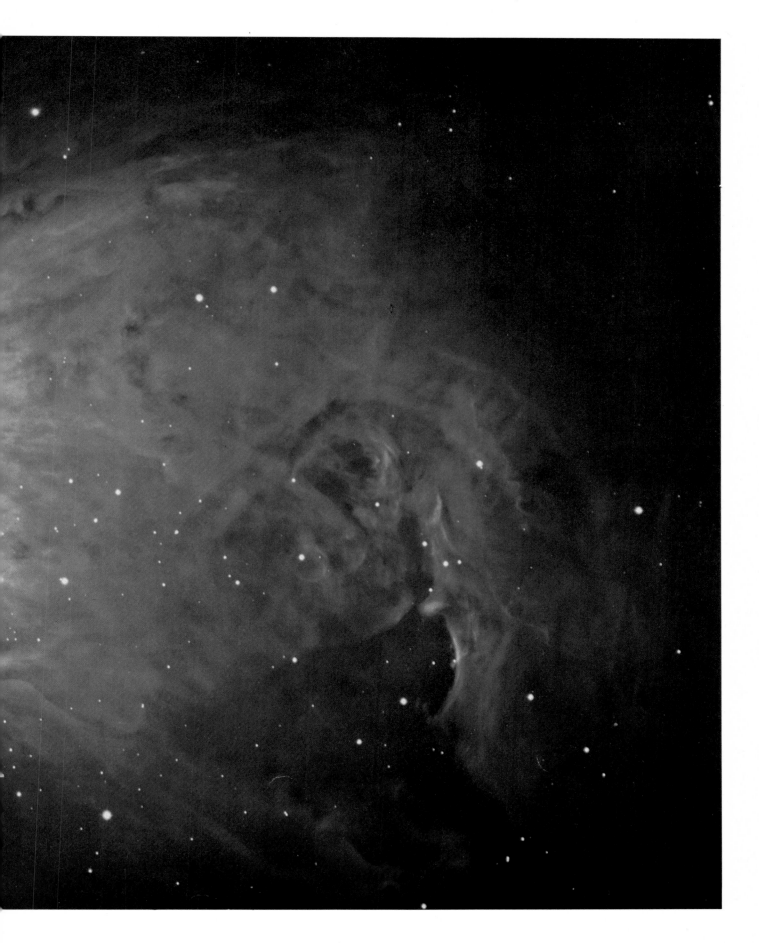

The newly-born stars in the Orion Nebula show up prominently at X-ray wavelengths. The Einstein X-ray Observatory found a total of over a hundred stars in the nebula. Its view of the central six arcminutes is here (**Fig. 4.15**) superimposed as a contour map on the optical photograph from the Lick Observatory (Fig. 4.12). Each 'peak' or closed contour line is a source detected by Einstein's High Resolution Imager. Almost all have been identified with stars, labelled here either by their conventional letter designation or by their 'π' number from P.P. Parenago's catalogue of stars in the nebula.

The lower picture (**Fig. 4.16**) is the central region of this Einstein view, to twice the scale. Here the X-ray intensity has been converted to photographic brightness, and only the most intense sources show. The brightest X-ray star, theta-1C, is the southernmost (lowest) of the four stars in the Trapezium cluster; theta-2A is the right-hand of the three stars visible on the optical photograph forming a line at the lower left of the Lick photograph. 'MT' and 'LQ' are relatively faint at optical wavelengths; they are two of the many young variable stars, whose brightness changes erratically as they stabilise into ordinary 'main sequence' stars.

The central star theta-1C is far more powerful than the other three Trapezium stars, which do not show up at all. The X-ray view underlines the fact that although the four Trapezium stars look rather similar at visible wavelengths, theta-1C is by far the most impressive – and important. With a mass of about 20 Suns, it is approximately twice as heavy as any of the others; it has a surface temperature some 10 000 degrees hotter, at 40 000 K; and it emits almost ten times as much radiation at visible and ultraviolet wavelengths. Its searing surface is surrounded by an even hotter corona (outer atmosphere). At a temperature of 20 million K, the corona shines a million times more brightly than does the Sun's corona in X-rays.

At radio wavelengths, the Orion Nebula's emission comes not from invidual stars but from the hot gas in the nebula. The two radio pictures (**Figs. 4.17** and **4.18**) were both made with the Very Large Array at Socorro, at a wavelength of 6 centimetres. The top picture (Fig. 4.17) covers the central region of the Orion Nebula on roughly the same scale as the X-ray/optical overlay opposite. It is coded so that red indicates the brightest regions, and yellow, green and blue successively less intense parts of the nebula. The most powerful radio-emitting region lies just to the right of the Trapezium, about a quarter of an arcminute (0.1 light years) away from theta-1C which supplies most of the energy. Radio observations indicate that

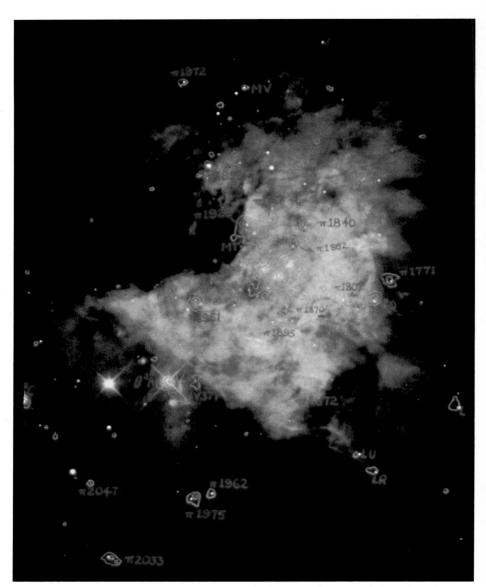

4.15 *X-ray contours from Einstein Observatory, overlaid on true colour optical photo, 3 m telescope, Lick Observatory*

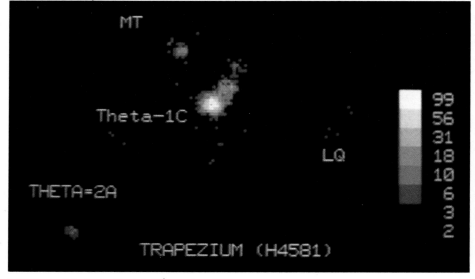

4.16 *X-ray, 0.4–8 nm, Einstein Observatory HRI*

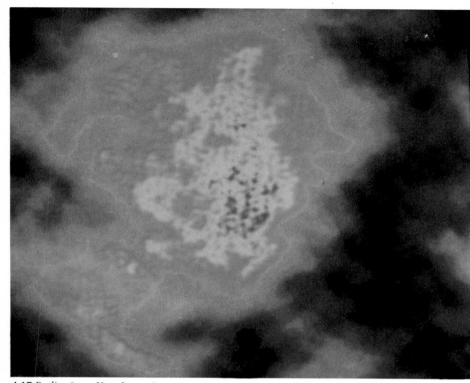

this central part of the nebula consists of about ten Sun-masses of hot gas spread over a couple of light years; the extended outskirts contain perhaps ten times as much again.

The lower radio picture (Fig 4.18) is to twice the scale, and shows finer details in the brightest regions. The most intense parts are purple, with red, yellow and blue indicating the lower brightness levels. The tiny bright spots to the right of centre are hot gas clouds around stars in the Trapezium cluster. This picture emphasises the 'bar' at lower left, which is faintly visible on the upper radio picture, and forms the lower boundary of the optical nebula in the Lick Observatory photograph used for the overlay opposite. Astronomers believe the bar is an ionisation front, marking the distance to which ultraviolet radiation from theta-1C travels before it uses up all its energy heating up the gas. To the lower left of the bar is undisturbed, cooler gas and dust.

**4.17** *Radio, 6 cm, Very Large Array*

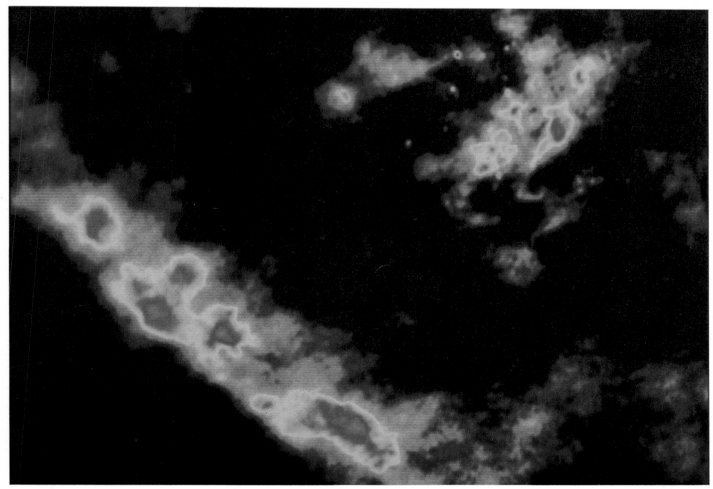

**4.18** *Radio, 6 cm, Very Large Array*

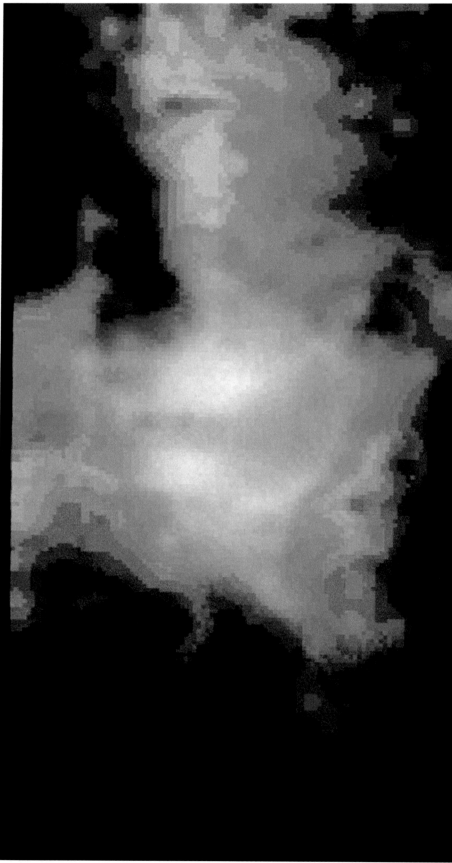

## Orion infrared cluster

Stars are being born right now in the dense dark cloud of gas and dust lying behind the Orion Nebula. The concentration of dust, however, blocks off the light from these very young stars, dimming them to less than a million-millionth of their original intensity – making them completely unobservable with optical telescopes. But the longer wavelength infrared and radio waves can cut through the dust, and reveal star birth in action.

Molecules in the gas clouds emit radio waves, and the powerful broadcaster, carbon monoxide, is a useful guide to the extent and density of the clouds. A 150 light year long molecular cloud runs down through the lower left of Orion (Figs. 4.4 and 4.5), but it is only condensing at its densest part, behind the Orion Nebula. The larger scale carbon monoxide picture (**Fig. 4.19**) covers the central region of the large optical photograph (Fig. 4.14), over the same 10 light year range from top-to-bottom, but much more restricted left-to-right. It is centred just to the right of the Trapezium. The brightness of the image represents the intensity of carbon monoxide emission, and the superimposed coloration indicates the velocity of the gas – receding from us at a few kilometres per second in the blue regions near the centre, and slightly faster in the green, orange and red regions towards the edge.

The intense central region, about 3 light years across, contains enough gas to make 10 000 Suns. It is called the Orion Molecular Cloud 1 (OMC1), and radio astronomers have discovered over thirty different kinds of molecule here, some consisting of as many as seven atoms. Molecules can only survive in space when dust protects them from ultraviolet radiation, so pictures like this reveal only the dense murky dust clouds. The nebulae M42 and M43, brilliant at other wavelengths, are too hot for molecules to survive, and so do not appear. (The former, the Orion Nebula itself, would stretch diagonally across, while M43 would lie in the dark patch on the upper left.)

Infrared astronomy has the advantage of resolving finer details than are revealed by conventional radio telescopes. The long wavelength far infrared from cool dust, however, does not penetrate Earth's lower atmosphere, and the far infrared picture (**Fig. 4.20**) is from a high-flying telescope, aboard the Kuiper Airborne Observatory. Michael Werner and colleagues observed Orion at wavelengths of 30, 50, 100 and 200 microns, covering just the 3 light year core of OMC1 (the white and blue region on the carbon monoxide map); and Allan Meyer has converted the results to a false-colour form and superimposed them on a black-and-white optical photograph of the region.

*4.19 Radio, 2.6 mm carbon monoxide line, 13 m telescope, Five College Radio Astronomy Observatory*

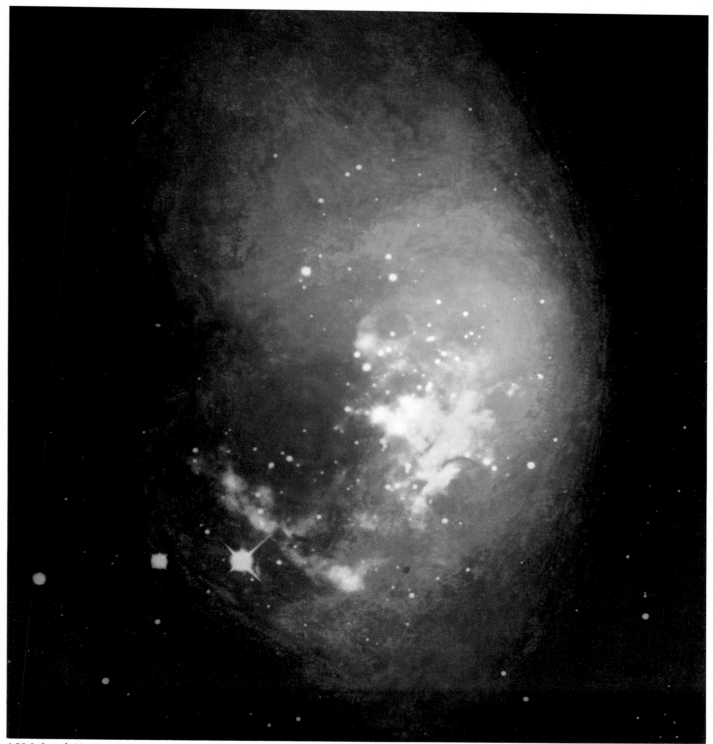

**4.20** *Infrared, 30, 50, 100 and 200 μm combined, 0.9 m telescope, Kuiper Airborne Observatory, overlaid on black-and-white optical photo*

The black-and-white optical images show the four close-set Trapezium stars just to the right of centre, and the diagonal 'bar' on the bottom left (compare with the optical picture Fig. 4.12); all these stars and nebulosities are, however, too hot to show up in the far infrared. Since longer wavelength infrared comes from cooler dust, and shorter wavelengths from warmer dust, these observations reveal the temperature of the dust as well as the region where it is densest. Here the colour coding indicates temperature. Red regions are coldest, at about 15 K; the scale runs through orange, green and blue to violet for 'warm' regions at 100 K. (The colours are chosen to show roughly how the region would actually look if our eyes were sensitive to radiation a hundred times longer than light.) The brightness of the colour indicates the intensity of the infrared, and hence the density of the dust.

The dust is most concentrated, and warmest, in the small region just to the upper right of the Trapezium. This region, half a light year in size, is the present 'star nursery' of Orion. Within it, a new cluster of stars is condensing from the dense gas, and shining – unseen to ordinary telescopes – as powerfully as 100 000 Suns.

**4.21** *Infrared, 1.2 (blue) and 2.2 (yellow) μm, 3.9 m Anglo-Australian Telescope*

Observations in the near infrared, at wavelengths of a few microns, reveal both the Orion Nebula and objects in the molecular cloud behind – including the region where stars are now forming. **Fig. 4.21** is a composite picture of observations at 1.2 microns (blue) and 2.2 microns (yellow). The bluish-white images here are the stars and gas of the Orion Nebula, detected at both wavelengths, and they look very similar to optical photographs like Fig. 4.12. The yellow-orange images are stars behind the nebula, in the dense Orion Molecular Cloud 1

(Fig. 4.19). The dust here blocks off their light and much of the infrared wavelengths as short as 1.2 microns, but their emission at 2.2 microns passes through. The colour-coding in Fig. 4.21 hence also represents a crude coding for distance: the white images are the nearest, and the reddest images the most distant stars, the deepest-embedded in the cloud.

The yellow nebulosity to the top right (at the edge of the bluish nebulosity) is a region where stars are forming. This region is most intense at longer infrared wavelengths, and it was discovered at a

wavelength of 22 microns by pioneer infrared astronomers Douglas Kleinmann and Frank Low in 1967. It is usually called the Kleinmann–Low Nebula, or just KL. The detailed false-colour photograph of KL (**Fig. 4.22**) combines views at three wavelengths. The longest, 20 microns, has been coded red, and shows regions of dust at temperatures of about 150 K. Wavelengths of 11 and 12 microns are coded blue and green, and these colours add to the red background to produce white spots where the temperature is highest. (The vertical streaks are not real, but are an artefact of the scanning system.)

The image-processed 'contour map' (**Fig. 4.23**) is based on observations made with the Goddard infrared camera at a wavelength of 12.4 microns. Different levels of brightness are shown by a series of colours (running from dark blue, through light blue, buff and green, to brown) repeated in succession to produce fringes like contour lines around the brightest spots.

The bright spot at the top of KL is the Becklin-Neugebauer object (BN), discovered by Eric Becklin and Gerry Neugebauer in 1967. BN emits most of its radiation at short infrared wavelengths (it is the brightest object in this region in Fig. 4.21), showing that it has a temperature of about 600 K. At first, astronomers thought it was a protostar – a small clump of gas and dust which is warming up as it condenses to become a star, but which does not have an energy-producing core. Astronomers have now, however, detected radiation which comes specifically from hydrogen atoms at a temperature of 10 000 K, showing that BN must contain a star which is already shining. This star is only a few thousand years old. The infrared source of BN is a surrounding ring of warm dust – about the size of the Solar System – left over from the star's formation.

Despite its prominence here, BN is probably not the most important object in KL. That distinction is reserved for another dusty ring surrounding a young star. This object, IRc2, looks fairly insignificant at long wavelengths – it appears in the middle of the Wyoming photograph (Fig. 4.22) and as the left-hand brown peak in the Goddard picture (Fig. 4.23) – but at short wavelengths it is half as bright as BN. Detailed investigation of the spectrum, moreover, shows that IRc2 is mostly dimmed by dust lying in front. In reality, IRc2 is probably ten times brighter than BL, its central star shining as brightly as 100 000 Suns.

Although KL must contain many other, fainter, young stars, the brilliance of IRc2 alone may be responsible for much of the nebula's activity. It is certainly capable of heating up the dust in the surrounding gas to a temperature of 150 K over a region the size of KL – notice how, in Fig. 4.22, IRc2 lies right in the centre of the red (20 micron) nebula – and to rather lower temperatures over the whole region seen at long infrared wavelengths (Fig. 4.20). The gas in KL is flowing outwards at speeds up to 100 kilometres a second; most probably as a result of the 'stellar wind' of gases from the surface of the superluminous IRc2. The larger, cooler sources in KL may be the very first stages of protostars, or perhaps just clumps of denser gas and dust which will eventually be disrupted by the energetic radiation and winds of IRc2.

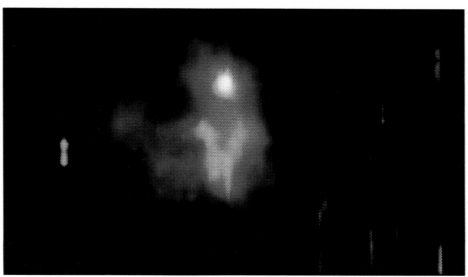

4.22 *Infrared, 11 (blue), 12 (green) and 20 (red) μm, 2.34 m Wyoming Infrared Telescope*

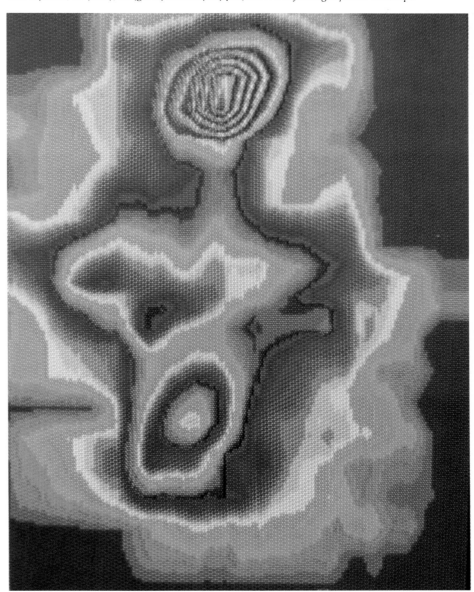

4.23 *Infrared, 12.4 μm, 3 m NASA Infrared Telescope Facility, Mauna Kea*

4.24 *Optical, true colour, 10 cm camera lens*

## W3

Just to the left to the well-known W shape of the constellation Cassiopeia lies a nursery of superstars – the birth place of some of the most brilliant stars in our region of the Galaxy. The stars are hidden within the dusty cloud from which they were formed, however, and astronomers only know them because they heat up the dust and gas around to produce infrared and radio emission. The region is known by its radio designation W3, the third source in Gart Westerhout's catalogue of extended radio sources.

The dark cloud stands out in silhouette in the upper right of the wide-angle view of the Milky Way (**Fig. 4.24**), made from exposures through three different coloured filters with a 10 centimetre lens. (Cassiopeia is out of the picture to the top right; at bottom centre is the Perseus double cluster.) Its left edge is rimmed by bright nebulae, shining in the red light of hydrogen atoms.

According to Hélène Dickel of Illinois University, the dark cloud originally extended as far to the left as the nebulosities now run, and contained enough matter to form 50 000 Suns. A few million years ago, it was swept up by the Perseus Arm of our Galaxy (Fig. 1.9), and stars began to form at its left-hand end. They heated up and dispersed the gas and dust around, and their radiation and outflow of gas compressed the adjacent region of the cloud into forming stars. The process has been spreading ever since. (A similar process has been occurring in Orion, Fig. 4.2.) The nebulae visible in Fig. 4.24 are heated by the penultimate generation of stars, while the youngest stars

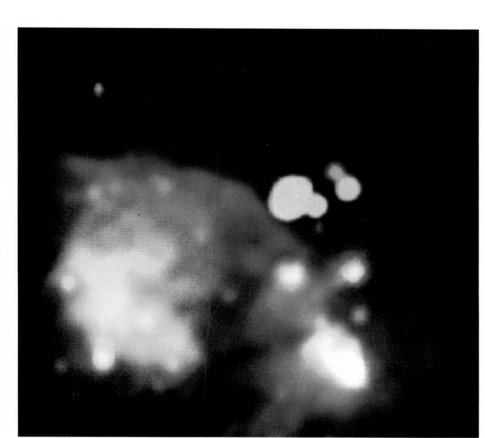

4.25 *Combined wavelengths: optical (green and red), 0.9 m telescope, Kitt Peak National Observatory, and radio (blue), 20 cm, Westerbork Synthesis Radio Telescope*

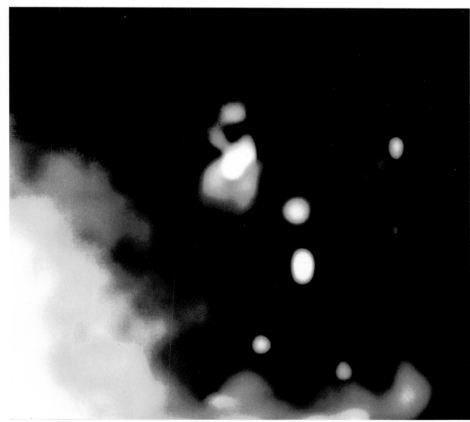

4.26 *Combined wavelengths: optical (green and red), 0.9 m telescope, Kitt Peak National Observatory, and radio (blue), 6 cm, Westerbork Synthesis Radio Telescope*

lie just within the edge of the dark cloud.

The false-colour pictures reveal the positions of some of these new stars, betrayed by their radio emission. The top picture (**Fig. 4.25**) shows in detail the region of the triangular nebula IC 1795, which is visible as a small bright red patch at the top centre of the optical photograph (Fig. 4.24). In Fig. 4.25 the green light from oxygen is coded green, and red light from sulphur red – so these two colours give a true-colour (but much enhanced) picture of IC 1795. Blue here represents the radio view, as observed at a wavelength of 20 centimetres with the Westerbork telescope. The four most intense radio sources, at the top right, are shells of hot gas, about a light year in size, surrounding new-born stars over 100 000 times more luminous than the Sun.

The lower false-colour photograph (**Fig. 4.26**) is a close-up of the central region, with oxygen light coded green, but with red here indicating the red light of hydrogen. The 'blue' radio map is at a wavelength of 6 centimetres, and shows finer details (although the observations cover a smaller field and the right-hand pair of radio sources does not appear). Here the largest and brightest radio source (W3A) appears as an inverted 'V' – the patchy shell around a central star as bright as half a million Suns.

This photograph also reveals a small wisp of optical nebulosity curling up from the top of W3A. It appears green and red, and includes the bright white spot where it overlies the 'blue' radio map. Radiation from the central star must be escaping to 'light up' this gas, so W3A must lie only just beneath the front surface of the cloud as we see it. The exciting deduction is that the 'hollow' cavity has only to expand a little more before it breaks through and we see the central star itself. In perhaps a thousand years time – a short time in human history as well as astronomy – our descendants will see an extra star in Cassiopeia, equal in brightness to the present five stars and extending their W shape to a zigzag.

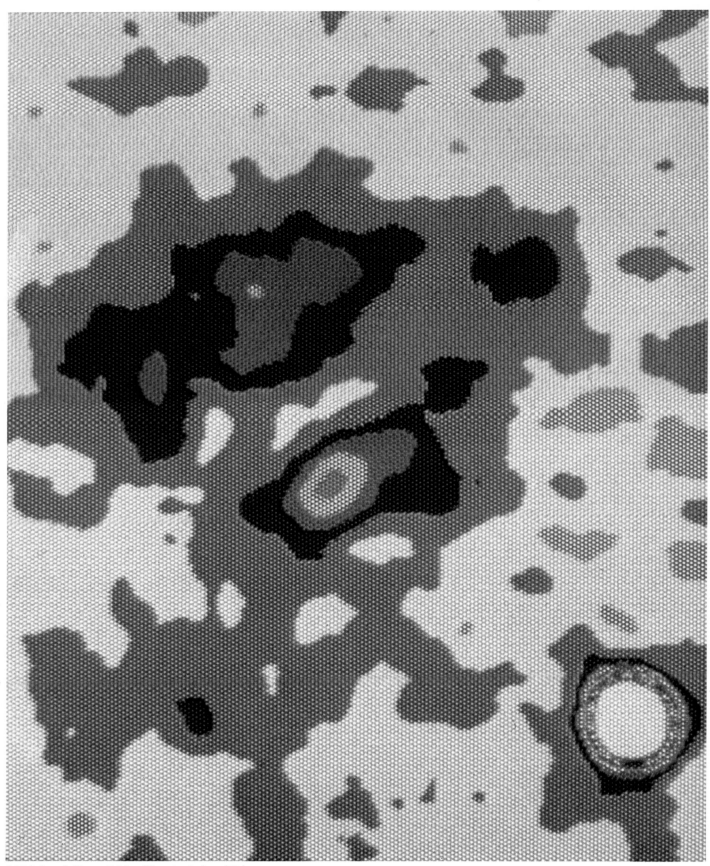

**4.27** *Infrared, 10 μm, 2.34 m Wyoming Infrared Telescope*

Many of the stars in W3 are slightly older than BN and IRc2 in Orion, and their intense radiation has had time to scour away the gas and dust, to form dense shells surrounding almost-empty cavities. Each shell intercepts its central star's radiation. The dust is warmed to a temperature of 100 to 200 K, and shines at middle infrared wavelengths. W3 is thus one of the most luminous infrared nebulae known in our Galaxy, and the third brightest as seen from Earth.

The infrared picture (**Fig. 4.27**) at 10 microns shows the largest of its dust shells – that around W3A (seen also in Fig. 4.26). The contour map shows the background dark sky as white, with successively brighter regions green, blue, brown, turquoise and red. The ring is 2 light years across, and contains a star 500 000 times brighter than the Sun; it appears to be just under 1 arcminute across. The small intense source (IRS5) at the lower right (pale blue) surrounds a younger star, or a compact star group like KL in Orion.

The radio picture (**Fig. 4.28**) from the Cambridge Five Kilometre Telescope shows a smaller hot gas shell, one-twentieth of a light year in diameter (1.3 arcseconds). The colour coding runs from blue for the dark background, through green, pink and blue to white for the brightest region. This object, W3 (OH), lies in the dark cloud just beyond the lower left corner of the triangular IC 1795 optical nebula (Fig. 4.24). As well as the radio emission at all wavelengths from hot gas, W3 (OH) shines intensely at the particular wavelengths (near 18 centimetres) which are emitted by the molecule OH (hydroxyl). The emission comes from natural masers, the radio equivalent of lasers. The masers are small regions of gas within the dust shell around the star, and they draw their energy from the hot dust's infrared output.

In 1976, Mark Reid and colleagues looked at W3 (OH) simultaneously with eight radio telescopes on sites ranging from Massachusetts to California, using the techniques of Very Long Baseline Interferometry (page 130) to resolve details as fine as could a radio telescope the size of the United States. The false-colour picture (**Fig. 4.29**) shows some of the 70 masers they found in W3 (OH). The frame here covers a region only 0.15 arcseconds across, the central one-twentieth of the region shown on the Cambridge map (Fig 4.28). The colour coding in Fig. 4.29 indicates speed (calculated from the Doppler shift of the masers' OH lines). All the masers are falling in towards the central star of W3 (OH), and all at speeds near 45 kilometres per second. The narrow range in speed, from 44 to 46.5 kilometres per second is coded by a wide range in colour, from red to violet. The positions, intensities and velocities of masers provide a new, highly sensitive probe of star-birth regions.

4.28 *Radio, 1 cm, Cambridge Five Kilometre Telescope*

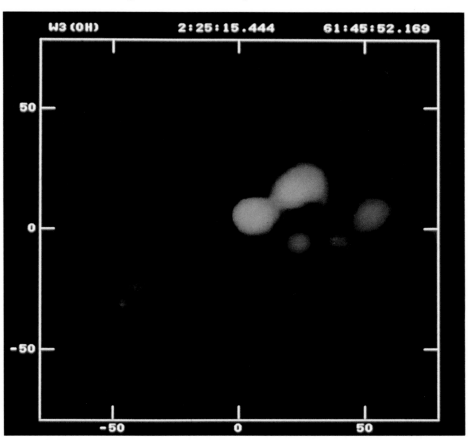

4.29 *Radio, 18 cm hydroxyl line, velocity coded, VLBI (Haystack, Maryland Point, Green Bank, Algonquin, Vermilion River, Fort Davis, Owens Valley, Hat Creek)*

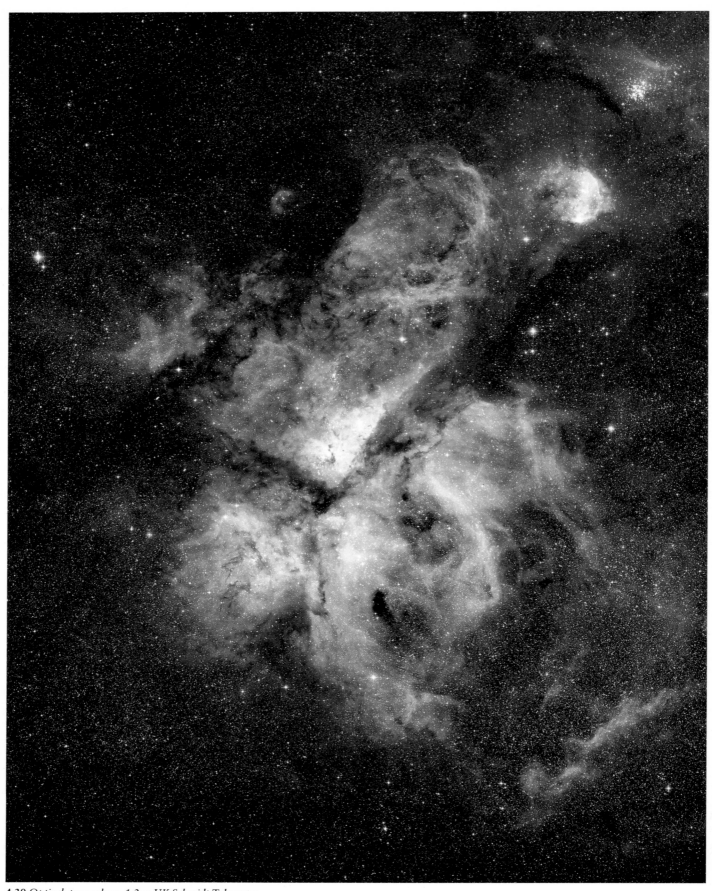

**4.30** *Optical, true colour, 1.2 m UK Schmidt Telescope*

4.31 *Optical, true colour, 3.9 m Anglo-Australian Telescope*

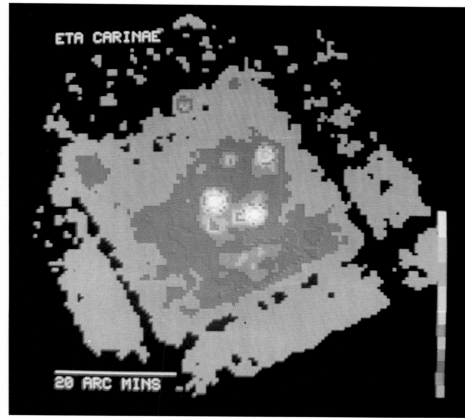

4.32 *X-ray, 0.3–2.5 nm, Einstein Observatory IPC*

# Carina Nebula

The huge Carina Nebula is visible from Earth's southern hemisphere as a misty glow in the band of the Milky Way late in the summer and autumn. The colour photograph (**Fig. 4.30**), taken by combining separate exposures through different coloured filters on the UK Schmidt Telescope in Australia, covers a region of sky about 3° by 4°. The Carina Nebula stretches over some 2½° (about five Moon breadths). At its distance of 9000 light years, the nebula is thus a staggering 300 light years across – twenty times the size of the Orion Nebula.

Two million years ago this nebula was a huge, dark, dense molecular cloud. Much of this matter has now condensed into stars, and the fierce radiation from the heaviest, hottest stars is lighting up the remaining gas. The tatters of the original cloud appear as the dark dust lanes which make a huge 'tick' across the nebula. In a star-formation region, stars of all masses are born, but the heaviest are very rare. Because it is so immense, the Carina Nebula is one of the few places in our Galaxy where we find many heavy stars.

The less-exposed photograph from the Anglo-Australian Telescope (**Fig. 4.31**) reveals the stars in the central 1° of the nebula. Many of the stars strewn over the picture lie well in front of the nebula, but there are two clusters of stars which do reside in the nebula and are largely responsible for heating it. The compact cluster (to top right of centre) is Trumpler 14, and the looser cluster in the centre Trumpler 16; both were first catalogued by Robert J. Trumpler of the Lick Observatory in the 1920s. Between them, the two clusters contain over a dozen stars heavier than twenty Suns (O type stars) – the kind of star (like W3A or theta-1C in Orion) which would be the outstanding member of an ordinary star cluster.

Amongst these awesome stars, one in each cluster contends for the title of the Galaxy's most massive and luminous star. The central star of Trumpler 14, catalogued HD 93129A, weighs at least 120 times as much as the Sun, and shines as brightly as five million Suns. Because of its high surface temperature of 52 000 K, HD 93129A emits most of its energy as ultraviolet radiation. Its rival, eta Carinae, lies in Trumpler 16, at the lower-left end of the small, banana-shaped dark cloud. It is surrounded by a shell of dust (Fig. 4.33), and here appears slightly oval. Eta Carinae is about as heavy and luminous as HD 93129A, but most of its radiation is emitted at infrared wavelengths.

The Einstein Observatory's view (**Fig. 4.32**) is to the same scale as the optical above. The colour coding runs from blue for faint regions, up through the scale of colours (on the right) to white for the most intense; the dark square is due to supporting struts in the instrument. The source in the centre of the square is eta Carinae, and the X-ray emitter to the top right is HD 93129A.

The X-rays come from gas in the stars' outer atmosphere (corona), at a temperature of 10 million K, and they make up only a ten-millionth of the stars' total output of radiation. The bright X-ray source to the right of eta Carinae (HD 93162) is a star only one-tenth as massive, but going through an active phase (during which it is classified as a Wolf Rayet star). It has an exceptionally high X-ray output for a star, amounting to over a millionth of its light output. The gases from these stars' coronae are blowing a tenuous bubble of multi-million degree gas, responsible for the background 'glow' (red, purple and pale blue in Fig. 4.32) within the cooler, 10 000 K, gas of the Carina Nebula itself.

# Eta Carinae

The massive star in the centre of the Carina Nebula is one of the most remarkable known – for its power, and for its instability. It was relatively faint (magnitude 4) in 1603 when Johannes Bayer catalogued the star and denoted it by the Greek letter 'eta'. But in the 1820s, the star became some twenty times brighter, and stayed that way for forty years before fading completely from sight (to magnitude 8). During its bright period, it fluctuated from year to year, with one spectacular outburst in March 1843, when the star reached magnatude -1. Eta Carinae then outshone every star in the sky, except for Sirius.

Since its rapid fading in the 1860s, eta Carinae has been very slowly brightening, although it is currently only just visible to the unaided eye (magnitude 6). But at wavelengths around 20 microns, eta Carinae is the brightest star in the sky. The infrared picture of eta Carinae (**Fig. 4.33**), made at a wavelength of 18 microns, shows that the radiation is coming from a shell of dust around the star. The colours here represent the infrared brightness, rising from faint red for the outer envelope through yellow and white to the intense blue centre. The infrared image appears 8 arcseconds wide, which (at eta Carinae's distance of 9000 light years) means the dust shell is about one-third of a light year across. Visible light from the central star heats up the dust from the 3 K of interstellar space to 250 K (just below the freezing point of water), and the dust re-radiates the energy as infrared radiation of 10 to 20 microns wavelength.

American astronomer Kris Davidson used a TV camera (SIT vidicon) on the 1.5 metre telescope at Cerro Tololo, Chile, to take the optical picture shown here in two different computer-processed versions. An orange filter (600 nanometres wavelength) blocked off the strongest spectral lines from hot gas around the star, and we are seeing the shell of dust lit up by the central star. In the straightforward colour-coded view (**Fig. 4.34**), the grey-blue and lime-green are the dark background, and successive contours show brighter regions, up to the star eta Carinae in the centre. (The contours are logarithmic, corresponding to equal ratios of brightness. The effect is to enhance the low-brightness areas and suppress the intensity of eta Carinae itself which is in fact thousands of times brighter.) The scale is approximately the same as the infrared picture above.

The small strangely shaped nebula around eta Carinae is called Homunculus – the 'little man', with his head to the upper right and legs to the lower left. Robert Gehrz and Ed Ney have measured the size of the Homunculus as sketched and photographed since its discovery in 1914, and have found that it is gradually

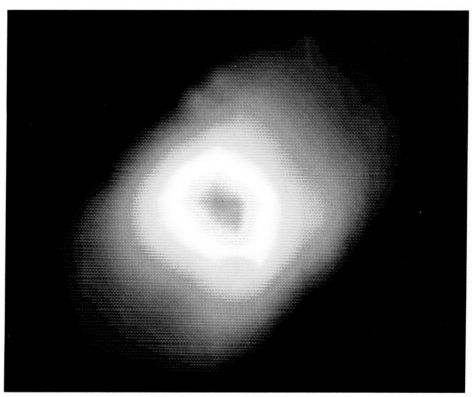

4.33 *Infrared, 18 μm, 2.34 m Wyoming Infrared Telescope*

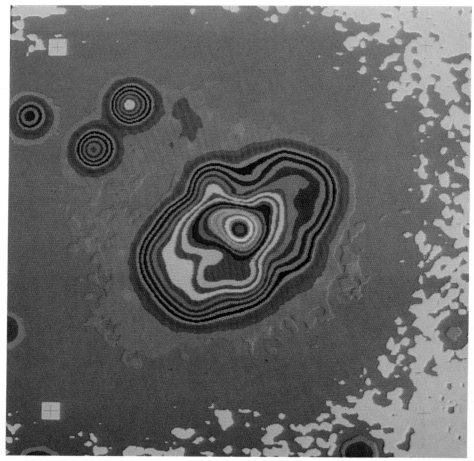

4.34 *Optical, 600 nm, intensity coded, SIT Vidicon, 1.5 m telescope, Cerro Tololo Inter-American Observatory*

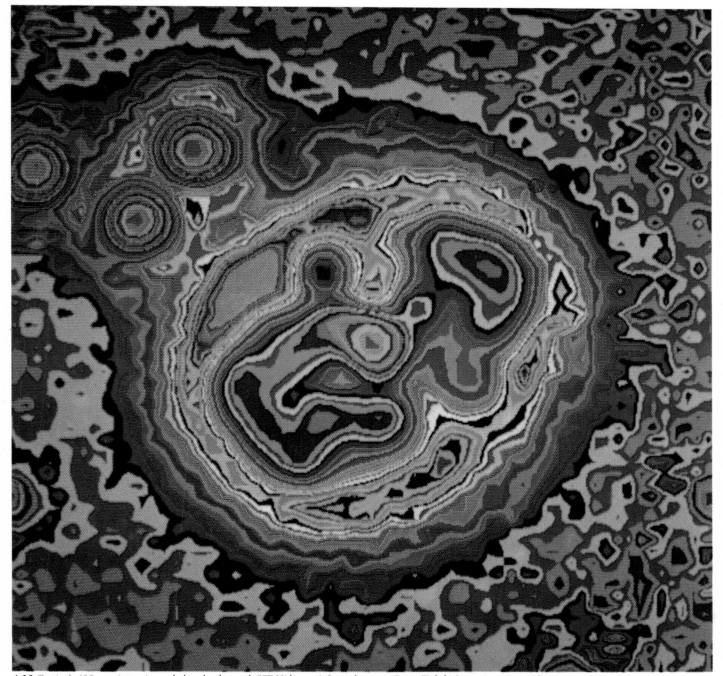

4.35 *Optical, 600 nm, intensity coded and enhanced, SIT Vidicon, 1.5 m telescope, Cerro Tololo Inter-American Observatory*

growing. Their measurements show that the dust (and gas) of the Homunculus were ejected from eta Carinae during its great upheavals of the 1840s.

The Homunculus contains as much matter as ten Suns, in a thick ring which we see at such an angle that it looks oval – with condensations in the ring appearing as the little man's head and limbs. Davidson has sharpened up his orange-light image by computer to reveal finer details in the Homunculus, seen as the green and red ridges to the top right and lower left of the central star in the 'slightly nightmarish' picture (**Fig. 4.35**). (The processing has

artificially lowered the brightness of the outer regions of the Homunculus, so viewed as a contour map the central regions are projecting from a surrounding 'moat'.)

Davidson emphasises that eta Carinae is not much fainter than it was in the 1840s; the ejection of dust from the Homunculus dimmed the central star in the 1860s, and its power equivalent to five million Suns now emerges as infrared from the dust. Stars are expected to become unstable and eject matter near the end of their lives. In 1981 Davidson and colleagues obtained ultraviolet spectra of gas in the Homunculus which had a composition

typical of gas ejected from stars near the end of their lives – comparatively rich in nitrogen, but with very little oxygen and carbon. How can an 'old' star lie in a 'star nursery' like the Carina Nebula? The answer is that a heavy star has a short lifetime, and a star like eta Carinae – as heavy as 120 Suns – lives for only a million years. It has lived its brief and spectacular life while the nebula is still leisurely spawning new stars. Within the next few thousand years, eta Carinae will end its life in a supernova explosion, and briefly shine in our skies more brilliantly than anything except the Sun and Moon.

## Hyades

This loose scattering of stars makes up the 'head' of Taurus (the bull), and it is the nearest cluster of stars. Here astronomers can study a variety of moderately bright stars of the same age.

The Hyades cluster lies 150 light years away, and consists of over 400 stars, spread over 20° of the sky (40 Moon-breadths). The pictures here cover just the central few degrees. In the blue-filtered photograph (**Fig. 4.36**) the brightest star is the red giant Aldebaran – brilliant even through the filter because it lies only 65 light years away. The others are true Hyades stars. The next brightest (near the centre) is theta-2, about twice as heavy as the Sun and twenty times brighter (a white A type star like Sirius). Slightly heavier Hyades stars have already become red giants. The Hyades cluster must thus be just about as old as the lifetime of a star like theta-2 *before* it becomes a red giant – according to theory, 500 million years. This makes the Hyades over a hundred times older than the Trapezium in Orion, but still only one-tenth as old as the Sun.

The younger the star, the more rapidly it rotates, and so the more powerful the magnetic field it generates. Since the magnetic field heats up the corona, young stars are stronger X-ray emitters. The X-ray view of the Hyades (**Fig. 4.37**) shows this powerful X-radiation from some thirty stars. The picture covers roughly the same region as the optical photograph, Fig. 4.36. Many of the X-ray stars correspond to optical images, but their different brightness make a detailed comparison quite a challenge. (Try it!)

Once a star is in the prime of life, however, almost all its radiation is in the form of light. To study stars in detail, astronomers' strongest tools are the traditional methods of optical astronomy. The analysis relies mainly on spectra, breaking the light up by wavelength and investigating the dark lines due to absorption by different elements in the star.

The spectra of dozens of Hyades stars (**Fig. 4.38**) have been obtained simultaneously by placing a thin objective prism in front of a wide-field Schmidt. The image of each star has been broadened horizontally, and stretched vertically into a spectrum. The stars here consist of basically the same mix of elements (mainly hydrogen, with some helium, and a little of the other elements), but their different temperatures cause different spectral lines to become prominent. Cool red stars have bland spectra; in bright Aldebaran (left) we can see that its overall appearance is due to a multitude of faint lines from many different elements. The hot stars like theta-2 (centre), however, show only a few strong lines from hydrogen and helium.

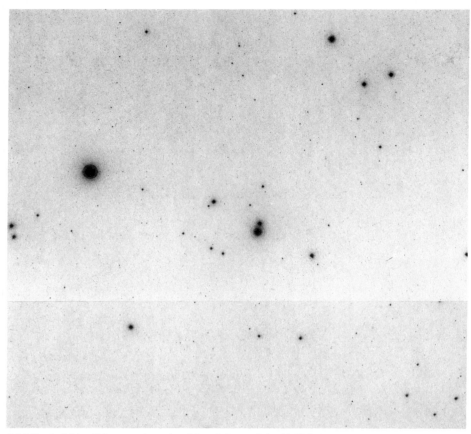

4.36 *Optical, blue light, negative print, 1.2 m Palomar Schmidt Telescope*

30 ARC-MIN:

4.37 *X-ray, 0.3–2.5 nm, mosaic, Einstein Observatory IPC*

**4.38** *Optical, objective prism spectra, 0.6 m Schmidt telescope, Cerro Tololo Inter-American Observatory*

# 5 Infrared Astronomy

In 1800, Sir William Herschel – best known for finding the planet Uranus – turned his attention to the amount of 'heat' carried by the different colours in sunlight. He spread sunlight out into a rainbow spectrum, and measured the temperature rise when a thermometer was placed in each colour in turn. Herschel also moved the thermometer beyond the red end of the spectrum – and found that here too the thermometer was receiving heat, carried by 'invisible' radiation. It was the first evidence for radiation other than light, and Herschel named the invisible rays 'infrared'.

As we define it today, the infrared region of the spectrum covers a wide range of wavelengths (Fig. 5.1), from waves just longer than red light (about 700 nanometres) up to one millimetre, which marks the boundary with radio waves. For infrared wavelengths, the most convenient unit is the *micron* ($\mu$m), one-thousandth of a millimetre. The infrared region thus stretches from 0.7 to 1000 microns. The longest infrared waves are, therefore, over a thousand times greater in wavelength than the shortest, and over this wavelength range the appearance of the sky changes far more than it does over the visible region of the spectrum, which spans only a factor of two in wavelength from blue-violet to red light.

Although the Sun produces so much infrared that we can feel its heat on our own skins, and it affects a simple thermometer like Herschel's, its strength is only due to the fact that the Sun is so near to us on the cosmic scale. Very sensitive detectors are needed to pick up the infrared radiation from any other star or from the planets, nebulae and galaxies. The only exception is the Moon, whose

infrared emission was detected in the mid-nineteenth century – again because the Moon is so close to the Earth. Otherwise, infrared astronomy languished for a century and a half after Herschel's discovery until physicists began producing sufficiently sensitive infrared detectors in the 1950s. This new branch of astronomy has made tremendous strides in the past thirty years, revealing aspects of the universe hidden to detectors of other wavelengths. But infrared astronomers face some difficult problems in their search for 'heat radiation' from deep space.

The easiest infrared rays to detect and measure are the shortest – those with wavelengths between 0.7 and 1.1 microns, just beyond the region of red light. Although we cannot see these rays, in most ways they are very like ordinary light, and can be dealt with in the same way. For example, some photographic emulsions are sensitive to radiation out to a wavelength of 1.1 microns, so infrared astronomers can photograph the sky in exactly the same way as taking a plate in visible light. For this reason, wavelengths from 0.7 to 1.1 microns are often called the *photographic infrared*. Astronomers can also use some types of electronic image-tube light-intensifiers to detect these wavelengths of infrared; and the new CCD silicon chip light-detectors (Figs. 3.17–3.19) are particularly sensitive to such short-wavelength infrared rays. The one type of light-detector which cannot pick up the photographic infrared is the human eye. So although we draw a line at about 0.7 microns to distinguish visible light from photographic infrared, to astronomers they are much the same: 'light', to them, extends through the invisible

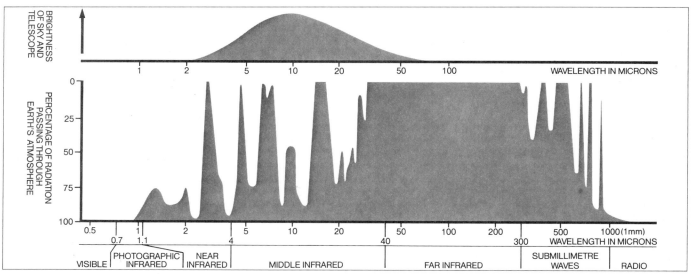

**Fig. 5.1** *Middle infrared observations suffer from heat radiation produced by the atmosphere and the telescope itself. Apart from the photographic infrared, all wavelengths are absorbed by atmospheric carbon dioxide and water vapour. There are reasonably transparent 'windows' at certain wavelengths in the near and middle infrared, and at submillimetre wavelengths, but far infrared is totally absorbed. These measurements were made from the 4200 metre peak of Mauna Kea, Hawaii; lower observatories suffer far worse absorption.*

rays up to a wavelength of 1.1 microns.

As seen in the photographic infrared, the sky does not look very different from the sky seen by the human eye. It is still studded with stars, although when viewed in this new waveband their brightness is altered. Bluish stars, like Rigel in Orion, are dimmer, while red stars shine more brightly. Rigel's neighbour in Orion, the intensely red Betelgeuse, becomes the brightest star in the sky, outshining at wavelengths of one to two microns all other stars, including the visually brightest star Sirius.

For astronomers, however, the rays of photographic infrared have one tremendous advantage over visible light. They can penetrate the dust in space that obscures our view of distant stars, and of stars embedded within dense clouds of gas and dust. The dust particles are about the same size as the wavelength of visible light, and they block it most efficiently. Even in the nearly empty space between the stars, dust particles dim a beam of light to less than half its original brightness over a distance of three thousand light years, a small fraction of our Galaxy's 100 000 light year extent. Dense clouds of dust can dim light so much that any stars within are totally invisible to optical telescopes. But infrared rays – even those as short as the photographic infrared – can penetrate this dust quite readily. The new CCD cameras are now photographing stars in regions of space which had previously been hidden from our view.

As we move to longer infrared wavelengths, the sky becomes more interesting. The familiar, light-emitting stars fade, and we begin to detect new objects. But the observational problems now become formidable. For a start, traditional light-detectors like photographic plates are of no use. They do not respond to the longer infrared wavelengths. Astronomers must use new, infrared-sensitive materials. They have chosen substances which change their electrical conductivity when infrared radiation falls on them. These work very like the exposure meters built into modern cameras, but are designed to measure infrared rays rather than light.

For the shorter wavelength infrared, from the limit of the photographic infrared at 1.1 microns up to four microns – the *near infrared* – astronomers originally used the substance lead sulphide. Modern detectors incorporate indium antimonide instead. This substance is ten times more sensitive than lead sulphide, but it must be kept very cold to achieve its maximum response to the very faint infrared rays from space. It can be chilled by solid nitrogen to a temperature of only 50 K (−223° C), or even better to a temperature just a few degrees above absolute zero with liquid helium.

The Earth's atmosphere causes major difficulties to astronomers working at all infrared wavelengths, but for those wavelengths of a few microns, the atmosphere has advantages as well as disadvantages. The air scatters sunlight, giving us such a bright sky during the day that our eyes cannot see any stars. This scattering affects shorter wavelengths more than longer wavelengths – which is why the daytime sky appears blue, rather than white or red. Near infrared waves are scattered so little that the sky appears almost black at these wavelengths, even during the day. Infrared astronomers can observe the sky during daytime just as happily as they can at night. Often they will mount infrared detectors on a large optical telescope, like the Hale 5 metre on Palomar Mountain or the Anglo-Australian Telescope, to make their observations during the day when the telescope would otherwise lie idle, and leave the telescope to the optical astronomers at dusk.

The atmosphere's curse is that molecules of water vapour and carbon dioxide absorb infrared rays very strongly, and at particular wavelengths (or *bands*) they can obscure the sky completely. At wavelengths between these absorption bands, however, there are 'windows' – wavebands where the air is quite transparent – and infrared astronomers have to choose these particular wavebands at which to investigate the sky (Fig. 5.1).

One clear window occurs at a wavelength of 2.2 microns. American astronomers Gerry Neugebauer and Bob Leighton surveyed the sky at this wavelength in the 1960s to build up the first comprehensive view of the sky in the infrared. For six years they scanned the heavens with a crudely built 1.5 metre telescope fitted with a lead sulphide cell cooled with liquid nitrogen, and they published the results as the Infrared Catalogue (IRC). This catalogue, known colloquially as the Two Micron Survey, contains 5612 infrared stars. This is roughly the number of stars that our eyes see at night, but they are generally not the same – apart from red stars like Betelgeuse which produce both light and infrared in copious amounts.

To infrared detectors sensitive at 2.2 microns, the brightest stars are distended red giants like Betelgeuse and Antares (which lies opposite in the sky in the constellation Scorpius). Some of the fainter sources, however, are objects which are quite invisible to optical telescopes, and astronomers turned to longer wavelength observations to work out what they are.

This required a different type of infrared detector. It was found that one of the best detectors for wavelengths longer than four microns is a crystal of the semiconductor germanium containing traces of the rare metal gallium. The germanium–gallium detector is extremely sensitive – but only when it is cooled to two degrees above absolute zero. The crystal must be refrigerated with liquid helium, at a temperature of 2 K (−271° C), and this must, in turn, be surrounded by a jacket of liquid nitrogen to prevent the helium from boiling away too quickly in the 'warmth' of an open telescope dome on a mountain top at night.

Our atmosphere has windows at wavelengths of about

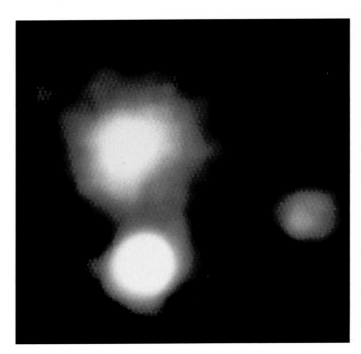

**Fig. 5.2** *Young stars near the nebula NGC 7538, their light hidden by dust, apear in this infrared picture from the Wyoming telescope (Fig. 5.3). It combines views at 5, 10 and 20 microns, coded blue, green and red. The hottest star is the white image; it warms up a cloud of dust (above). To the right lies a cooler star (red).*

ten and twenty microns – sufficiently different from visible light to show new objects in the sky. But astronomers have to struggle to see through these windows. All objects produce radiation, and they emit most strongly at some particular wavelength region which depends entirely on their temperature. The Sun, at 5800 K, produces mainly visible light; an electric fire, at about 1000 K, emits infrared of around two to three microns. The sky above a mountain-top observatory has an average temperature of around 270 K – near the freezing point of water at 273 K (0° C). Anything at that temperature produces strong infrared radiation at around ten microns wavelength.

So the ten-micron-sensitive infrared detector looks up to see the whole sky glowing brightly – day and night – with its own heat energy. The telescope itself is at roughly the same temperature, so it too is glowing (Fig. 5.1). It is as if an optical astronomer had to observe the sky in daytime, through a telescope coated inside and out with luminous paint! Infrared astronomers have tried hard to overcome these problems. They design telescopes so that no stray radiation from struts or supports can make its way to the infrared detector at the focus. In addition, they reduce the problem of sky brightness by the technique of 'nodding'.

The telescope is pointed at the object of interest – perhaps a star selected from the Two Micron Survey – and its brightness is measured. The infrared radiation picked up includes radiation both from the source and from the Earth's atmosphere. The telescope is then moved slightly so it is no longer 'seeing' the source, but is picking up only radiation from the sky. By subtracting the second measurement from the first, it is possible to discover how much radiation the source is producing. In practice, the technique gives more accurate results if the telescope is continuously 'nodded' back and forth between the infrared star and nearby empty sky. And rather than nod a bulky telescope, it is easier to wobble the secondary mirror at the top of the telescope, which moves the images back and forth over the fixed detector below, to give the same effect.

By nodding between the source and blank sky, an infrared astronomer can measure the brightness of a star, but this technique does not produce an image of an extended source like a nebula. Only recently have techniques in the infrared been developed to produce images of such sources, and the infrared images in this book have been obtained within the past three or four years. The simplest method is to let just a small region of the nebula fall on the detector, and measure its brightness in the usual way; by scanning the telescope backwards and forwards over the nebula, an image can be built up. This is stored in a computer, and can be processed by any of the standard programmes now used in optical astronomy (Chapter 3). The scanning must, however, be very precise, because any inaccuracy in the telescope motion will blur the resulting image. The Anglo-Australian Telescope in

New South Wales is highly suitable for such work. Even though it is essentially an optical telescope, infrared astronomers have used this outstandingly precise computer-controlled instrument equipped with an extremely sensitive indium antimonide detector to build up detailed infrared images of the outer planets (Figs. 2.23 and 2.36) and the centre of our Galaxy (Fig. 8.15) at wavelengths of a few microns.

Astronomers at the University of Wyoming have developed another imaging technique for use on their 2.3 metre infrared telescope (Fig. 5.3), which involves a modification of the wobbling secondary mirror system. In the Wyoming telescope, the secondary mirror shifts the detector's field of view up and down, from source to sky and back, ten times per second. In normal use, the view shifts back to the same small region of the nebula each time. But the wobbling can be modified so that on each wobble the detector's view is returned to the region just below. As the wobbling progresses, the detector is thus made to scan a vertical strip of the nebula. In only 6.4 seconds it scans a strip of 64 pixels. The telescope is then moved slightly sideways, and an adjacent vertical strip is scanned in the same way. In seven minutes, a two-dimensional image of 64 strips each containing 64 pixels can be built up – although in practice most sources are so faint that the observation must be repeated several times to increase the sensitivity. The Wyoming astronomers have produced unique images at longer wavelengths, up to 20 microns, of objects like NGC 7538 (Fig. 5.2), the centre of our Galaxy (Fig. 8.14) and the infrared sources in the Orion Nebula (Fig. 4.22).

The future, however, lies with infrared cameras which can record an image in a single exposure. Researchers at the Goddard Space Flight Center in Maryland have developed a prototype to work at ten microns wavelength (Fig. 5.4). In principle, the device is very like the CCDs now widely used in optical astronomy. The detector is a single chip of silicon containing a small proportion of bismuth, and its surface is made up of 32 rows containing

**Fig. 5.3** *The Wyoming Infrared Telescope is amongst the largest, with a 2.34 metre main mirror. Sited at an altitude of almost 3000 metres, on Jelm Mountain near Laramie, it has produced pioneering images at middle infrared wavelengths.*

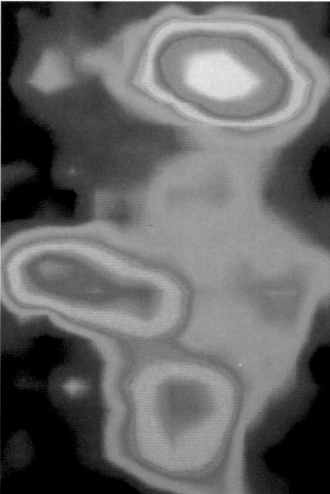

32 sensitive elements, each acting as a separate infrared detector. The main difference from an optical CCD camera is that the telescope's secondary mirror is wobbled to subtract the sky signal from the picture that the camera is building up. The prototype camera has produced superbly detailed views of the Orion Nebula infrared sources (Figs. 4.23 and 5.5) and of planetary nebulae like NGC 7027 (Fig. 6.5).

Despite all the effort and complexity involved in helium-cooled detectors, specially designed telescopes and wobbling secondary mirrors, infrared astronomers can still see relatively little through the ten and twenty micron windows if they have to look up from sea-level through the total thickness of atmosphere; and, from sea-level, they have no hope at all of detecting any of the longer wavelength infrared. The only solution is to raise the telescopes as high as possible, above the worst of the obscuring carbon dioxide and water vapour.

There are now several purpose-built infrared observatories around the world, perched on high mountains, sometimes in arid deserts. The best site of all is the peak of the Hawaiian mountain Mauna Kea, towering 4200 metres above the Pacific Ocean (described more fully in Chapter 3). As well as its optical telescopes, Mauna Kea boasts the world's two largest infrared telescopes. The American space agency NASA operates a 3 metre reflector built specifically to study the infrared radiation from the planets (Fig. 5.6). The huge United Kingdom Infrared Telescope (UKIRT) takes pride of place amongst Mauna Kea's giants, with a main mirror 3.8 metres across (Fig. 5.7).

The longer wavelength view gives astronomers a new perspective on the infrared stars from the Two Micron Survey. Some are just ordinary, very bright stars, which happen to lie behind dense clouds of dust. They actually produce far more light than 'heat' – infrared – but the dust dims their light to a thousandth or even a millionth of its original intensity. The infrared, on the other hand, passes through relatively unscathed.

Many infrared sources, however, are clouds of dust which shine most brightly at infrared wavelengths because they are at a temperature of a few hundred degrees above absolute zero, basking in the warmth of a nearby star. As a rule of thumb, infrared 'stars' are dust clouds associated either with very young stars or with very old stars – stars in the prime of life have no surrounding shell of dust.

The young infrared stars lie in dark dense clouds in space, where globules of gas and dust are contracting to form new stars. When a star first 'switches on' its nuclear reactions, the newly created energy warms up the surrounding dust to around room temperature, and the dust radiates away this energy as infrared waves. Eventually a new star disperses the dust 'cocoon' in which it was born, and shines uninhibitedly.

**Figs. 5.4–5.5** *The first true infrared camera (*Fig. 5.4*) uses the square lens (right) to refocus an infrared telescope's image onto a square array of 1024 infrared-sensitive pixels (hidden behind the wheel). The wavelength is selected by interposing a filter from among the eight on the wheel. The large cylinder (below) is part of a liquid helium system chilling the array and its electronics to 10 K. The picture of the Orion infrared cluster (*Fig. 5.5*) was taken with this camera on the IRTF (*Fig. 5.6*), using a 12.75 micron filter. The colour coding for intensity shows brightest regions white, with dimmer parts pink, red, yellow, green, pale and dark blue. (See also Figs. 4.21–4.23).*

**Fig. 5.6** *The dome sheltering the second largest single-mirror infrared telescope – NASA's Infrared Telescope Facility (IRTF) – is perched on 4200 metre high Mauna Kea, Hawaii. Although built to study the planets, IRTF also observes star birth regions, galaxies and quasars.*

**Fig. 5.7** *The full Moon rises behind the dome of the largest single-mirror infrared telescope, UKIRT, on Mauna Kea. The dark triangle at the horizon is the shadow of Mauna Kea on the distant atmosphere.*

Towards the end of its life, a star begins to develop its own dust cocoon. It swells up in size, to become a red giant star a hundred or a thousand times larger than the Sun. Gas from its cool outer layers seeps out into space, and some of the atoms inside it clump together to form tiny dust grains of graphite and silicates: the star wreathes itself in its own soot and smoke, and provides enough heat to make the dust cloud shine brightly in the infrared.

Even from the heights of a mountain like Mauna Kea, the sky is permanently bright as seen in the middle infrared wavelengths around ten microns. Although the 'tricks of the trade' enable astronomers to study the infrared glow from particular objects, detectors must be placed above the atmosphere if they are to see the infrared sky as it really is. The US Air Force undertook just such a survey in the early 1970s. Ten different rocket flights carried small liquid-helium-cooled infrared telescopes above the atmosphere, spinning them round to scan the sky at wavelengths of 4, 11, 20 and 27 microns. In a total observing time of only half an hour, these flights scanned nine-tenths of the sky, and picked out about two thousand infrared stars and nebulae.

At these longer wavelengths, the Air Force survey was picking out cooler objects than the earlier Two Micron Survey had found. At 27 microns, in particular, the scanning telescope was seeing mainly the regions in the depths of the thick dark clouds where stars are born. The longer wavelength infrared shows us progressively cooler dust clouds: while ten micron radiation comes from dust at room temperature (300 K), the 27 micron infrared is radiation from dust at 100 K ($-173°$ C). As we go to longer and longer wavelengths, we are seeing the progressively younger embryos of stars.

To understand the formation of a star, astronomers must work at wavelengths of a hundred or several hundred microns – in the *far infrared*. As seen from Earth, the sky is quite dark at these wavelengths – the atmosphere does not emit much far infrared radiation of its own. But the other bane of the infrared astronomers' life is a major problem: the molecules of water vapour and carbon dioxide absorb practically all the far infrared coming from space; the radiation cannot even penetrate down as far as the highest mountain peaks.

Astronomers must position their infrared detectors to intercept this radiation before the atmosphere can absorb it. One very effective method seems at first sight surprisingly primitive: the telescope and its detector can be hung from a balloon. By making the infrared equipment automatic, balloons that carry it up to heights well above the limit for manned balloon flights can be used. Such infrared telescopes at altitudes of 30 kilometres or more are very useful for looking at the radiation from particular objects. It often helps to have an astronomer at the telescope's controls, however, and with this in mind, the

American space agency NASA has converted a C-141 *Starlifter* transport plane into a flying observatory. Named after leading planetary astronomer Gerard P. Kuiper, the Kuiper Airborne Observatory (Fig. 5.8) carries a 0.9 metre telescope specially designed to observe at long infrared wavelengths.

But the best way to see the far infrared is to get above the Earth's atmosphere altogether. For this purpose, infrared astronomers from the United States, the Netherlands and the United Kingdom constructed the Infrared Astronomy Satellite (IRAS) which, during 1983, studied the sky from its orbit at an altitude of 900 kilometres. But even IRAS could not escape the problem of having a 'luminous telescope'. The satellite was basking in sunlight, and at some wavelengths its own heat radiation would have swamped the feeble infrared radiation from space. The designers had to enclose the 0.6 metre diameter IRAS telescope (Fig. 5.9) in a large jacket filled with 70 kg of liquid helium at a temperature of only 16 K, and with the detectors themselves cooled to 2 K.

Although astronomers had flown satellites to measure other radiations from space (X-rays, for example), IRAS was one of the most difficult missions ever attempted because of the difficulties of keeping its huge amount of helium coolant – about a man's weight of liquid – near the absolute zero of temperature, out of reach of human intervention. But the returns have repaid the effort handsomely. IRAS has surveyed the sky at the whole range of infrared wavelengths from 8 to 120 microns – the middle and far infrared wavelengths, which are difficult or impossible to study from the ground – with a sensitivity hundreds of times better than the balloon-borne detectors. As a result, its catalogue should contain a third of a million infrared stars, nebulae and galaxies. To put that into perspective: the most complete middle infrared catalogue before IRAS, the Air Force Survey, contains two thousand sources, whereas IRAS could pick up two thousand sources *per day*.

Both the shorter wavelengths of visible light, and the longer wavelength radio sky have now been well-explored, but the intermediate wavelengths of the infrared are still frontier territory in the study of the Universe.

**Fig. 5.9** *The Infrared Astronomical Satellite (IRAS) — seen here undergoing prelaunch tests — has a 0.6 metre reflecting telescope. The mirror is to the left (hidden within the satellite); the open end (right) is protected here by a convex cover, ejected once in orbit. The large cowl shades the telescope from the Sun's infrared; the bulky middle section is a tank of liquid helium for cooling the telescope and detectors; the satellite's controls are in the short cylindrical section (far left).*

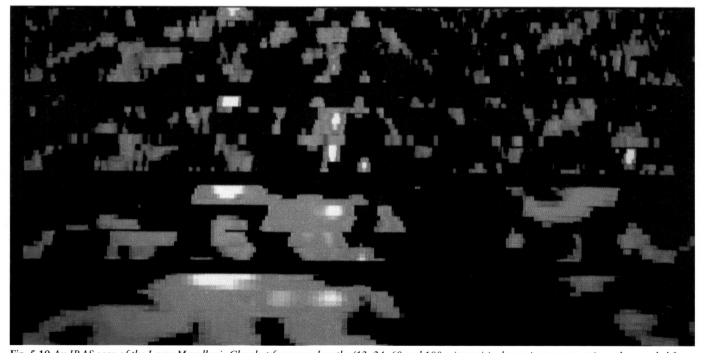

**Fig. 5.10** *An IRAS scan of the Large Magellanic Cloud at four wavelengths (12, 24, 60 and 100 microns) is shown in separate strips colour-coded for intensity. Large cool clouds (bottom) have small warm cores (top) where stars are forming.*

# 6 Stardeath

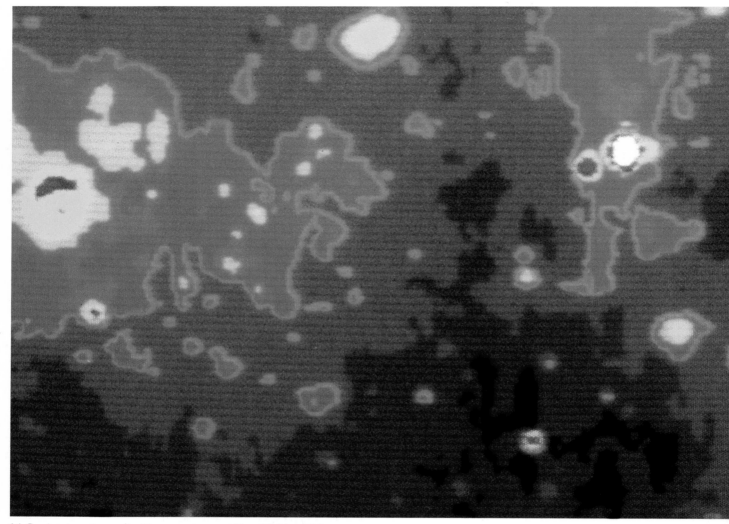

**6.1** *Cassiopeia region, radio, 21 cm continuum, 100 m Effelsberg Telescope*

The manner of a star's death depends on its mass. A relatively lightweight star like the Sun dies quite quietly. After it has expanded to become a red giant, it puffs off its gaseous envelope, as a huge, slowly expanding shell. The dense core is left as a tiny white dwarf star, with a temperature of around 100 000 K, emitting most of its radiation as ultraviolet. This heats up the surrounding shell, making its gases emit light and radio waves, while the rather cooler dust in the shell shines brightly in the infrared. Through a small optical telescope, such round shells look rather like the planet Uranus or Neptune, and they are called planetary nebulae.

A star more than six times heavier than

the Sun has a more eventful life and a spectacular death. As a red giant star, its core becomes hot enough to convert helium into a multitude of heavier elements. Within the helium core, nuclei combine to make carbon and oxygen; later, the hottest central regions convert carbon and oxygen nuclei to heavier elements like silicon; and the central temperatures and pressures within the silicon core eventually convert these nuclei to iron. At this point, the star's structure resembles an onion, with concentric shells of different composition from the external unchanged layer consisting mainly of hydrogen, through interior shells of helium, carbon and oxygen, and silicon, to the central iron

core. At each stage in the giant star's evolution, nuclear fusion reactions in the very centre have provided energy to support the core against the pressures generated by the weight of the star's overlying layers. But iron has the most stable of all nuclei: iron nuclei will not take part in further nuclear fusion reactions. The upshot is dramatic. The central iron core collapses into a ball of neutrons only 20 kilometres across. This neutron star is so dense that a pinhead of its material would weigh a million tonnes.

The energy from this central collapse heads outwards, hurling the star's outer layers into space at a tremendous rate – around 7000 kilometres a second. In this

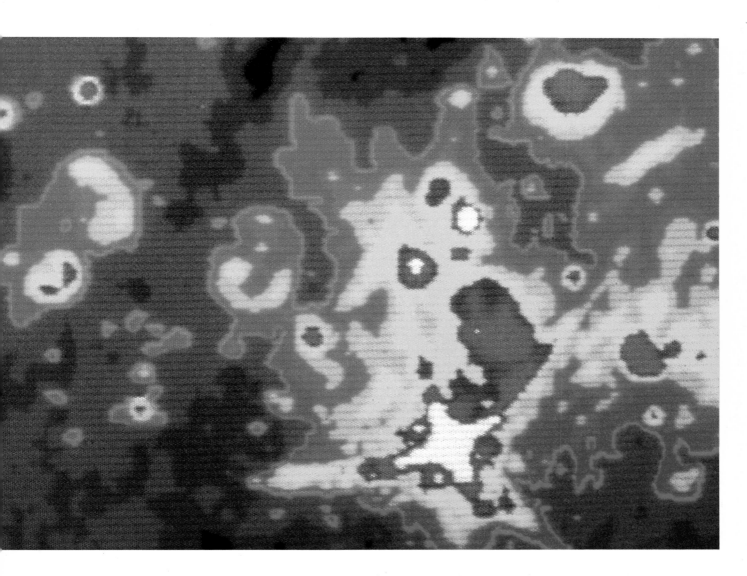

colossal explosion, a type II supernova, the star brightens up to become temporarily equivalent to 500 million Suns.

The less-understood type I supernovae are even more brilliant, and they throw out their gases at an even higher speed of around 11 000 kilometres a second. A type I supernova may be an old white dwarf, drawing gases from a companion star until it becomes unstable and starts to collapse uncontrollably. Conditions in a collapsing white dwarf make its contents of carbon and oxygen fuse directly into iron, in a reaction so rapid that the collapse is halted, and the white dwarf is blown apart.

The high-speed gases from either type of supernova sweep up the tenuous interstellar gas, and form huge shells – the supernova remnants. These shells of gas are too hot to emit light, but they shine brightly at X-ray and radio wavelengths. Young remnants appear as small, intense rings to X-ray and radio telescopes; older ones are larger and fainter.

The radio view of part of the Milky Way which lies in Cassiopeia (**Fig. 6.1**) is dominated by supernova remnants. It covers 12° from left to right (24 Moon breadths), with the Milky Way running horizontally. It was made at a wavelength of 21 centimetres (but not at the precise wavelength of the hydrogen line at 21.106 centimetres). Black represents the dark background and blue the general Milky Way emission, with more intense regions green, yellow, red and white. On the lower right is Cassiopeia A (Figs. 6.23–6.27), the strongest radio source in the sky (the cross shape is an artefact of the telescope). It is the remains of a supernova of the seventeenth century. Tycho's supernova remnant (Figs. 6.20–6.22), seen as a white spot above the centre, is a century older. Very ancient remnants have expanded more and are visible as 'rings' on this scale: for example, the remnant just to the right of centre, CTB 1, is over 10 000 years old and a degree across. The red semicircle on the extreme right is the strange remnant G109.1-1.0, which contains an active neutron star (Figs. 6.39–6.41).

# NGC 7027

This planetary nebula is one of the youngest and smallest known. It is a shell of gas and dust shed some 2000 years ago by a red giant star, and since then it has grown to only one-fifth of a light year in diameter. At its distance of 5000 light years, the nebula is only a few arcseconds in apparent size. Although NGC 7027 is bright enough to be seen in binoculars (at magnitude 8.5), its small size led to the nebula being mistaken for a star up until 1879, when an amateur astronomer, the Reverend T. W. Webb, made out its slightly extended shape – in time for it to be included in the definitive New General Catalogue of Nebulae and Clusters of Stars, published in 1888.

**Fig. 6.2** shows gas which has been shed from the old red giant star, and observations of Doppler shifts in the spectral lines from this gas reveal that it is moving outwards at 22 kilometres per second – quite a modest speed in astronomical terms. (The nebula's age is calculated by dividing its present radius by this speed.) The expansion of the red giant's outer layers has left bare its extremely hot core, as a new white dwarf star, in the centre of the nebula. This star is at a temperature of about 200 000 K: it is faint optically, and is not visible in Fig. 6.2, but its strong ultraviolet radiation was easily detected by the IUE satellite. **Fig. 6.3** is the spectrum obtained by IUE. The radiation has been spread out by wavelength along a band from the upper left to the lower right, and the intensity coded by colours which run from grey for the dark background, through white, blue and green, to red for the most intense features. The strong spectral lines, seen as red blobs, are emission from magnesium atoms which have been stripped of electrons by the intense heat of the white dwarf's surface.

The nebula seen in Fig. 6.2 is heated up by the ultraviolet radiation from the central white dwarf, but the light from the glowing gases is also absorbed by dust grains in the nebula. Where the gas is dense and bright, the dust is also thick and opaque, and so the optical nebula looks rather featureless and certainly does not reveal the nebula's true shape. Observations at longer wavelengths can, however, show the gas and dust of the nebula separately.

The hot gas generates radio waves, and these long wavelengths are completely unaffected by dust. The radio view (**Fig. 6.4**), to a slightly larger scale than Fig. 6.2, reveals the true distribution of hot gas – a symmetrical oval around the central star 0.2 light years long and 0.17 light years across (the central star is undetectable at radio wavelengths). Here the dimmest regions are coloured blue, and brighter regions red, pale green, pale blue and

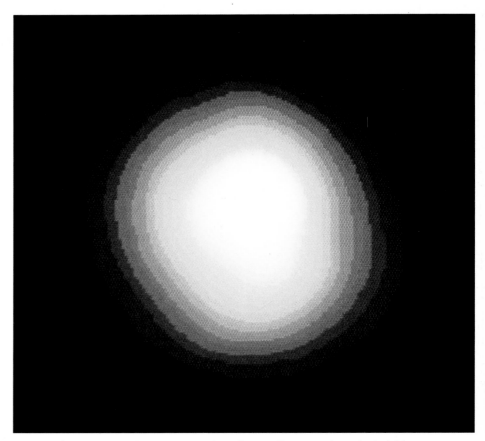

6.2 *Optical, 650–660 nm, Reticon array, 1.8 m telescope, Dominion Astrophysical Observatory*

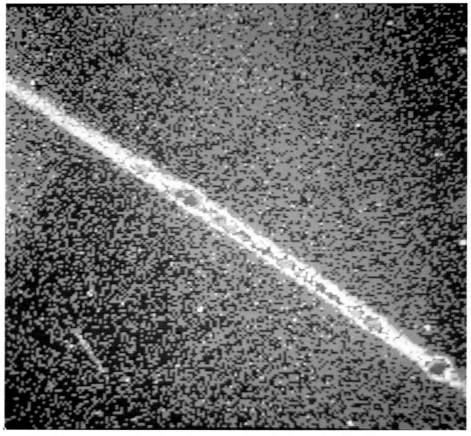

6.3 *Ultraviolet, spectrum 180–320 nm, International Ultraviolet Explorer Satellite*

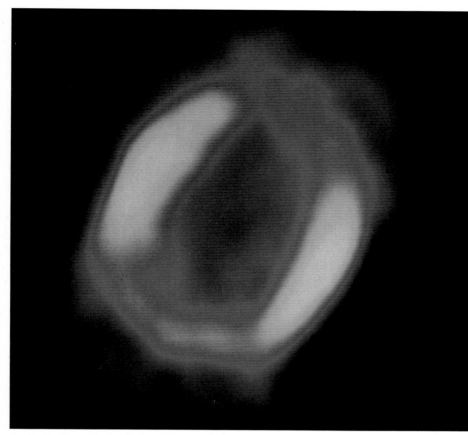

6.4 *Radio, 2 cm, Cambridge Five Kilometre Telescope*

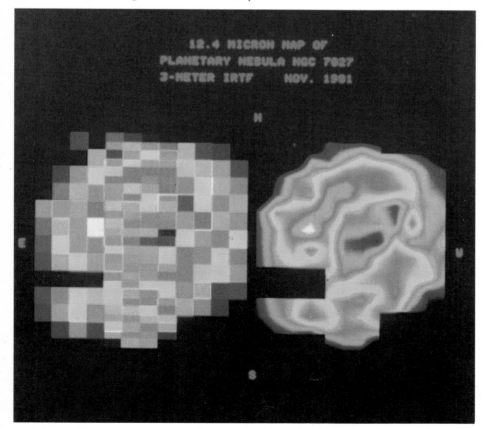

6.5 *Infrared, 12.4 microns, raw data (left) and smoothed (right), 3 m NASA Infrared Telescope Facility, Mauna Kea*

white. There is a largely empty region around the central star itself. The gas is brightest at the sides of the oval, probably because here it is closest to the hot central star. The nebula's radio brightness shows that the gas is at a temperature of 14 000 K, rather hotter than the 10 000 K gas in nebulae surrounding young stars (like the Orion Nebula, Fig. 4.17). The total emission from the gas makes NGC 7027 the brightest planetary nebula in the sky at radio wavelengths.

As the shell of gas has expanded, some of the heavier elements – like silicon and carbon – have condensed to form small grains of dust. The grains are at a temperature of 90 K, and they naturally emit radiation at infrared wavelengths. The structure of dust in the nebula thus shows up in the infrared picture (**Fig. 6.5**). These observations were made with a camera developed by John Arens and colleagues at the Goddard Space Flight Center, which records the brightness in individual small squares, each 0.7 arcseconds across (smaller than the detail that an optical telescope can resolve). The two original sets of observations are combined in the left-hand picture of Fig. 6.5, and they have been smoothed by computer to produce the more realistic version on the right. In both representations, the faintest levels are coded blue, and brighter regions green, yellow, blue and white (the black region on the left is due to defective elements in the camera).

The dust clearly forms a shell similar to the hot gas shell revealed in Fig. 6.4, although the dust shell is rather thicker, and the distribution of dust differs in detail from that of the gas. In particular, the dust does not show the strong brightening to either side of the oval shell, because even at the ends of the shell the dust can be heated to the low temperatures required to emit infrared, and its distribution is rather more clumpy. NGC 7027 is the brightest planetary nebula at infrared wavelengths as well as radio, for its dust absorbs most of the central star's ultraviolet radiation and emits it in the infrared with a power equal to the total output of 20 000 Suns.

6.6 *Optical, true colour, 5 m Hale Telescope*

## Crab Nebula

In the early morning of 4 July 1054, watchful Chinese astrologers saw a bright new star rising in the east just before the Sun. Over succeeding days, the 'guest star' brightened until it outshone all the other stars in the sky (reaching magnitude −4). For a period of three weeks it was brilliant enough to be visible in daylight; then it faded, until by April 1056 it had disappeared from sight even on the darkest night.

The 'guest star' has since been identified as a type II supernova: the explosion of a heavyweight star at the end of its life. The wreckage of the star is still expanding outwards, as the Crab Nebula (**Fig. 6.6**). It is some 6500 light years away from us, and over the nine centuries since the explosion has expanded into an egg shape about 15 light years long and 10 light years across.

English astronomer John Bevis first noticed this dim, nebulous patch in the constellation Taurus in 1731. Twenty seven years later, it was rediscovered by the French comet-hunter Charles Messier, who put it first in his famous catalogue of nebulae. Hence the Crab is also known as M1, or from its listing in the 1888 New General Catalogue as NGC 1952. The great nineteenth century amateur astronomer, the third Earl of Rosse, saw for the first time 'resolvable filaments . . . springing principally from its southern extremity'. These reminded him of the legs and pincers of a crab – hence the nebula's popular name.

The Crab is a moderately bright nebula in visible light, but it is far from being a conspicuous object. At magnitude 8, the Crab is too faint to be seen with the unaided eye. When observed at other wavelengths, however, it ranks with the most brilliant celestial objects. As a result, it has always been one of the first objects identified when astronomers have explored new wavelengths. In 1949, Australian radio astronomers pinned down the position of the intense radio source 'Taurus A', and identified it as the Crab Nebula. The story was repeated in the early years of X-ray astronomy, by a rocket flight in 1964 which discovered and identified the Crab Nebula as an X-ray source.

Four years later, radio astronomers found that the 'power-house' of the Crab Nebula is a pulsar at its centre. This is the youngest pulsar known, one of the fastest pulsing and it was the first to be associated with a known supernova event. The Crab Pulsar is visible in this photograph as the lower of the two central stars. Its light is, in fact, flashing thirty times per second. It was the first pulsar to be detected flashing at optical and gamma ray wavelengths. At the latter wavelengths, the pulsar outshines the rest of the nebula, and is the third brightest gamma ray 'star' in the sky.

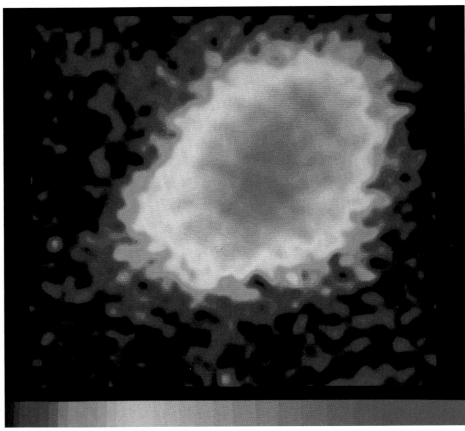

6.7 *Radio, 0.9 cm, 100 m Effelsberg Telescope*

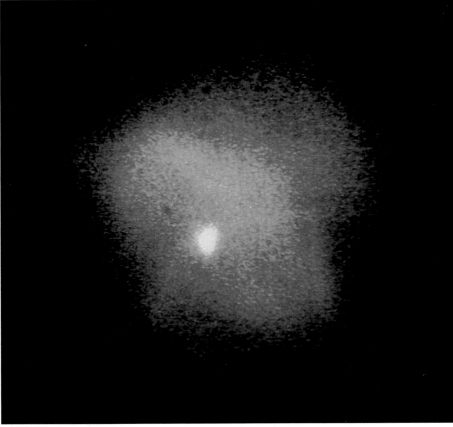

6.8 *X-ray, 0.4–8 nm, Einstein Observatory HRI*

The Crab Nebula is in fact two nebulae in one. The reddish filaments prominent in optical photographs (**Fig. 6.9**) form the twisted bars of a 'cage' which encloses a much smoother, bluish-white *synchrotron nebula*. This nebula's magnetic field is twisting the paths of fast electrons streaming outwards from the Crab Pulsar, (Figs. 6.17 and 6.18), causing them to broadcast radiation at wavelengths corresponding to their speeds. The fastest electrons produce the shortest (X-ray) wavelengths and the slowest the long radio waves. The Crab Nebula's bluish-white glow (some four-fifths of its light output) is the only source of synchrotron light that can be seen using only a moderate telescope or even good binoculars.

The radio view (**Fig. 6.7**) shows the full extent of the nebula's magnetic field, stretching over an oval region some 5 by 7 arcminutes in size – as large as the 'cage' of filaments seen optically. The synchrotron nebula is decidedly smaller when observed at shorter wavelengths. The optical picture taken with a CCD detector (**Fig. 6.10**) shows contrasts of 1000:1 in brightness, and reveals that the brightest part of the synchrotron nebula is only 3 by 5 arcminutes across – well within the filaments. In X-rays, the Crab Nebula is little more than 1 arcminute in size: the picture here (**Fig. 6.8**) is magnified three times relative to the radio map.

The Crab Nebula's magnetic field is itself responsible for this apparent change in size with wavelength. The torrents of electrons from the pulsar cover a wide range of speeds, and near the centre of the nebula the electrons generate the whole range of wavelengths of radiation. The electrons are, however, detained by the contortions of the magnetic field and remain in the nebula for a couple of centuries. During this time, the energy of each electron is sapped by the continuous emission of radiation, and so the electron gradually slows down.

The slower the speed of the electron the slower the energy loss. The radio wave emitters can percolate right through the field, 'lighting up' the whole nebula at radio wavelengths. A faster light-emitting electron will traverse only three-quarters of the magnetic field before it is slowed down to a speed at which it produces infrared instead. Hence the visible nebula appears smaller than the radio one. The very fast electrons which produce the nebula's X-rays penetrate and 'light up' a little more than a tenth of the magnetic field before losing their inordinate energy. The X-ray nebula (Fig. 6.8) is actually not symmetrical around the pulsar (the small bright circular patch), suggesting that the pulsar is ejecting the electrons mainly towards the top right of the nebula.

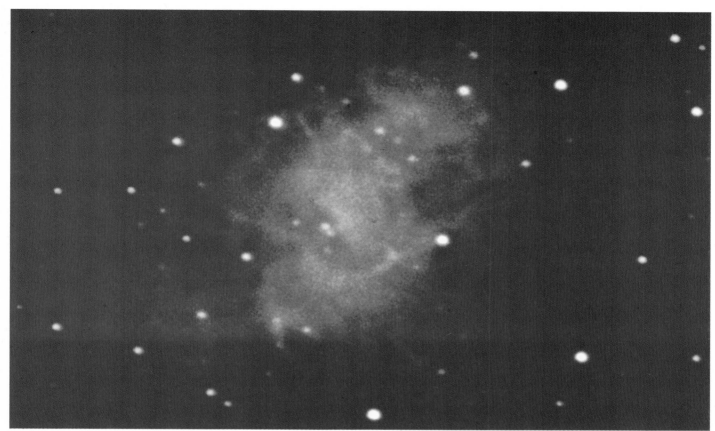

**6.9** *Optical, enhanced true colour, 1 m telescope, US Naval Observatory*

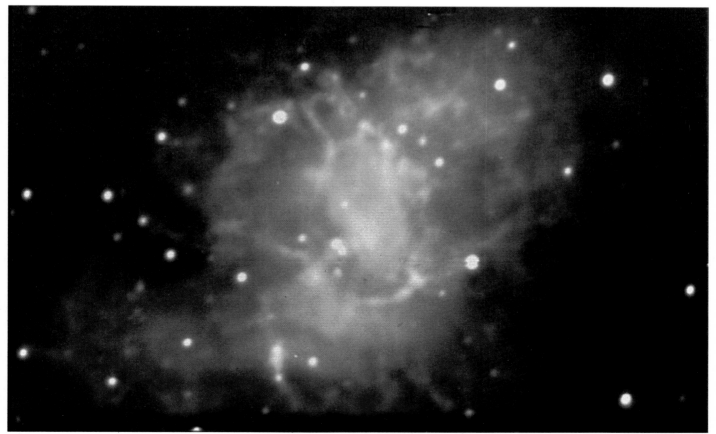

**6.10** *Optical, 570–720 nm, CCD, 0.6 m telescope, Smithsonian Astrophysical Observatory*

Optical photographs taken through polarising filters show dramatically that the Crab Nebula's central regions shine by synchrotron radiation. A polarising filter, like the 'lens' in a pair of polaroid sunglasses, passes light waves vibrating in only one direction – say 'up and down' or 'left and right'. Whenever an electron produces light, the wave is polarised in the direction of the electron's motion. In an ordinary hot gas nebula, electrons moving in different directions produce light waves of all possible orientations. But if a nebula's light comes from synchrotron emission, the magnetic field is controlling the electrons' motions. In a region of the nebula where the field is uniform, all the electrons are moving in the same direction, and so produce light with the same direction of polarisation. As seen through a polarising filter set to pass this light, the emitting region looks bright. On turning the filter through 90°, it blocks off all this light, and this part of the nebula disappears.

These two photographs were taken through polarising filters set to pass only light vibrating up and down (**Fig. 6.11**) and left and right (**Fig. 6.12**) – directions indicated by the arrow marked EV (electric vector of the radiation). The startling difference between such polarised pictures first proved in the 1950s that the bluish-white light from the Crab Nebula is indeed synchrotron emission, as had been predicted by the Russian theorist Iosef Shklovsky.

The polarised appearance of the Crab reveals the structure of its magnetic field. Each electron is following a tightly coiled path round a magnetic field line, and as it travels round and round many times while moving only a short way along, its motion is primarily at right angles to the direction of the magnetic field line. The radiation from any part of the nebula is thus polarised at right angles to the magnetic field direction there.

Regions which are bright in the upper photograph (Fig. 6.11) are parts of the nebula where the magnetic fields runs left and right (East–West) – for example, the regions just below the centre and towards the top left. Conversely, the lower photograph (Fig. 6.12) emphasises regions of the synchrotron nebula where the field lines are up and down (North and South). Note the projection to the right of the centre, which is invisible in the top picture.

The polarisation photographs also emphasise the synchrotron nebula's fine wisps – 'ropes' of more intense magnetic field. Since the magnetic field lines run along the wisps, the lower picture emphasises the wisps which run up and down the nebula, while the upper shows up the wisps going right and left. On ordinary unpolarised photographs, these wisps are superimposed and are more difficult to discern.

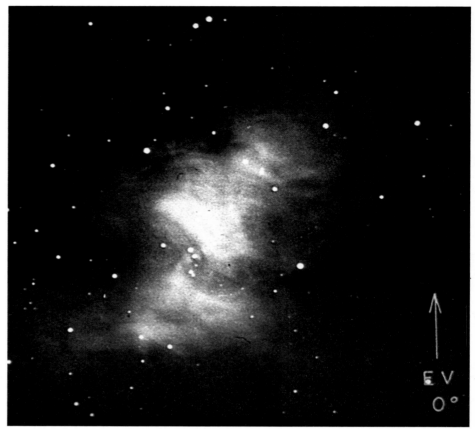

6.11 *Optical, polarised (0°), 5 m Hale Telescope*

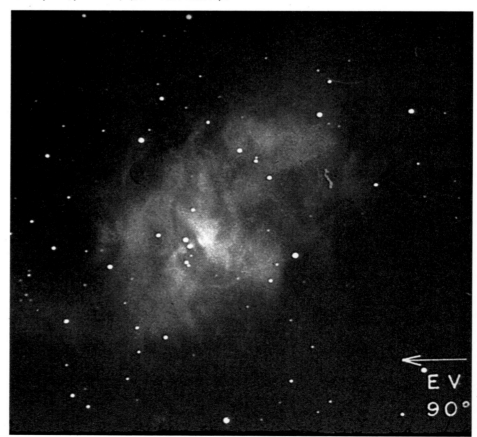

6.12 *Optical, polarised (90°), 5 m Hale Telescope*

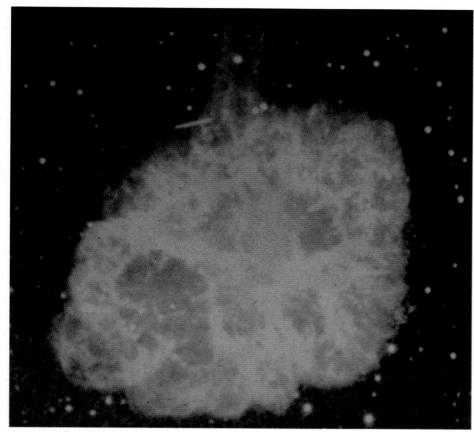

6.13 *Optical, 501 nm oxygen line, 0.9 m telescope, Kitt Peak National Observatory*

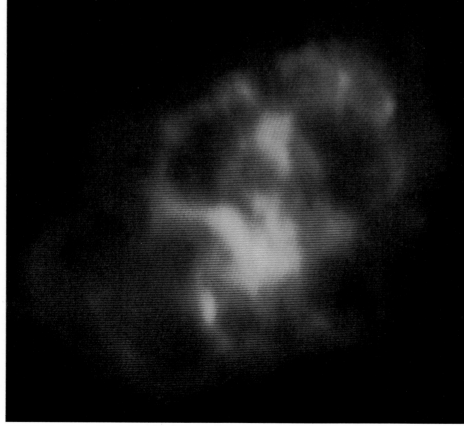

6.14 *Radio, 11 cm, Cambridge Five Kilometre Telescope*

The glowing gas filaments appear prominently in this photograph (**Fig. 6.13**), taken through a very narrow-band filter. It passes only light with a wavelength close to the green emission line from oxygen at 501 nanometres. The synchrotron nebula is almost invisible, and we see the full extent of the filaments. Compare this with Fig. 6.10, for example, which covers the optical synchrotron nebula but only the central half of the filamentary nebula. This picture also shows a strange double jet projecting upwards (North) from the nebula's edge. It could be a tube of gases seen sideways-on. Whatever the structure, it is difficult to explain the straight, parallel pair of jets which contrasts so markedly with the twisted main filaments.

The filaments contain two-and-a-half times as much matter as the Sun, spewed out by the supernova explosion. Nuclear reactions within the original star have converted much of its hydrogen to helium, so the filaments are unusually rich in helium. But there are normal proportions of oxygen – producing the light seen here – and carbon, which produces spectral lines in the ultraviolet. Japanese theorist Ken'ichi Nomoto has calculated that this particular mix of elements would be produced by a star eight times heavier than the Sun. Half this matter is now in the filaments and the central neutron star – the Crab Pulsar (Figs. 6.17–6.18). The outer four solar masses of the original star lie in a large halo surrounding the Crab Nebula (too faint to show on direct pictures like this).

Detailed radio maps also reveal the Crab's filaments. Here (**Fig. 6.14**), the dimmer outer regions of the nebula are coded blue, with brighter regions from red to white at the most intense spots. The filaments show up prominently near the centre, at the same locations but rather thicker than the main optical filaments. This radio emission is, however, synchrotron radiation. Since the Crab Pulsar 'pumps up' the synchrotron nebula with electrons and magnetic field, it expands faster than the cage of filaments. The magnetic field can expand freely between the filaments, but it is obstructed where it runs into one of them: it is compressed behind and around the filaments, until its strength is doubled. The gas filaments thus have magnetic casings which radiate more powerfully than the rest of the synchrotron nebula, and appear as the broad 'radio filaments' running along and around the gas filaments themselves.

Because the expanding synchrotron nebula is pushing outwards on them, the filaments are travelling slightly faster than they would if they had been moving at a constant speed since 1054. As a result, a simple division of the nebula's size by the filaments' speed gives a wrong age for the Crab – 90 years younger than it actually is.

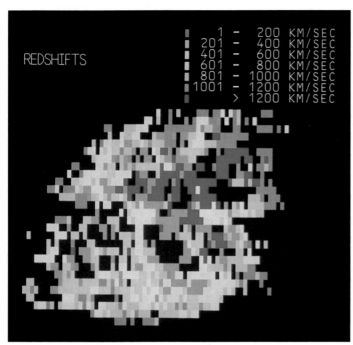

| | 1 | - | 200 | KM/SEC |
| | 201 | - | 400 | KM/SEC |
| REDSHIFTS | 401 | - | 600 | KM/SEC |
| | 601 | - | 800 | KM/SEC |
| | 801 | - | 1000 | KM/SEC |
| | 1001 | - | 1200 | KM/SEC |
| | > | | 1200 | KM/SEC |

| | 1 | - | 200 | KM/SEC |
| | 201 | - | 400 | KM/SEC |
| BLUESHIFTS | 401 | - | 600 | KM/SEC |
| | 601 | - | 800 | KM/SEC |
| | 801 | - | 1000 | KM/SEC |
| | 1001 | - | 1200 | KM/SEC |
| | > | | 1200 | KM/SEC |

*6.15 Optical, 501 nm oxygen line, velocity coded (redshifts), IPCS, 3.9 m Anglo-Australian Telescope*

*6.16 Optical, 501 nm oxygen line, velocity coded (blueshifts), IPCS, 3.9 m Anglo-Australian Telescope*

The false-colour photographs (**Figs. 6.15** and **6.16**) are colour-coded velocity maps, showing the speed at which the filaments are expanding – and, incidentally, revealing the structure of the Crab in three dimensions.

Paul Murdin, David Clark and John Danziger made these maps by scanning the 3.9 metre Anglo-Australian Telescope across the Crab Nebula. They measured the spectrum at 1800 points with a spectrograph fitted with the very sensitive IPCS light detector (p. 51), concentrating on the spectral line emitted by oxygen atoms at a wavelength of 501 nanometres (in the green part of the spectrum). At each point, the oxygen line appears split into two lines closely spaced in wavelength. One component is the light from filaments at the back of the nebula, shifted to a slightly longer wavelength (redshifted) by the Doppler effect, as they are moving away from us. The other is a blueshifted line from filaments on the near side coming towards us.

To make these photographs, the values of the redshift and the blueshift at each point have first been converted to the velocity of the receding and approaching gases. The receding and approaching velocities have then been treated separately, to make two velocity maps. Fig. 6.15 shows the receding filaments on the far side of the nebula, with the colour coding indicating the speed. The scale runs from grey (for velocities less than 200 kilometres a second) through white, yellow and pink to red. The red regions show gas that is racing away at over 1200 kilometres per second.

A similar velocity-coded map for the blueshifted gases (Fig. 6.16) shows the filaments on the nearside speeding towards us. The colour coding runs similarly, from green for gases travelling at less than 200 kilometres per second through shades of pale blue to the dark blue which represents filaments coming towards us at over 1200 kilometres per second.

The two pictures together thus show the entire expanding cage of filaments. Since the time of the explosion, the filaments at the edge of the nebula have moved predominantly sideways rather than towards and away from us – such filaments display only small velocities in these maps and they have also moved comparatively little in 'depth'. Conversely, the filaments seen near the centre of the Crab have the highest speeds along the line of sight, and they have moved furthest in depth. Hence these colour maps also represent the relative distances of the filaments from us.

Think of the left picture (Fig. 6.15) as a geographical contour map of a basin, with its deepest point where the red is most intense, to visualise the far side of the filamentary network in three dimensions. Fig. 6.16 is a 'hill', with dark blue marking the nearest part of the network. The Crab's shape thus resembles an egg (viewed sideways-on), slightly 'waisted' around the middle.

The collapsed core of the exploded star remains in the centre of the nebula as the Crab Pulsar. This neutron star is fifty per cent more massive than the Sun, but only 20 kilometres across, and it spins around thirty times every second.

The Crab Pulsar has a magnetic field a million million times more intense than the Earth's magnetism. The magnetic field and the star's rapid rotation accelerate electrons to speeds of almost the velocity of light, and as a result the neutron star generates synchrotron radiation of all wavelengths. This radiation shines out as two beams from above the star's magnetic poles. Because the magnetic poles are not at the neutron star's poles of rotation, the two beams are swept around during the star's rotation like the beams of a lighthouse. As the beams sweep past the Earth, we pick up regular pulses of radiation – hence the term *pulsar* for the neutron star.

American radio astronomers picked up radio pulses from the Crab Nebula in November 1968. The following year, optical astronomers at the Steward Observatory, Arizona, discovered that the optical radiation from the central star is also pulsing. The pulses recur so rapidly that they are smoothed out in normal photographic exposures. But stroboscopic techniques can show the star 'pulsing'. When it is 'on' (**Fig. 6.17**, left), we are looking straight down one of the pulsar's intense radiation beams; when it is 'off' (Fig. 6.17, right), the beams are pointing away and we see nothing.

The Crab Pulsar is even more evident in X-rays, and stroboscopic pictures at these wavelengths also show the pulsar 'on' (**Fig. 6.18**, left) and 'off' (Fig. 6.18, right). Detailed analysis of the images reveals the pulsar faintly even when the beams are pointing away. The neutron star's surface is thus a weak source of X-rays itself, indicating that it is incredibly hot, with a temperature of two million degrees.

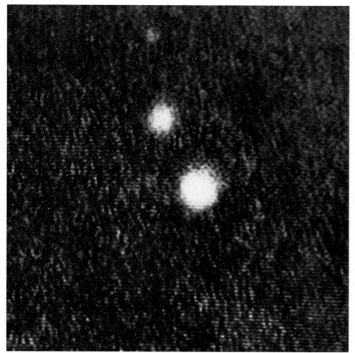

**6.17** *Optical, 3 m telescope, Lick Observatory*

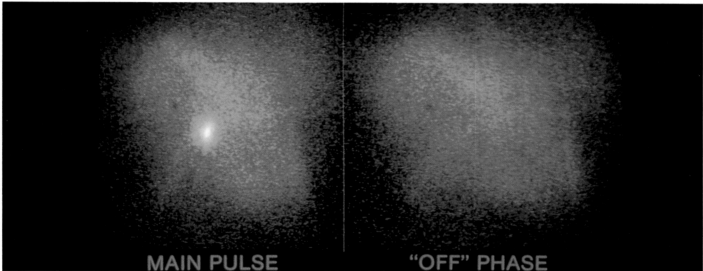

MAIN PULSE      "OFF" PHASE

**6.18** *X-ray, 0.3–2.5 nm, Einstein Observatory IPC*

The Crab Pulsar is one of the fastest rotating neuton stars known, but it is gradually slowing down. The pulse interval of about 1/30 second is increasing by a hundred-thousandth of a second each year. This indicates that the neutron star is losing rotational energy at a rate equal to the luminosity of 100 000 Suns. Its magnetic field is transferring this energy to the Crab's synchrotron nebula (Figs. 6.7–6.12), and accelerating the electrons which make it shine. In effect, the pulsar's loss of rotational energy is lighting up the nebula in the same way that the rotational energy of a heavy flywheel attached to a dynamo can be used to power a light bulb, and is slowed down as a result.

This outflow of energy causes noticeable changes in the synchrotron nebula in the space of just a few years. American astronomer Jeffrey Scargle has emphasised these by superimposing two optical photographs taken nine years apart (**Fig. 6.19**) – one a negative and the other a positive. Regions of the nebula which have stayed the same brightness turn out grey in the composite, as does the sky background. But most of the synchrotron nebula has changed in intensity, with some regions becoming brighter (white on the composite) and some darker (black). The nebula's brightness is changing in great waves, about a light year in extent, sweeping out from the pulsar.

**6.19** *Optical, superposition of 1955 (negative) and 1964 (positive) plates, 5 m Hale Telescope*

## Tycho's supernova remnant

The great Danish astronomer Tycho Brahe was in for a surprise as he took his early evening stroll on 11 November 1572. As he looked up, 'directly overhead, a certain strange star was seen, flashing its light with a radiant gleam'. The new star was in the constellation Cassiopeia, and night after night Tycho carefully measured its position and its brightness compared to the stars and planets visible at the same time. A few days later, the star reached its maximum brightness, shining as brilliantly as Venus with a magnitude of −4. Then it gradually faded until it became invisible some fifteen months later.

From Tycho's careful measurement of the star's fading, we can deduce that it was a type I supernova. The remnant it has left is extremely faint at optical wavelengths. Canadian astronomer Sidney van den Bergh took this optical photograph (**Fig. 6.20**) with the Hale 5 metre telescope in 1970. Even with such a large telescope and a two-hour exposure, the wisps of glowing gas hardly show. The brightest of these long, extremely thin filaments lies near the left-hand edge of the picture and runs vertically. Above and to the right, a more diffuse patch of wisps heads towards the top right, and beyond these is another isolated wisp at right angles. In this photograph through a red filter, we are seeing the light emitted by hydrogen atoms.

Most of the gas in the supernova remnant is too hot to produce light, but at a temperature of 40 000 000 K it shines brilliantly in X-rays. The picture from the Einstein X-ray observatory (**Fig. 6.21**) shows the remnant as a complete, almost circular shell of gas – appearing brightest at the outer edge because we are here looking through a greater thickness of shining gas. This celestial ring is about 8 arcminutes in diameter, one-quarter of the Moon's apparent size. Since Tycho's remnant is about 7500 light years away from us, we can calculate that the gas shell has grown to a size of 17 light years during the past four centuries.

The radio photograph from the Cambridge Five Kilometre Telescope (**Fig. 6.22**) shows the remnant in even finer detail. In this picture, the dimmest parts – mainly in the centre – are coded blue, with brighter regions green and red, shading to purple and white for the most intense radio-emitting regions at the top left of the gas shell. Unlike the X-rays, which come from the atoms of hot gas in the shell, the radio waves are synchrotron radiation caused by high-speed electrons whirling in the magnetic field trapped within the shell. In the radio view, we can see that the outer edge of the remnant appears extremely sharp: this edge is the shock front where the shell is sweeping up fresh interstellar gas to add to its growing bulk.

**6.20** *Optical red light, negative print, 5 m Hale Telescope*

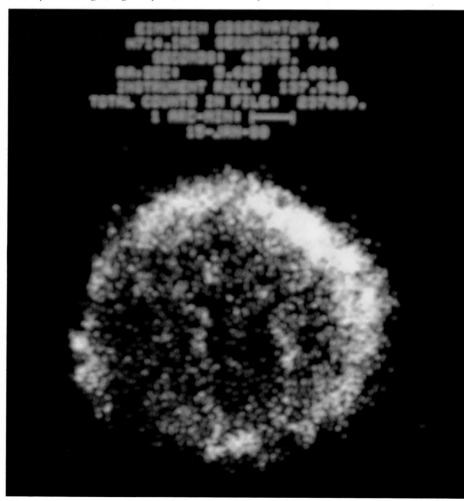

**6.21** *X-ray, 0.4–8 nm, Einstein Observatory HRI*

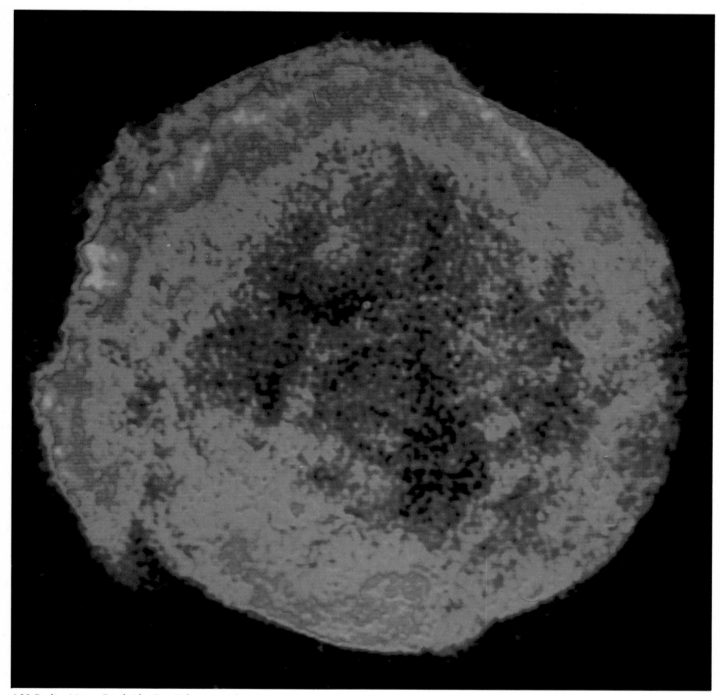

**6.22** *Radio, 11 cm, Cambridge Five Kilometre Telescope*

The remnant's outline, seen at X-ray and radio wavelengths, is slightly 'dented', where the shell has run into small, denser clouds of interstellar gas which have slowed its expansion. The dents at the left and at the top right coincide with the two most prominent optical filaments (Fig. 6.20) – those lying exactly along the edge of the shell that we see at radio wavelengths. Here the swept-up interstellar gas glows for a few years at optical wavelengths, as it is heated through a temperature of a few thousand degrees on its way to the multi-million degree temperature it will reach within the shock front.

Tycho's supernova remnant is clearly very different from the Crab Nebula (Figs. 6.6–6.19). Not only is it very faint when photographed at optical wavelengths, but the X-ray and radio emission come from a hollow shell of gases, while the Crab's radiation of all wavelengths is strongest at the centre. Whereas the Crab Pulsar powers the Crab's bright synchrotron nebula, there is no powerful radiation and energy source at the centre of Tycho's remnant. A young neutron star would be hot enough to show

up as a bright point in the X-ray picture, even if its beamed radiation were emitted in a direction which missed us.

6.23 *Optical, red light, 5 m Hale Telescope*

## Cassiopeia A

The most recent star explosion in our part of the Galaxy shows up on optical photographs merely as a few tatters of dimly glowing gas; but at radio wavelengths it is the brightest object in the whole sky. In this photograph (**Fig. 6.23**) a large patch of gas clumps is visible just above the centre; a few other isolated wisps can be seen around the middle. The faint filaments have been emphasised by taking the picture through a red filter which dims all the stars but has little effect on the filaments glowing in the red light of hydrogen and sulphur.

Optical astronomers would not have noticed this faint patch of nebulosity at all if radio astronomers had not called their attention to it. Early radio astronomers detected Cygnus A (Figs. 12.13–12.16) in 1946; two years later, the Cambridge team

set up the first double radio telescope interferometer to study Cygnus A, and immediately discovered an even more powerful source in Cassiopeia. This object still bears the name Cassiopeia A given by radio astronomers to denote it as the strongest radio source in the constellation.

After identifying Cassiopeia A with this optical nebulosity, Walter Baade and Rudolph Minkowski used the Hale 5 metre telescope to break up the light from the brighter blobs into a spectrum. They found that the gases glow in the spectral lines of hydrogen, oxygen and sulphur. The wavelength shift of these lines due to the Doppler effect reveals the gas cloud's motion towards or away from us. Baade and Minkowski discovered the strange fact that some of the gas 'blobs' (quasi-stationary flocculi) are hardly moving at all, while others (fast-moving knots) are speeding along at around 6000 kilometres

per second.

Only a supernova explosion could have flung out the fast-moving knots at such speeds. A series of optical photographs of Cassiopeia A over the past forty years show that the knots are indeed moving, heading outwards from a point below the brightest filaments (more or less in the centre of Fig. 6.23). Canadian astronomer Sidney van den Bergh has backtracked the motions of the fast-moving knots, and found that they all coincided at the central point around the year 1660.

Cassiopeia A lies some 10 000 light years away, and we can calculate that the supernova should have appeared at least as brilliant as the brightest stars. Yet seventeenth century astronomers recorded no new bright star in Cassiopeia. American historian William B. Ashworth has, however, recently discovered that the first Astronomer Royal, John Flamsteed,

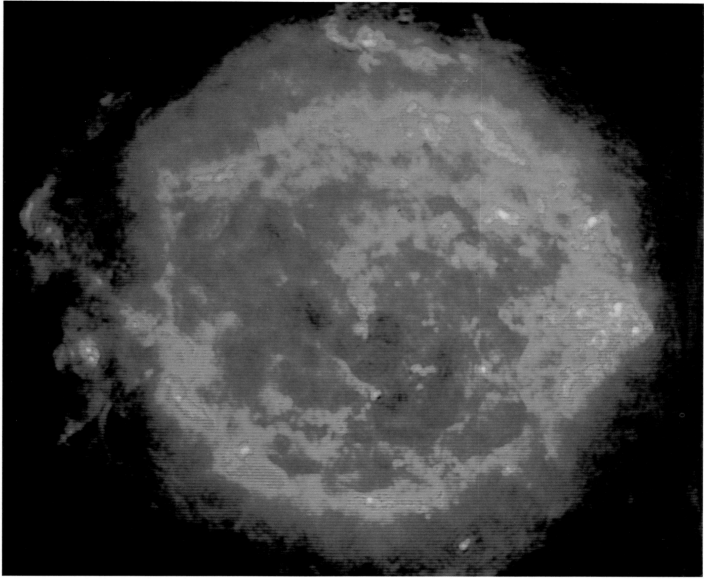

**6.24** *Radio, 6 cm, Cambridge Five Kilometre Telescope*

recorded a star very near the present position of Cassiopeia A. Flamsteed saw the star in August 1680; and there is no such star there now, but it was only just visible to the naked eye, at magnitude 6. If Ashworth is right, then the most powerful supernova remnant at radio wavelengths is the remains of an unusually dim supernova!

The colour-coded picture (**Fig. 6.24**) from the Cambridge Five Kilometre Telescope shows in detail the radio emission from Cassiopeia A at a wavelength of 6 centimetres. The colours run from blue for the faintest regions, at the centre and around the rim, through green, red and purple to white for the most intense regions of this brilliant radio source. The radio photograph covers the same region of sky as the optical photograph opposite, and at radio wavelengths the exploding shell of gas is

instantly obvious. The tiny brilliant ring of Cassiopeia A appears 5 arcminutes across – one-sixth the apparent size of the Moon.

Cassiopeia A is evidently a complex maelstrom of activity, but the radio photograph shows that it has two distinct regions. The outer region is a relatively faint (blue) rim around the brighter (green) inner ring. In three dimensions, we are seeing a faint outer shell surrounding a bright inner shell of radio emission – unlike Tycho's remnant where there is only one shell which gets brighter towards the outer edge.

Astronomers generally believe that the outer shell shows where the supernova's ejected gas is sweeping up the interstellar gas. This outer shell is the equivalent of the shell of Tycho's remnant, although it is about half the size (13 light years across). Cassiopeia A's outer shell is, in fact, rather brighter than the older Tycho remnant; it

appears faint here in contrast with the brilliance of the inner shell, which is about sixty times brighter in radio waves than Tycho's remnant.

The inner, intense shell lies in the same region as the fast-moving knots of gas seen in the optical photographs, although the brightest swirls seen in the radio picture do not coincide exactly with the optical knots. The fast-moving knots may, in fact, be responsible for generating the radio emission. As they plough through the dense gas, the knots may stir and twist its weak magnetic field. The writhing magnetic field lines could accelerate electrons up to high speeds, and as these electrons were whirled through the magnetic field, they would emit powerful synchrotron radiation at radio wavelengths.

The subtle shades in colour in the composite optical photograph (**Fig. 6.25**) show that different parts of the Cassiopeia A remnant are made of different mixtures of gases, and prove that the exploding star has thrown out newly created elements into space. Sidney van den Bergh and Karl Kamper have prepared the false-colour composite by combining three photographs, taken in the mid-1970s through different coloured filters, to isolate the light from only one or two types of atom on each photograph. The original negatives have been printed with different coloured illumination onto the same positive print, in colours corresponding to wavelengths shorter than those of the original filters.

The extensive 'reddish' nebulosities visible here are emitting light from sulphur or oxygen atoms at very long red wavelengths, around 670 nanometres. The 'green' patches of nebulosity are producing light from hydrogen and nitrogen, in reality at shorter, but still red, wavelengths of 656 nanometres. The faint 'bluish' filaments (on the left) shine in the green light of oxygen at 501 nanometres.

The different coloured filaments are moving at different speeds, and this gives valuable clues to the history of the supernova explosion. The 'greenish' ones are the slowly moving 'quasi-stationary flocculi'. Some thirty of these can be seen on the original plate, and they travel so slowly and alter so little that these flocculi can be recognised at almost the same positions in the photograph taken some twenty years earlier (Fig. 6.23).

The more intense patches of 'reddish' nebulosity, and the few 'blue' filaments are fast-moving knots. About a hundred distinct knots are visible on the original photographs. They are moving and changing so rapidly, however, that it is difficult to recognise any individual filament after a few years have passed. For example, the short arc of 'reddish' nebulosity on the lower right of the photograph has appeared only recently. The earlier photograph (Fig. 6.23) does not show it at all, although the 'green' quasi-stationary flocculus next to it can be easily picked out.

These fast-moving knots show no sign of light from hydrogen (which would appear green in this representation), the most common element in the Universe and the gas which makes up most of the outer layers of ordinary stars. Instead, they shine in the light of oxygen, sulphur and argon. The unusual composition of these knots suggests that they are 'lumps' of gas from

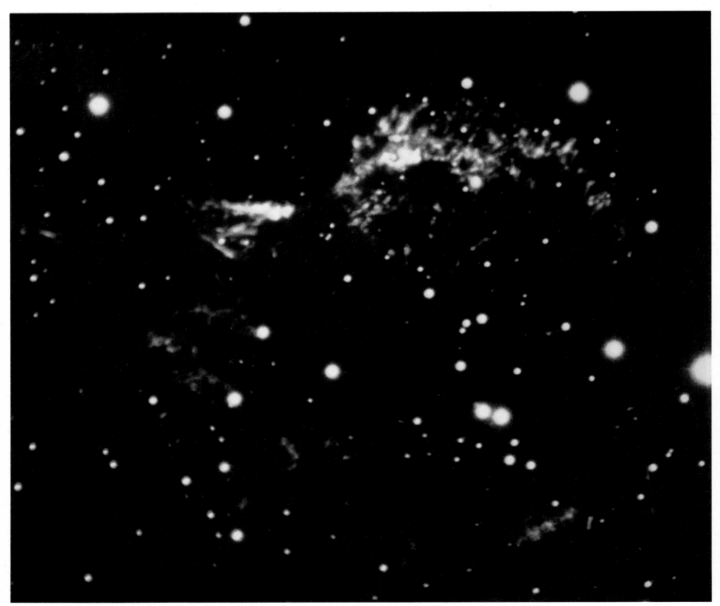

**6.25** *Optical, 501 nm oxygen line (blue), 656 nm hydrogen line (green), and 670 nm sulphur line (red), 5 m Hale Telescope*

deep inside the original star, thrown out into space in the supernova explosion. During its lifetime, a star converts hydrogen at its core into helium, then the helium into heavier elements like carbon, oxygen, sulphur and argon. The knots in Cassiopeia A consist of exactly the same gases we would expect to find near the centre of a star many times heavier than the Sun – and theoretical studies tell us that such a heavy star should end its life as a supernova.

The gases emitting visible light are, however, only a small fraction of the total thrown out by the star, and swept up from interstellar space by the expanding remnant. The shock of this sweeping-up process heats the gases up to temperatures of millions of degrees, and such gas emits not light, but X-rays. The X-ray view of Cassiopeia A from the Einstein Observatory (**Fig. 6.26**) thus shows where most of the matter lies. Although it looks very similar to the radio photograph (Fig. 6.24) the X-ray structure is a result not of magnetic fields creating synchrotron radiation, but of radiation from the dense, superhot gases.

The observatory's High Resolution Imager was used to resolve the fine details of this small X-ray source. In the colour-coded map, the dimmest regions are light blue, and the colours progress up the series in the scale on the right, to white for the most intense regions where the gas is densest.

X-ray detectors see two concentric shells making up this supernova remnant. The outer shell (light blue) is formed by the swept-up interstellar gases. These gases are at a similar temperature to those in the swept-up shell of Tycho's supernova remnant (Fig. 6.21), around 50 million degrees. The inner, more intense shell on the X-ray photograph is rather cooler – 'cool' here meaning some ten million degrees – and the gas is also much denser. This inner shell contains most of the matter thrown off in the star explosion – about fifteen solar masses of gas. The gas has been travelling outwards at a slower speed than the small amount of gas which has shot ahead to sweep up the gases forming the outer shell. The latter is, however, slowing down, and the inner shell is catching up with the outer shell. The collision, or 'reverse shock' is heating up the inner gas shell, and because it is so massive it emits X-rays copiously and dominates the X-ray view.

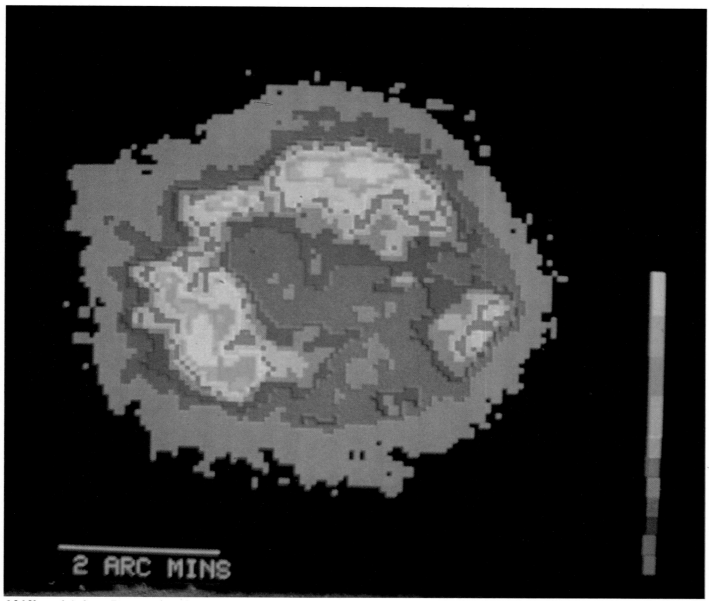

2 ARC MINS

**6.26** *X-ray, 0.4–8 nm, Einstein Observatory HRI*

This 'complete view' of Cassiopeia A (**Fig. 6.27**) combines three pictures which span the entire wavelength range from X-rays (green), through optical (red) to the long-wavelength radio waves (blue). The 'green' X-ray image is the most fundamental, revealing most of the matter from the deceased star – some fifteen solar masses. The 'red' optical view shows many stars; the faint knots and flocculi of Cassiopeia A itself contain less than one per cent of its gases. The 'blue' radio image reveals the tangled magnetic fields.

The composite reveals the relations – and contrasts – between these aspects of the explosion. On the broad scale, the radio (blue) image is brightest to the right, and the X-ray (green) to the left. American astronomer John Dickel, who made this composite, suggests two possible causes. Perhaps the magnetic field in interstellar space is slightly stronger to the right, and when swept up its interaction with the electrons has produced stronger radio synchrotron emission. Or the gas in interstellar space between Cassiopeia A and the Sun is uneven, and denser gas to the right absorbs X-rays more efficiently.

Most of the optically bright fast-moving knots lie at the North (top) of Cassiopeia A. The radio and X-ray images are intense here too, but their brightest parts do not coincide exactly with the optical knots. Nonetheless, they are probably related. The knots are dense blobs of gas, travelling outwards from the original explosion. As they enter the inner shell of gas, they churn up its magnetic field and so create stronger synchrotron radio waves from the region around.

The knots are comparatively cool, however, and the gas shell at a temperature of 10 000 000 K rapidly heats them up. The outer parts of the knots thus 'evaporate' to join the rest of the gas in the shell. This recently evaporated gas is still fairly dense, and hence emits X-rays powerfully. The result is that many of the optical knots are surrounded by a 'halo' of intense X-ray emission. Small, bright X-ray clouds without an associated optical knot may be regions where one of them has recently evaporated completely.

The quasi-stationary flocculi, seen at optical wavelengths, lie further out, in the interstellar gases just beyond the shock front (marked by the outermost edge of the remnant as seen in radio and X-rays). Many astronomers believe that the flocculi are small dense gas clouds which drifted away from the original star centuries before it exploded. Now the shock wave from the explosion is catching up with them. As the shock engulfs a flocculus, it glows at optical wavelengths. Parts of an engulfed flocculus are heated to multi-million degree temperatures. They shine as brighter X-ray spots in the generally faint outer shell, appearing adjacent to optical flocculi.

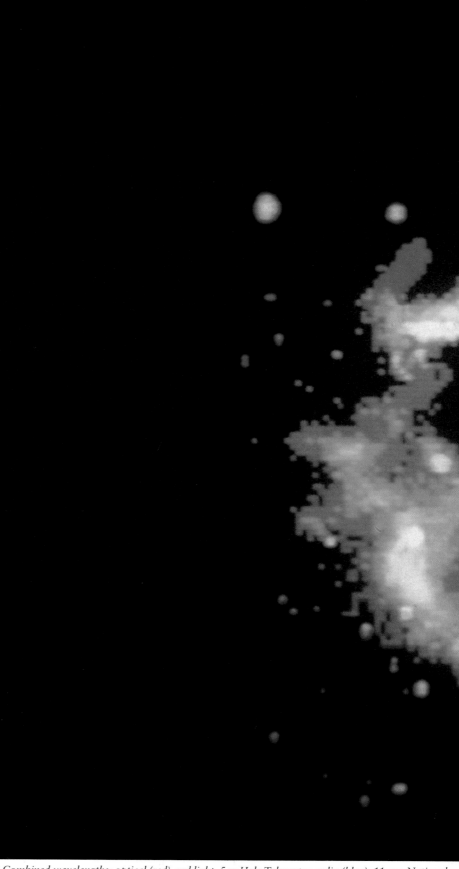

6.27 *Combined wavelengths: optical (red), red light, 5 m Hale Telescope; radio (blue), 11 cm, National*

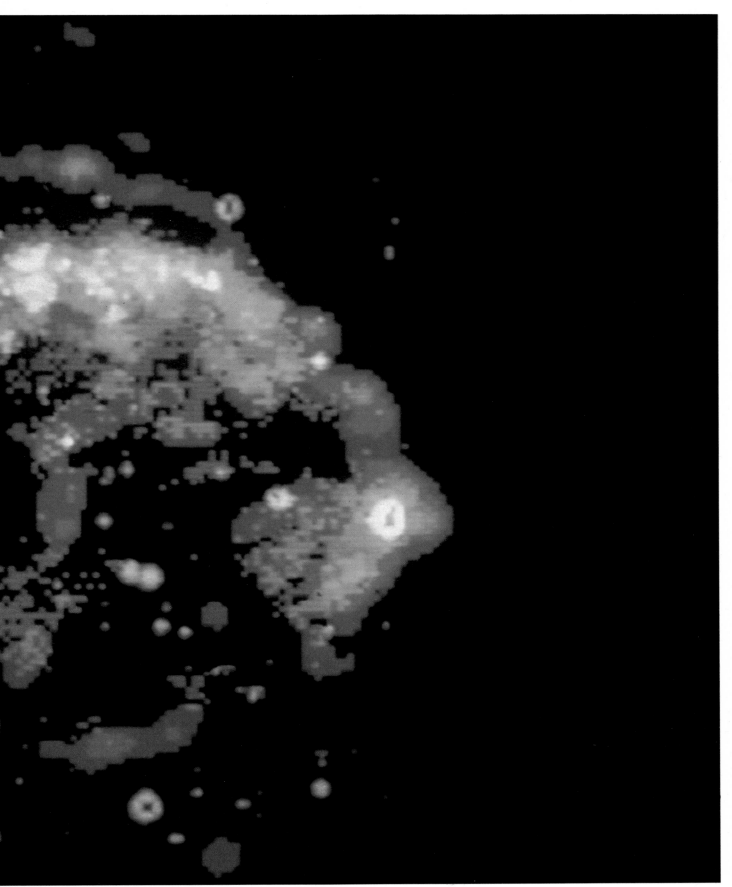

*Radio Astronomy Observatory interferometer; X-ray (green), 0.4–8 nm, Einstein Observatory HRI*

# Vela supernova remnant

Straggly wisps of gas, stretching over several degrees of sky – almost ten Moon widths – mark the site of a supernova which exploded long before human records were written down. Fig. **6.28** was taken through a green filter to emphasise the filaments which glow in the green light of oxygen gas. These delicate traceries are thin sheets of glowing gas, draped in folds across the sky. Where a sheet is running across in front of us, its light is so spread out that we cannot easily see it; but where it bends round in a fold, we are seeing the sheet edge-on and its light is concentrated into a narrow bright filament. The thin sheets mark where the shock wave from the explosion is driving outwards through the interstellar gases, and lighting them up as it passes.

The Vela supernova remnant is one of the closest we know, lying only 1500 light years away. Because it is much older than remnants like the Crab Nebula and Cassiopeia A, the filaments are moving very slowly and we cannot see them shifting from year to year. As a result, astronomers cannot backtrack their motions to find out when the explosion occurred. Fortunately, the Vela remnant is not just a shell of expanding gases; the original star's core has collapsed to leave a pulsar – too faint to appear on the photograph – and radio astronomers have calculated its age to be 11 000 years. Around 9000 BC, prehistoric people of the southern hemisphere must have witnessed a brilliant 'new star' about a hundred times brighter than Venus, at magnitude −9.

Like other supernova remnants, the Vela remnant is far more striking when observed at radio and X-ray wavelengths. At radio wavelengths it is one of the brightest sources in the sky, as strong as the Crab Nebula. The radio picture (**Fig. 6.29**) covers the same area of sky as the optical photograph (Fig. 6.28), and is colour coded so that the faintest outer regions are pink, with successively brighter parts in shades of blue, green, orange and red. (The pulsar is too weak to show up here; its position is shown by the black spot.) Fig. 6.29 shows the total extent of the gases and shockwaves from the explosion much more clearly. The radio-emitting remnant is 4° (about 100 light years) across.

The radio emission does not come from a neat ring, as it does in the younger supernova remnants. As the supernova gas has expanded, different parts of the shell have run into interstellar gas of different densities; and consequently the shell has broken up into uneven clumps. The first radio astronomers to map the remnant thought that there were three different souces; Vela X (the intense region on the right); Vela Y (to the left) and Vela Z (the top left). Later observations revealed that

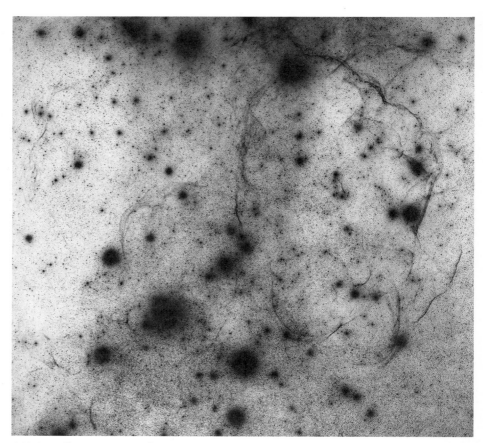

**6.28** *Optical, 501 nm oxygen line, negative print, 1.2 m UK Schmidt Telescope*

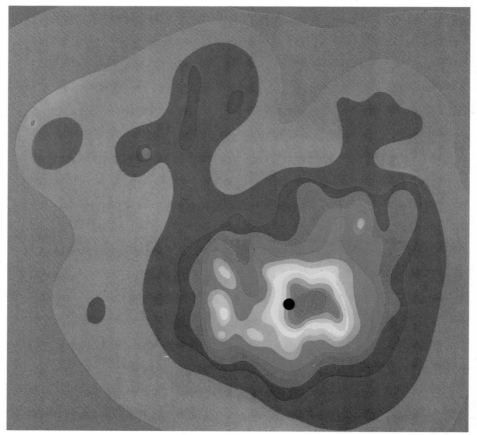

**6.29** *Radio, 11 cm, 64 m Parkes Telescope*

6.30 *X-ray, 0.3–2.5 nm, mosaic, Einstein Observatory IPC*

these are just brighter portions of one supernova shell. The appearance of detailed maps has led radio astronomers to dub Vela X,Y,Z the Donald Duck Nebula!

The Vela remnant is so large that the Einstein X-ray Observatory could not see it all at once, even with its wide-field Imaging Proportional Counter (IPC). In the mosaic

(Fig. 6.30) covering the same region as the optical and radio pictures, each small square is one view from the IPC (the bordering dark square within each is a shadow cast by the supports for the IPC window). The colour coding runs from red for the dimmest regions through green to blue and white for the brightest parts.

At X-ray wavelengths, the pulsar is very bright, and it appears as the tiny intense white point within the square just below the centre. The rest of the X-ray emission comes from the hot gas ejected by the explosion.

6.31 *Optical, true colour, 1.2 m UK Schmidt Telescope*

The colour photograph (**Fig. 6.31**) shows details of intricate optical filaments in the Vela remnant, from the region at the top right of the previous photograph (Fig. 6.28). The composite was made from plates taken through three different colour filters on the UK Schmidt Telescope. Most of the filaments glow bright red with light from hydrogen atoms; other elements produce a tinge of different colours like the green of oxygen.

Right at the heart of the remnant lies the Vela Pulsar, the collapsed core of the original supernova. Like the Crab Pulsar (Figs. 6.17 and 6.18), it is a compact neutron star, which appears to pulse as it spins around. Australian radio astronomers discovered the Vela Pulsar in 1968. It is spinning thirteen times per second, but its rotation is gradually slowing down so that its period increases by four millionths of a second in a year. Working backwards, astronomers calculate that the pulsar was born about 11 000 years ago.

The Vela Pulsar is the second youngest known, after the Crab Pulsar. Since the Crab Pulsar emits light, a team of astronomers at the Anglo-Australian Telescope looked for flashes of light from the Vela Pulsar at precisely the period of the radio pulses, using the very sensitive Image Photon Counting System (p. 51). In 1977 they found the pulsar, so faint that it was right at the limit of the telescope's light grasp. The average brightness of the Vela Pulsar is only 1/400 that of the Crab Pulsar. At magnitude 24, it is one of the faintest stars whose brightness has been measured.

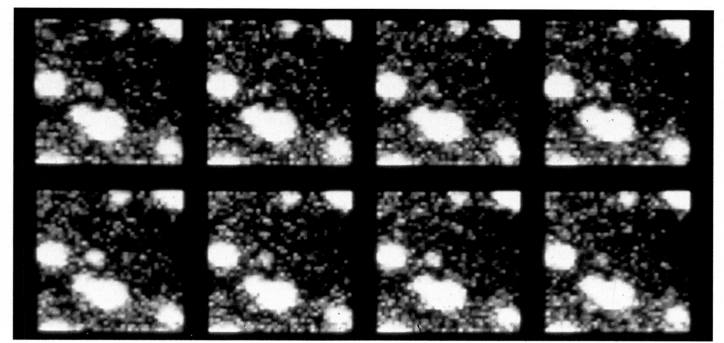

6.32 *Optical, IPCS, 3.9 m Anglo-Australian Telescope*

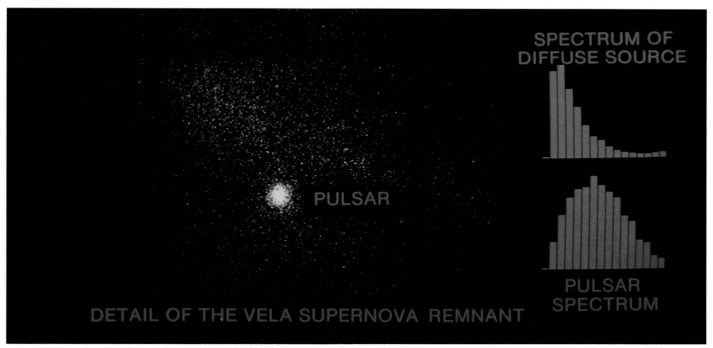

SPECTRUM OF
DIFFUSE SOURCE

PULSAR

DETAIL OF THE VELA SUPERNOVA REMNANT

PULSAR
SPECTRUM

6.33 *X-ray, 0.3–2.5 nm, Einstein Observatory IPC*

The sequence of eight pictures (**Fig. 6.32**) shows the pulsar (at the centre) running through one complete cycle: the series runs horizontally from the top left to right and then the bottom left to right. The bright spots are ordinary stars. The pulsar flashes twice in each cycle (in the first and third frames of the lower row). Oddly enough, the flashes are not equally spaced, and the main pulse of radio waves occurs in the third frame of the top row, when the pulsar is producing hardly any light.

The Einstein Observatory's Imaging Proportional Counter (IPC) detected the Vela Pulsar shining brightly within the remnant (**Fig. 6.33**), but its X-ray output does not pulse. We are probably picking up not the pulsar's beamed radiation but X-rays from its hot surface, at about 1 500 000 K.

The pulsar is surrounded by a small X-ray emitting synchrotron nebula, one-fifth the size of and one-thousandth as bright in X-rays as the Crab Nebula. The X-ray spectrum of the pulsar plus small nebula (Fig. 6.33 bottom right) differs markedly from that of the rest of the Vela remnant (Fig. 6.33 top right) which is due to hot gas. (The bars in the spectra represent the amount of energy present at each wavelength with the wavelength increasing to the left). The former resembles the X-ray spectrum of the Crab Nebula, so the small Vela nebula is probably also a magnetic halo surrounding the pulsar and powered by the transfer of energy from the pulsar as it gradually slows down.

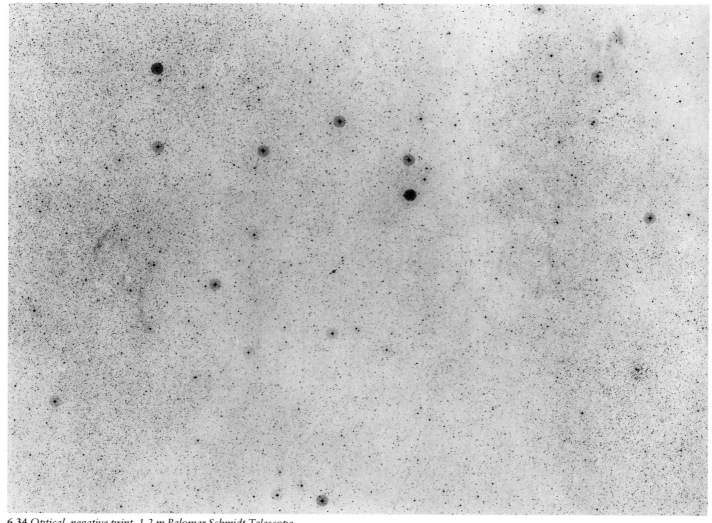

6.34 *Optical, negative print, 1.2 m Palomar Schmidt Telescope*

# SS 433

The arrowed star in the photograph (**Fig. 6.34**) looks misleadingly unimpressive – similar to the hundreds of others here. In fact, it is the notorious SS 433, one of the most bizarre objects in our Galaxy.

SS 433 achieved fame in 1978, but it had been noticed in the 1960s, when American astronomers Bruce Stephenson and Nicholas Sanduleak were looking for stars with peculiar spectra. This faint (magnitude 14) star in Aquila had bright spectral lines of hydrogen and duly went down as number 433 in the Stephenson-Sanduleak ('SS') catalogue. In 1978, Canadian astronomer Ernie Seaquist searched for radio waves from SS stars, and found that SS 433 is a radio emitter. But Seaquist did not realise that SS 433 is also surrounded by a huge ring of radio emission, known as W50 from its cataloguing by Dutch radio astronomer Gart Westerhout in the 1950s.

Radio astronomers in Australia and Cambridge were investigating W50 at the time, and they independently discovered the intense radio source at the centre. It turned out to be an X-ray source too. Intrigued by these results, optical astronomers David Clark and Paul Murdin pointed the Anglo-Australian telescope at this position – and rediscovered the star with the strong hydrogen lines. Later studies of SS 433's hydrogen lines showed that they are moving regularly backwards and forwards in wavelength by a small amount, due to the Doppler effect, indicating that this is a double star system, where the stars orbit one another with a period of 13 days.

One of the pair is a fairly normal star, whose outer gases are falling onto a compact companion star (a neutron star or black hole) in a spiralling accretion disc at a rate of a hundred Earth masses per year. This whirling disc's hot gases generate the X-rays, and its magnetic field generates the radio waves which we pick up from SS 433. Its light comes from the cooler, outer regions of the accretion disc and from the normal star. So far, the system resembles other X-ray binaries, like Scorpius X-1. Clark and Murdin had, however, noticed

other bright lines in the spectrum of SS 433, at wavelengths which are not emitted by any of the common elements. American astronomer Bruce Margon followed up these observations, and discovered that these unusual lines were changing dramatically in wavelength, first one way then the other – marching up and down the spectrum in a regular rhythm which repeats every five-and-a-half months.

The lines are moving back and forth around the wavelengths of the ordinary hydrogen lines, so they are simply due to clouds of hydrogen, and the wavelength change must be a tremendous Doppler shift. What we are seeing is gas in two 'jets' 'boiled off' either side of the hot accretion disc, and swinging round with a period of five and a half months as the disc precesses, like an unstable top.

The jets themselves keep a constant speed. SS 433's great surprise was the enormous size of this speed: just over a quarter of the speed of light – some 80 000 kilometres per second. Although electrons are probably travelling at such high speeds in many cosmic radio sources, the

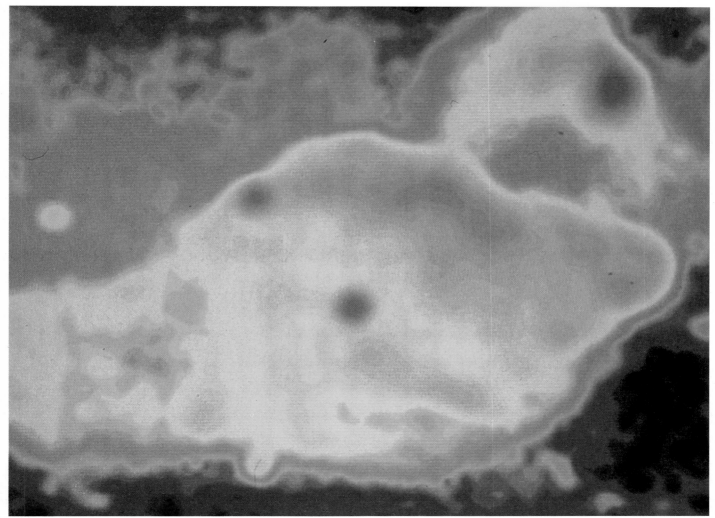

**6.35** *Radio, 11 cm, 100 m Effelsberg Telescope*

discovery of the speed of these jets of ordinary gas was totally unexpected.

In the radio picture (**Fig. 6.35**), covering a slightly larger area than the optical photograph, the most intense radio regions are colour coded red; successively fainter regions yellow and green, while the background sky is coded dark blue.

SS 433 itself is the red spot at the centre, while W50 is the horizontal oval filling most of the picture. The curved 'cloud' (top right) is not part of W50, but is a separate hot nebula. Such nebulae are the dispersing remains of the dense, dark and cold interstellar clouds where stars are born. SS 433 is itself relatively young, so this nebula may well be the outskirts of the dark cloud in which SS 433 was born. If so, we are seeing, in one picture, signs of both the birth and death of the same star.

The radio waves from W50 come from a thin, hollow egg-shaped shell, where fast electrons, whirled around in compressed magnetic fields, generate synchrotron radiation. It stretches about $\frac{2}{3}°$ North-South and 2° East-West (four Moon breadths). W50 is about 18 000 light years away, which means that it is huge for a supernova remnant: 200 light years by 600 light years in extent.

Apart from its size, W50 differs from other supernova remnants in its shape. It is neither circular in outline, like Tycho's remnant, nor irregular like the Vela remnant. The easiest way to describe W50 is as a circle with 'ears'. More detailed radio observations (Fig. 6.36), and X-ray pictures (Fig. 6.38) reveal that the jets from SS 433 are heading out East and West (left and right), and it seems that they are pushing back the interstellar gas there with more force than the expanding supernova shell can muster, and so are inflating the radio ears of W50.

The radio map shows no distinct traces of the 'inner edges' of the ears, where the circular rim of the supernova remnant would have run in their absence, but these do show in the optical photograph (Fig. 6.34) as thin wisps of gas near the left and right edges.

The simplest explanation of the formation of the SS 433/W50 complex is that two very similar stars were born together just over five million years ago, each with a mass of as much as 25 Suns. The slightly heavier, shorter-lived, star exploded as a supernova after only five million years, and its core was left as the neutron star or black hole in SS 433. Now, 100 000 years later, we detect this supernova's remnant as W50, while the overspilling gases from the other star are powering the radiation and fast jets of SS 433.

Dutch astronomer E.P.J. van den Heuvel, however, suggests an alternative scenario: the two original stars were about twenty and eight times heavier than the Sun, and they were born ten million years ago. The heavy star exploded as a supernova three million years ago, and its supernova remnant has long since dissipated into space. Some 40 000 years ago, however, the lightweight star started to lose gases to the compact star left over from the explosion, which then began to squirt out its high-speed jets of gas. The jets are entirely responsible for blowing up the 'balloon' of W50, which is thus not a genuine supernova remnant at all.

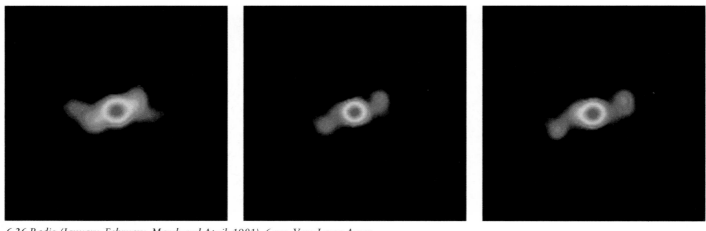

**6.36** *Radio (January, February, March and April, 1981), 6 cm, Very Large Array*

**6.38** *X-ray, 0.3–2.5 nm, mosaic, Einstein Observatory IPC*

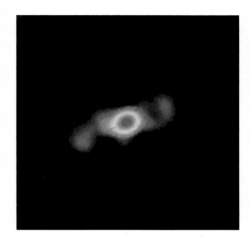

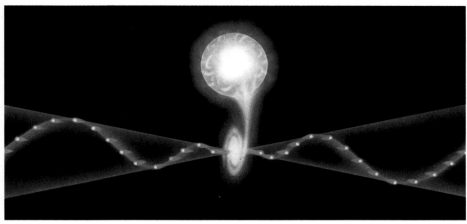

6.37 *Diagram, formation of SS 433 jets*

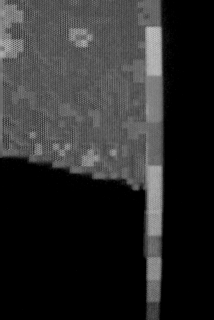

Radio observations reveal that the first interpretations of the optical spectrum were astonishingly successful in interpreting the three-dimensional structure of SS 433 and its jets. Israeli astronomer Mordehai Milgrom, and George Abell and Bruce Margon in the United States had worked out from the spectrum that there are two oppositely directed jets, swinging around like the spray from an eccentric lawn sprinkler, tipped up as shown in the diagram (**Fig. 6.37**). The 'sprinkler's' axis runs more or less across the sky, at an angle of about 80° to our line of sight. Each jet swings around in such a way that it is always at an angle of 20° to this axis, so the jet is pointing along one of the two imaginary cones.

The jets consist of successive 'blobs' of gas shot out by SS 433's accretion disc, each blob following its own path in a straight line along the direction in which it was originally ejected. If we were to draw a line to connect the successive blobs in each jet, it should make an unwinding spiral along one of the imaginary cones.

With the Very Large Array radio telescope in New Mexico, astronomers have detected the blobs making up SS 433's jets – and by observing them month after month they have found that the jets behave exactly as Milgrom, Abell and Margon had predicted. The four observations here (**Fig. 6.36**) were made in 1981, at intervals of about a month from mid-January to mid-April. The pictures are colour coded so that the brightest regions are red, with successively fainter regions yellow and green, and the dimmest parts blue.

Three-quarters of the radio emission comes from SS 433 itself, the central red blob. But the two jets are clearly weak radio emitters too. They stretch out about one-sixth of a light year on either side of SS 433 (these maps are at a scale of 250 times larger than the previous radio photograph of SS 433 and W50, Fig. 6.35). The first picture shows a pair of outer blobs (blue) at about positions of 'three o'clock–nine o'clock'; by the second view these

have faded from sight. In the first picture, there is also an inner, more intense pair of blobs (green), emitted at a different angle ('two o'clock–eight o'clock') because the accretion disc has swung around. In the later pictures we can see this pair of blobs moving steadily outwards along the same directions, and gradually fading (to blue) as they go. By the last frame, they have almost disappeared. Now we can see the next pair of blobs, closer in at 'three o'clock–nine o'clock'.

The composite of three views from the Einstein Observatory (**Fig. 6.38**) is about 1° across, half the extent of the radio photograph of W50 (Fig. 6.35). The colours run from white for the most intense X-ray emission at the centre (from SS 433 itself) down through green, yellow, purple and red, to blue for the faintest regions.

The X-ray jets here stretch out a hundred light years in each direction, about six hundred times further than the radio jets above. They show how the 'cones' fan out from SS 433, and reach the inner edge of the radio ears of W50, where the optical nebulosity lies (Fig. 6.34). The X-rays may come from hot gas in the blobs, or from synchrotron radiation like the radio jets; in either case they show that the jets have been travelling outwards for a thousand years or more.

As well as the main jets running roughly East–West, there are other X-ray blobs around SS 433 in this view. Some of them form pairs that straddle SS 433 – at 'seven o'clock–one o'clock', for example, and particularly the line of compact blobs at 'eight o'clock–two o'clock'. Has SS 433 shot out its two jets at completely different orientations in the remote past, to create these other pairs of blobs? This is one of many questions which still haunt astronomers caught up in the mystery of SS 433.

# G109.1-1.0

Is this the missing link between the bizarre SS 433 and ordinary supernova remnants? G109.1-1.0 is a large semicircular ring of gases in the constellation Cassiopeia (Fig. 6.1), shining brightly in radio waves and X-rays, but only faintly as seen in optical telescopes.

It is a rather aged, misshapen supernova remnant. But a 'star' in the centre is apparently ejecting jets of gas highly reminiscent of SS 433's jets. Canadian astronomers Phil Gregory and Greg Fahlman discovered this extraordinary supernova remnant by its X-ray emission, when they observed this part of the sky with the Einstein Observatory in 1979. In the X-ray image (**Fig. 6.40**) we see radiation from hot gases in the remnant. The left-hand rim shines brightly (coded yellow) while the right-hand side is much fainter (blue). But the real surprise is the very bright (red) compact X-ray source to the right of the centre.

The X-rays from this compact source pulse regularly with a period of seven seconds. Relatively slow X-ray pulsars like this are invariably part of a double star system where one partner is a rotating neutron star, feeding on gases from the other, normal star. The neutron star is presumably the core of the supernova which created the surrounding remnant.

In the false-colour radio map (**Fig. 6.41**), the brightest regions are coded red and fainter parts yellow and green, while the background sky is blue. The remnant is about 33 by 25 arcminutes in extent (about the same apparent size as the Moon). Its radio emission is synchrotron radiation from fast electrons moving through a magnetic field. G109.1-1.0 lies 12 000 light years from us, so that its shell is, in fact, 100 light years across, suggesting that G109.1-1.0 is the remnant of a star which exploded some 20 000 years ago.

The original supernova exploded at the edge of a dense cloud of gas (a molecular cloud), which extends from the middle of these images towards the right. The small intense radio source to the right of the semicircular remnant is a nebula within the cloud, where stars have just been born. This nebula also shows up as the brightest wisp of gas on the optical photograph (**Fig. 6.39**), which covers the same region of sky. The dense cloud has prevented the ejected supernova gases from travelling far to the right; but they have billowed out to the left to create a hemispherical supernova remnant – which we can see in projection as a semicircle.

Gregory and Fahlman suggest that the pulsar can account for the appearance of the radio emission which resembles two intersecting ovals (ellipses), the main ring of emission and a smaller ring of radio emission lying in the lower-right part of the remnant.

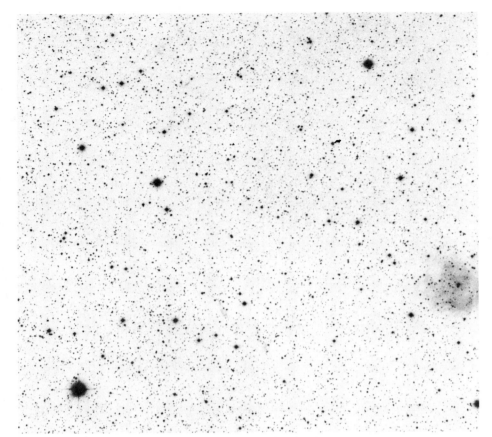

6.39 *Optical, red light, negative print, 1.2 m Palomar Schmidt Telescope*

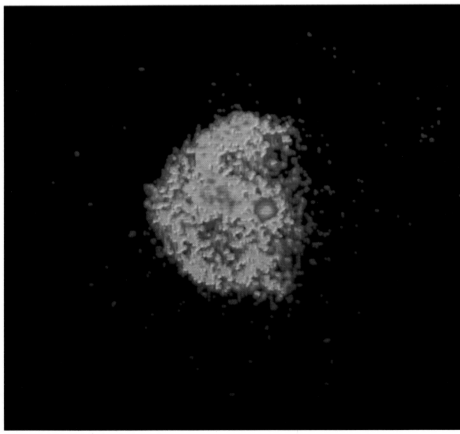

6.40 *X-ray, 0.3–2.5 nm, Einstein Observatory IPC*

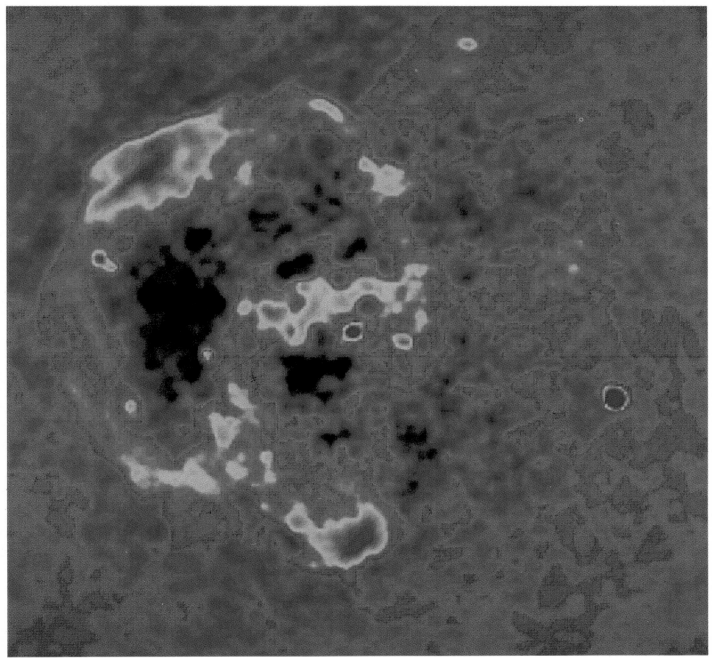

**6.41** *Radio, 20 cm, Very Large Array*

According to their model, the central pulsar is emitting two jets of matter in the opposite directions, like a weaker SS 433. The jets travel freely within the relatively empty cavity of the remnant's interior, but where they hit the shell of the remnant the jets generate radio waves. They swing round, like those of SS 433 but through a larger angle, and so each jet traces out a loop on the shell, over and over again. The jet travelling to the upper left draws out a large ellipse on the hemispherical part of the remnant, which happens to coincide mainly with the left-hand edge of the remnant as seen from Earth. The second jet cannot travel far to the right before it meets

the flattened wall of the remnant on this side, so it draws out a small ellipse (made to appear smaller still by projection effects, because we view the flat wall almost edge-on). The small, bright radio source here, incidentally, is not the pulsar, but lies close to the upper edge of the smaller ellipse.

To support this interpretation Gregory and Fahlman point out that the X-ray picture (Fig. 6.40) shows a broad jet heading from the pulsar towards the top left – the predicted direction for one of their jets – just as the jets of SS 433 can be seen as broad X-ray emitting features a long way out from the central source.

G109.1-1.0 is, however, weak compared

to SS 433. Although the central sources emit about the same power in X-rays, SS 433 shines a thousand times more brightly than the pulsar in G109.1-1.0 in visible light, and it is over 3000 times brighter at radio wavelengths. In addition, there is no *direct* evidence in G109.1-1.0 for the kind of extremely high speeds found in SS 433's jets. Nonetheless, further study of this source will undoubtedly help us understand how these strange swinging-jet star systems operate, and how they arise in the process of star death.

# 7 Radio Astronomy

In an America hit by the Depression of the early 1930s, scientists achieved a major breakthrough in the study of the Universe: they acquired the first view of the skies at wavelengths of invisible radiation. The new instrument was strange by any standard, and it was certainly a far cry from anyone's idea of a 'telescope'. Sited in the midst of the potato fields of New Jersey, the first radio telescope consisted of eight large metal hoops, supported on a wooden frame which rotated slowly on a set of Model T Ford wheels. It was as important a step forward as Galileo's first optical telescope of three centuries earlier. Its descendants have not only revealed weird and unexpected radio sources in the sky – pulsars and quasars, to name but two – but have shown astronomers that they must turn away from ordinary optical astronomy if they are to investigate some of the strangest, and some of the most important, objects in the sky. The spectacular growth of the 'new astronomies' – X-ray, infrared, ultraviolet and gamma ray – has been inspired by the unpredicted successes of radio astronomy.

The first instrument to detect radio waves from space was not in fact a radio telescope. The Bell Telephone Laboratories were studying sources of radio 'static' or 'hiss' which interfered with ship-to-shore communications, and a young engineer called Karl Jansky was sent out to New Jersey to investigate. His big rotating radio aerial could pin down the direction from which the static was coming. After a year, he could distinguish hiss from local thunderstorms, distant thunderstorms, and background static which seemed to come from space. We now know that Jansky had detected radio waves generated out in the regions between the stars of our Galaxy by electrons travelling through the extended magnetic fields of the Milky Way.

Radio astronomers have been depicted by cartoonists as 'listening in' to the crackle and hiss picked up by radio telescopes. Although the early pioneers like Jansky did

listen to the static, radio astronomers no longer regard their radio telescope as 'ears', but more as 'eyes'. They pick up radio waves in much the same way as we all do at home – a radio telescope is really just a very sensitive version of a domestic radio set. Radio waves reach our radios from a whole variety of radio stations, in many different directions, all broadcasting at different wavelengths. We tune the receiver to a particular wavelength, and the radio set extracts the superimposed 'message' of voices or music from the radio waves which carry it from the transmitter to our aerial.

The distant natural radio sources in the Universe, however, simply emit a cacophony of hiss and crackle; 'listening in' tells us little about *where* the individual radio sources are in the sky, or what their size and shape are. For this, a 'radio picture' is needed, and that is what a modern radio telescope provides. If we operated a radio telescope at the wavelengths used for terrestrial broadcasting, the result would not be the latest news bulletin, but a 'photograph' showing the aerials of the radio transmitters.

In practice, radio astronomers must take care *not* to observe at broadcasting wavelengths, because the signals from artificial transmitters would swamp the faint radio waves from the depths of the Universe, just as the brilliance of the Sun makes optical astronomy impossible during daytime. International conventions have allocated specific wavelengths to radio astronomy (Fig. 7.1). No one is permitted to broadcast at any of these two dozen or so wavelengths, so radio telescopes tuned to them can observe the Universe without the foreground 'glare' of radio transmitters on Earth.

As well as these artificial limitations on radio astronomy, the Earth's atmosphere places restrictions on the wavelengths radio telescopes can use. In theory, radio waves include all radiation of wavelength longer than about one millimetre. The longest wavelengths, however, are reflected by the Earth's upper atmosphere, the

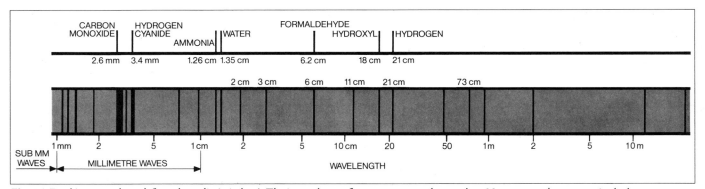

**Fig. 7.1** *Earth's atmosphere defines the radio 'window'. The ionosphere reflects away waves longer than 30 metres; water vapour in the lower atmosphere absorbs radiation shorter than two centimetres. Certain wavelengths are kept free from artificial broadcasts, for radio astronomy observations. These allow multi-wavelength studies of continuum sources (emitters of the whole range of wavelengths), the most commonly used wavelengths being labelled. Some are allocated at the specific wavelengths emitted by atoms and molecules in space, the most important being shown.*

**Fig. 7.2** *The 76 metre Mark IA telescope at Jodrell Bank is the second largest fully steerable radio dish. The original Mark I telescope, completed in 1957, has been given a new, more accurate, surface. On the right is the smaller, elliptical, Mark II telescope.*

*ionosphere*. This was a great advantage to early radio broadcasting pioneers, because artificial transmissions could be bounced around the world by the ionosphere. But long wavelength radiation coming from space is also reflected back out into space. So radio astronomers on the ground cannot pick up radio waves with wavelengths much greater than thirty metres.

At the other end of the scale, radio waves shorter than around one millimetre merge into the far infrared regions. The exact boundary between short radio and far infrared is really only a matter of convention, and in practice astronomers do not usually make the distinction at a particular wavelength, but instead relate the radiation to the kind of detector they are using. An astronomer using a germanium–gallium detector (Chapter 5) to look at 0.5 millimetre radiation would call it infrared; a colleague, detecting the same wavelength with radio astronomy techniques, would regard it as radio. Partly because of this confusion, special names have been given to these borderline regions of the spectrum. *Millimetre wave* radiation comprises waves ranging from one to a few millimetres wavelength, while the *submillimetre waves* are those with wavelengths that are between one-third and one millimetre. Water vapour in the Earth's lower atmosphere absorbs radiation of these wavelengths. From a radio observatory at sea-level, it is very difficult to pick up cosmic radiation of wavelength less than one or two centimetres.

Radio waves with wavelengths in the wide range from two centimetres to thirty metres, can penetrate both the ionosphere and lower atmosphere, but the air around is, in fact, surprisingly opaque to most radiation. The radio wavelengths are the only natural 'window' apart from the optical, which allows visible light through, so radio astronomy is the only branch of invisible astronomy whose telescopes can be situated, to our convenience, at sea level. As a result, radio observatories have been built around the world to pick up the natural cosmic 'broadcasts' – in the deserts of New Mexico, the fens of Cambridgeshire, the Caribbean and the Australian outback.

The main components of a radio telescope parallel those of a domestic radio set. At the front there is an aerial or antenna. The energy of radio waves passing the antenna is converted into electrical signals. These are very weak voltage fluctuations, and they are boosted by the second component, the amplifier – in practice, a series of amplifiers which increase the strength of the incoming signal about a thousand million million times. Finally, there is a computer to store the output, and display the signal in some convenient form.

Although we usually think of radio telescopes as large saucer-shaped dishes – like the giant dishes at Jodrell Bank (Fig. 7.2) or Effelsberg in Germany – they can actually be of several different forms. A single wire will act as an antenna to pick up radio waves from space, but its signal will naturally be weak. Some consist of thousands of simple wire *dipole* aerials next to one another, so that the radio waves can be collected over a much larger area and their combined output is considerably stronger. British radio astronomer Tony Hewish constructed one such

Fig. 7.3 *The Cambridge Four Hectare telescope (foreground) consists of thousands of simple aerials (dipoles) strung between wooden posts. Pulsars were discovered by the original half of this very sensitive array. In the background is a dish of the One Mile Telescope.*

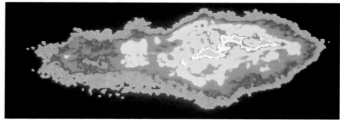

Figs. 7.4–7.5 *Two ways of displaying the same radio view of the supernova remnant 3C 58. Fig. 7.4 shows the radio intensity converted into brightness, similar to an optical photograph. In Fig. 7.5, the intensity is colour coded. The faintest regions are pale blue, and brighter parts dark blue, green, yellow, red, white and pale blue (again) for the intense central regions.*

array near Cambridge in the 1960s; it had 2048 individual wire dipoles spread over a two-hectare field (Fig. 7.3). This 'telescope' was so sensitive that Hewish could look at rapid fluctuations ('twinkling') in the radio emissions from sources in the sky. His student, Jocelyn Bell, noticed that some of the 'twinkling' was actually arriving in the form of regular pulses of radiation, at intervals of a second or so. A month's study convinced the Cambridge astronomer that these *pulsars* were not intelligent broadcasts from an alien civilisation. They are the natural radio emission from the tiny, collapsed cores of old stars.

Such a star is only about 20 kilometres in diameter, but contains rather more matter than the Sun; it is composed of subatomic particles called neutrons, packed so tightly together that a pinhead of matter from one of these *neutron stars* would weigh a million tonnes! The radio waves are emitted in beams from opposite sides of the star. As the neutron star spins round with a period of about a second, one of the beams regularly sweeps past the Earth, and a radio telescope picks up a 'pulse' of radiation. In most pulsars, the second beam points in such a direction that it misses the Earth, but in some cases its radiation is detected as a weak 'interpulse' between the main pulses.

Unusual telescopes like Hewish's are the exception. Although they are cheap, they are fixed, pointing more-or-less straight upwards. A dish-shaped telescope is much more versatile. It operates very like an optical reflecting telescope: radio waves from space hitting the inside of the bowl are reflected up to a focus, and the telescope can be swung round to look at any part of the sky. But a radio telescope cannot 'photograph' directly the image at its focus. A dipole antenna at the focus picks up and measures the strength of the radio waves, which come from just a small region of sky directly in front of the telescope dish. To build up a 'picture' of the source, the telescope scans back and forth across it, so that its brightness can be measured at every point. The telescope's computer stores the measurements, and builds up a 'map' of the source. This is a two-dimensional array of numbers, the number stored at each point indicating the brightness of the

corresponding region of the sky. The map can then be displayed in various ways. Traditionally, different intensity levels have been represented by contour lines, like a geographical contour map. An isolated bright source appears as a series of concentric circles, the smallest interior one marking the 'peak' of intensity.

The most direct way, however, is to project it onto a black-and-white TV screen, with the brightness of each point representing the number stored in the computer – hence showing the brightness of the radio sky. This method produces 'radio photographs' which show realistically how the sky would look through 'radio eyes' (Fig. 7.4). The numbers stored in the computer are equivalent to the array of numbers that represent an optical picture once an ordinary astronomical photograph has been scanned (Chapter 3). So the techniques of false-colouring and other computer-generated representations can also be used on radio maps (Fig. 7.5).

Such maps can be made with any radio telescope, but astronomers generally want to build the biggest radio telescope they possibly can. Their reasons are similar to the arguments of 'light grasp' and 'resolution' which have driven optical astronomers to build larger telescopes. A bigger radio telescope dish can collect radio waves over a larger area, and resolve more detail, responding to the small-scale structure, rather than blurring it out.

The question of resolution is one of the radio astronomers' main headaches. The finest detail which any telescope can reveal depends on the diameter of the telescope mirror (lens, or dish) *relative to the wavelength* it is detecting. Radio waves are almost a million times longer than light radiation, so a radio telescope must be roughly a million times bigger than an optical telescope if it is to resolve the same kind of detail in the sky.

The increase in scale is quite staggering. It means that when the world's largest steerable dish, the 100 metre diameter Effelsberg radio telescope (Figs. 7.6–7.9), observes at a typical wavelength of 11 centimetres, it actually 'sees' a more blurred view of the sky than does the human eye. An effort to build bigger telescopes encounters

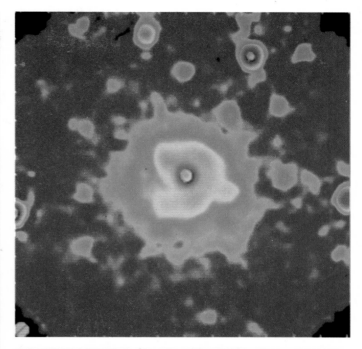

**Figs. 7.6–7.9** *The world's largest fully steerable radio telescope, opened in 1971, is the 100 metre diameter dish at Effelsberg, near Bonn in West Germany (**Fig. 7.6**). The outer part of the dish is made of mesh, to reduce wind pressure which distorts the telescope's shape. The valley location shields it from radio interference. The control building (left) has a grandstand view of the telescope (**Fig. 7.9**). A typical radio map, of the galaxy IC 342 (**Fig. 7.7**), is colour coded for intensity; the background sky is dark blue, brighter regions are green, yellow red, then blue, yellow and red again. **Fig. 7.8** shows similar intensity colour coding applied to a photograph of the Effelsberg telescope itself, with the most brilliant parts coded white.*

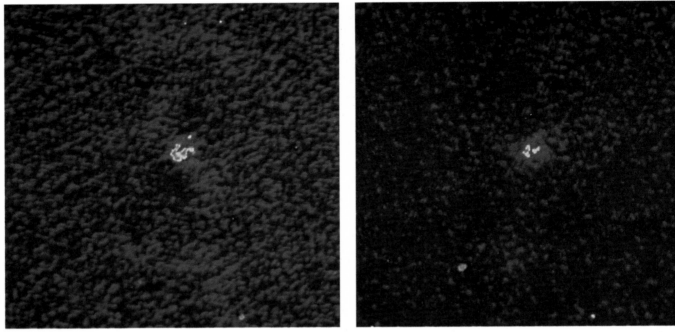

Figs. 7.10–7.11 *A supernova in the spiral galaxy M100 does not appear in* **Fig. 7.10** *(left), a radio map taken in April 1979 at the time of the outburst; radio emission is from the galaxy's nucleus. Colour coding indicates the brightest regions in red, with fainter parts yellow, light blue and dark blue.* **Fig. 7.11** *(right) shows the galaxy a year later, when the supernova's expanding gas does appear at radio wavelengths (red spot at lower left).*

a problem with the engineering: how to keep the enormous bowl true to shape as it is tilted from the horizontal to the vertical and buffeted by winds. The dish at Effelsberg, near Bonn, represents the limit in size for a fully steerable dish; although it was built fifteen years later, it is only slightly larger than the Jodrell Bank 76 metre radio telescope.

Astronomers from Cornell University, New York State, accepting that a larger dish can be built but not tilted, have constructed the word's largest radio dish in a natural hollow in the limestone hills near Arecibo, on the Caribbean island of Puerto Rico. The Arecibo dish is an immense bowl of wire-netting, 305 metres across. The dish reflects radio waves up to a receiving antenna strung on a girder 130 metres above it – the height of a fifty-storey building. The Arecibo telescope's huge area means that it can detect some very faint radio sources – but even this leviathan can barely resolve detail as fine as the human eye can perceive at visible wavelengths.

However, radio telescopes with such low resolution can perform many useful tasks. The steerable dishes can maintain observation of an individual radio source, timing the pulses from a neutron star, or watching a flare star for one of its unpredictable bursts of radio waves.

Such radio telescopes are also invaluable for surveying the sky. Most stars emit radio waves so faintly that even the most sensitive radio telescopes cannot detect them; the strongest radio sources in the sky are mostly objects that emit very little visible light, and a powerful optical telescope is needed to reveal their appearance at visible wavelengths. So surveys at radio wavelengths are used to locate these sources. A large radio telescope with a resolving power rather poorer than the human eye – a few arcminutes in angular scale – is about right for such work; a telescope with a finer beam would take an inordinately long time to scan the entire sky.

The earliest surveys showed that the brightest object in the radio sky is the band of the Milky Way – as the pioneer Jansky had suggested. The Milky Way outshines even the Sun at radio wavelengths – except when the Sun's surface is broken by the explosion of a solar flare, which shines in radio waves a million times more brightly than the whole of the rest of the Sun. As well as the diffuse band of the Milky Way, the radio sky is dotted with individual, small radio sources (Fig. 8.3). These are not stars, however. Many of them are so extended that their diffuse, 'woolly' shapes would be seen easily with radio detectors as sharp as the human eye. Some are the rings of *supernova remnants*, the exploding gas shells produced at the death of a star as a supernova; others are the glowing gases of nebulae – like the Orion Nebula – which surround relatively young stars.

Many of the brightest sources actually lie millions of light years away, well outside our own Galaxy. There are titanic explosions at the centres of some distant galaxies, with compact radio sources called *quasars* marking the site of the explosion in the most violent of these galaxies. The emissions from *radio galaxies* mark the aftermath: the energy blown out in the central explosion creates huge radio-emitting lobes on either side of the galaxy itself (Fig. 7.12). Radio detectors, blind to the millions of stars making up the galaxy, see a dumb-bell in the sky, sometimes stretching out millions of light years on either side of the galaxy which has spawned the colossal clouds.

Early radio astronomers named individual sources after the constellations they appear to lie in. The strongest source in the constellation Cygnus was called Cygnus A (Figs. 12.13–12.16), and so on through the alphabet. Later surveys at observatories around the world have listed too many sources for this simple system to be used. Instead, each observatory has produced its own catalogue

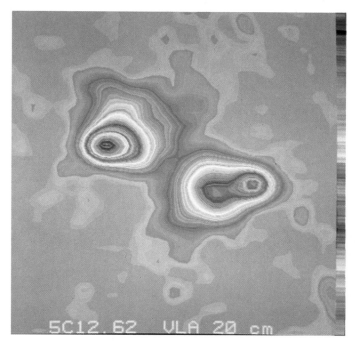

**Fig. 7.12** *The radio-emitting lobes of 5C 12.62 undoubtedly straddle a massive galaxy, but it is too distant for optical telescopes to detect. The colour coding for intensity follows the sequence on the right (blue–green for background sky; dark blue for the brightest region).*

– so the same source can appear under a variety of different catalogue numbers. The first major catalogue of the northern skies, the Third Cambridge Survey, is the most widely quoted. The nearest quasar, for example, is known as 3C 273 (Figs. 12.33–12.38): it was originally entered as the two hundred and seventy-third radio source in the catalogue. The most distant quasar currently known, PKS 2000-330, on the other hand, was too faint and too far South to have shown up in the 3C survey. It was picked up in a later survey by the Parkes radio telescope in Australia, hence the prefix PKS. The figures give its position in the sky in an abbreviated form.

In their more recent surveys, in addition to using more sensitive telescopes to look for fainter radio sources, astronomers have scanned the sky at different wavelengths, to pick out different kinds of sources. The hot gas in nebulae, for example, shines much more brightly at the shorter wavelengths, as do the cores of the distant quasars. A survey at short wavelengths picks out objects such as these as the brightest in the sky. Most other kinds of radio source consist of high-speed electrons travelling through magnetic fields. They produce radio waves of a type called *synchrotron radiation* which is strongest at the longer wavelengths. A long-wavelength survey is dominated by the synchrotron radiation, the supernova remnants and the distant magnetic lobes of the radio galaxies.

By looking at a radio source's relative strength at different wavelengths, something can be said about what is producing its radiation: whether it is a cloud of hot gas, or a magnetic 'bag' full of electrons. But to find out exactly what is going on, a higher resolution detector is needed to probe the finer details of the source.

Since a telescope's resolution depends both on its diameter and the wavelength that it is observing, there are two ways to improve your view of the fine structure in radio sources: either build a bigger telescope, or use existing telescopes at a shorter wavelength. The second alternative certainly seems easier – and cheaper.

When the giant Effelsberg dish is tuned to radio waves of three centimetre wavelength, it can resolve details down to about an arcminute in size – as good as the unaided human eye. If its antenna and receiver are replaced with a set which picks up radio waves of one centimetre wavelength, then the telescope can resolve details only one-third of an arcminute across: unlike the human eye, it can reveal the planets Jupiter and Venus as definite discs, rather than mere 'points'.

But short-wavelength research has its own problems. Water vapour in the atmosphere is a serious one; the amount of radiation it absorbs changes all the time with the changes in cloud and humidity. In addition, the surface of the telescope must be smoother and more accurate. Bumps a centimetre high have little effect on radiation of,

say, 20 centimetres wavelength; but if such a dish were used to pick up radiation of one centimetre wavelength, the bumps would scatter the radiation so badly that only a small fraction would reach the focus – it would be like trying to see a reflection in the unpolished surface of an aluminium saucepan. Apart from these instrumental problems, the synchrotron radio sources produce very little short-wavelength radiation anyway, so their strong, long-wavelength emission must be studied instead.

Radio astronomers have therefore been driven to inventing ingenious techniques to make radio telescopes that are effectively several kilometres in size – far larger than the size at which a single dish can be built. The most successful method – *Earth-rotation synthesis* – was pioneered in the 1960s by the Cambridge radio astronomer Martin Ryle. It relies quite simply on two small radio telescopes, a powerful computer, and the fact that the Earth rotates on its axis. It uses these ingredients to build up what is in effect a huge radio telescope mirror.

Any mirror forms an image at the focus by the merging of radiation reflected from each part of its surface, in such a way that the waves of radiation 'interfere' with one another – the 'crest' of one wave can reinforce the crest of another, or be 'damped' by another wave's 'trough'. In principle, it should be possible to create the effect of a very large optical telescope mirror with just two small mirrors. Keeping one mirror fixed in the centre, the other one can be moved to successive positions around it until it has covered the total area of an imaginary large mirror which is in effect being synthesised. When the mobile mirror is at each position, the image at the focus can be recorded; and eventually all these images can be combined to make the image which the larger imaginary mirror would have seen.

That is the theory; unfortunately it does not work for optical telescopes. You have to keep track of the phases of each image (the parts of the waveform at each point in the mirror) in order to add them all correctly at the end; and light waves are so short that this is impossible with current equipment. Radio astronomy has the advantage here: the

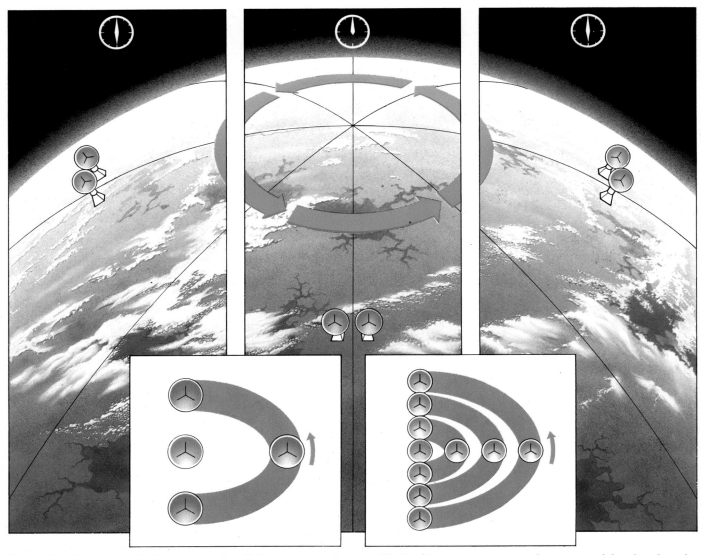

**Fig. 7.13** *Earth-rotation synthesis uses a pair of small dishes to mimic a large one. The Earth's rotation carries one telescope around the other, through 180° in 12 hours (left inset). By changing their separation daily, radio astronomers build up a complete synthesised large dish (right inset) – the missing half is added by computer (it contains no new information).*

phase of the incoming radio waves is preserved in the electrical signal that forms the radio telescope's output, and it is at a sufficiently low frequency that electronic circuits can record it. In addition, the two radio telescopes do not actually have to reflect radiation to a distant mutual focus. This too can be achieved electronically. Each small dish collects radio waves in the normal way, and their electrical outputs are merged to mimic the merging of radio radiation at the focus.

So a large radio telescope *can* be synthesised by two small dishes and some electronic wizardry. There is obviously a practical problem in moving a radio telescope dish continually from place to place to cover an area the size of the large imaginary dish, but for this we can make use of the Earth's spin (Fig. 7.13). Imagine looking down from a radio source which happens to lie above the Earth's North Pole. A radio astronomer in middle-latitudes – say England or the United States – has set up two small radio dishes on an East–West line. The Earth rotates anticlockwise and the dishes travel with it. As they do so, their *orientation relative to each other in space* changes

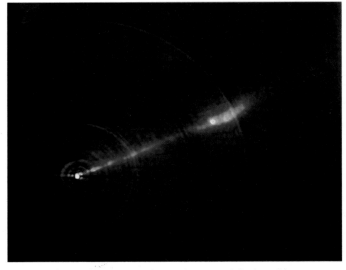

**Fig. 7.14** *The Earth-rotation technique has caused the faint false rings in this detailed view of the jet of electrons from the galaxy NGC 6251, taken by the Cambridge Five Kilometre Telescope.*

**Fig. 7.15** *Five of the eight dishes of the Cambridge Five Kilometre Telescope are seen here, spread out along their railway track.*

too. We see this effect from our position above the North Pole. Relative to the dish which was originally on the left, the other circles it in an anticlockwise direction; starting on the right, it moves round continuously, until after twelve hours this dish is now on the left. If we regard the first dish as the centre of a synthesised mirror, the second has travelled in a semicircle, around it, to trace out half a ring of mirror surface, some distance out from the centre.

The radio astronomer then moves the second mirror slightly closer to the first, and lets the Earth's rotation create another, slightly smaller semicircular ring of mirror. On moving it nearer for another twelve-hour run, a still smaller ring of the mirror is 'filled in'. Eventually, an entire half-mirror has been synthesised. Theory shows that the information from the 'missing' half of the mirror is mathematically related to that obtained from the synthesised half, so the computer can add in the missing information automatically. The technique puts into the computer all the information that would have been gathered by a complete dish equal in size to the largest separation of the small dishes.

This information is stored as a set of electronic intensities and phases, each measurement corresponding to one relative position of the dishes. A mathematical technique called Fourier analysis can be used to convert this mass of figures directly into a view of the radio sky – stored neatly as an array of numbers, just like the output of a scanning single dish (or the computer-memory version of an optical photograph). The Earth-rotation synthesis telescopes do not even need to scan back and forth. Each dish points constantly at a particular radio source, and the Fourier analysis effectively does the scanning within the computer.

Ryle's first Earth-rotation synthesis telescope was 1.6 kilometres long; it was followed in 1972 by another 5 kilometres in length – effectively fifty times larger than the world's largest steerable dish at Effelsberg. Each of these telescopes in fact has more than two dishes. The Five Kilometre Telescope (Fig. 7.15) has eight, and by connecting them in pairs sixteen rings of the mirror can actually be synthesised simultaneously. Four of the dishes can be moved along railway tracks to alter their separations. The Dutch have built a similar telescope, the Westerbork Synthesis Radio Telescope, which has fourteen dishes to synthesise a three kilometre telescope.

These telescopes have revolutionised our knowledge of radio sources, particularly of radio galaxies and quasars. Their fine resolution of detail has shown why a galaxy's central explosion can produce the distended radio-emitting lobes. They reveal narrow glowing 'jets' of matter (Fig. 7.14) stretching out from either side of the core, leading to the lobes. These jets are evidently beams of fast electrons, shot out at high speed from the core like the electron beam from the back of a television tube. Where these beams collide with the tenuous gas outside the galaxy, they are reflected back in disorder, like the jet of water from a garden hose splashing back from a wall. The electrons' motion constitutes a disorderly electric current which generates a tangled magnetic field in these regions. As the high-speed electrons move through this self-generated field, they emit synchrotron radiation at radio wavelengths.

The first Earth-rotation synthesis telescopes had problems looking at radio sources far from the North Pole of the sky. As seen from a radio source above the Earth's North Pole – in the direction of the celestial North Pole – the dishes in these arrays trace out semicircular rings as the Earth rotates. But the view from a radio source away from the pole is foreshortened; each dish traces out an oval path relative to the others. The synthesised dish is thus not a circular mirror, but an oval one. This shape resolves details of a radio source less well in the North–South direction. For a very small radio source whose size is at the resolution limit of the telescopes, the image is stretched from being a circular blur to an elongated one. The stretching gets worse for sources further from the celestial North Pole, and becomes so extreme near the celestial Equator (the imaginary line above the Earth's equator) that no North–South detail can be resolved at all.

The answer is to include more dishes in the radio telescope array, in lines that extend some distance North–South, rather than exactly East–West. As seen from the radio source, each dish now traces out a much more complicated arc relative to the other: the synthesised mirror is made from many differently shaped arcs of mirror surface. The telescope computer can, however, calculate the Fourier transform from the measurements made along these arcs, to produce the appearance of the radio sky as reflected in this large 'partial mirror'. Just as an optical mirror with pieces missing cannot form an accurate image, so the partial mirror of the radio telescope produces distortions in the image it makes. The worst distortions are now routinely removed using computer programs designed specifically to 'clean up' maps of the radio sky.

The most ambitious of these radio telescopes is the Very Large Array (Fig. 7.16), sited in the desert near Socorro in New Mexico. The VLA has 27 dish aerials each 25 metres across, which is large by the standards of a radio telescope anyway. They can move to various positions along the

Fig. 7.16 *Rainbows are a rare sight over the desert location of the most sophisticated purpose-built radio telescope array, the Very Large Array in New Mexico. The dishes are seen here unusually close together. They can be spaced out on railway tracks 21 kilometres long to synthesise a correspondingly large telescope.*

arms of a Y-shaped railway network. The whole array effectively makes a radio dish 27 kilometres in diameter. When the VLA is observing at its shortest wavelength of 1.3 centimetres, it can resolve details 0.13 arcseconds in scale – almost a thousand times better than the human eye, and nearly ten times better than any Earth-based optical telescope.

Arrays like the Five Kilometre Telescope and the VLA have their dishes connected electronically by buried cables or waveguide tubes to the central control computer building. The VLA represents about the largest size possible for directly linked telescopes, but a larger array can be built if instead the signals from each dish are amplified, then fed into an ordinary radio communication receiver which beams the signal to the control centre. Astronomers at Jodrell Bank have connected several isolated dishes in England by radio links, to make a dish effectively 133 kilometres across. This network is called MERLIN – the Multi-Element Radio-Linked Interferometer Network. It can perceive details as small as 0.02 arcseconds in size when it observes at 1.3 centimetres wavelength. In practice, though, longer wavelengths are often used which give MERLIN about the same resolution as the VLA. As a result, the variation in the structure of a radio source at different wavelengths can be observed.

The size of radio telescopes can, however, be pushed even further to mimic the performance of a radio telescope the size of the Earth! In the technique of Very Long Baseline Interferometry (VLBI), astronomers at widely separated radio observatories look at the same radio source simultaneously, and record the output of their telescopes, along with the signals from an atomic clock which keeps an extremely accurate track of time. The tapes are flown to a common centre, where they are played back, with the atomic time signals keeping them exactly synchronised. The outputs from the telescopes can then be added together electronically just as if the telescopes had actually been connected while they were observing.

With VLBI, details 0.001 arcseconds in size can be resolved – a thousandth the scale of the smallest detail an optical telescope can see, and equivalent to making out a pinhead at a distance of 200 kilometres. But there is a price to pay. Two radio telescopes separated by almost the diameter of the world form only two very tiny portions of the mirror they are trying to synthesise. In fact, they cannot form a proper image at all, although they can detect whether or not a radio source does have very fine details within it. Additional radio telescopes need to be involved to synthesise a more complete VLBI mirror. The combination of the results from, typically, half a dozen telescopes can produce fine-scale images – the most detailed of all the images formed at any wavelength.

The nearest quasar, 3C 273, for example, is a powerful explosion at the core of a galaxy lying some 2000 million

light years away. It is a compact blaze of light and other radiations, less than a light year in size. When we look at the galaxy with telescopes that are sensitive to radiations at wavelengths from X-ray to infrared, however, our telescopes or the Earth's atmosphere blur it out to a fuzzy blob which obscures all detail within 5000 light years of the quasar's centre. But a radio VLBI network can probe a thousand times smaller to reveal that the core is ejecting a jet of matter only a few dozen light years long (Fig. 12.38).

The quasars and their elderly relatives the radio galaxies, together with the supernova remnants and the hot gas clouds in our Galaxy, all radiate over the whole range of radio waves, and so can be observed at any chosen wavelength.

But there is also a different kind of radio emitter in the sky which radiates at just one or a few specific wavelengths. These are atoms or molecules of gas, in the almost empty space between the stars of a galaxy. To detect them, radio telescopes must be 'tuned-in' precisely to the relevant wavelength.

An accurately tuned radio telescope can thus probe the gases in interstellar space – both the gas surrounding us in our own Galaxy, and the gas in other distant galaxies. Conveniently enough, the most abundant kind of atom in space broadcasts at radio wavelengths. Hydrogen atoms naturally produce radiation at a wavelength of 21.106 centimetres. A radio telescope tuned to this wavelength sees a bright glow all along the band of the Milky Way, from the large amount of hydrogen between the stars in our Galaxy. It sees nearby spiral galaxies as bright whirlpools of hydrogen, the interstellar gas bunched up into the shape of a Catherine wheel like the stars we see with an optical telescope. Some galaxies are surrounded by streams of hydrogen stretching far beyond the distribution of stars making up the galaxy itself.

Radio telescopes tuned to the hydrogen wavelength can tell us not only where the hydrogen lies, but how fast it is moving towards or away from us. The *Doppler effect* alters the wavelengths we pick up from a moving radiation source, in the same way as for the wavelengths of visible light (p. 50). It moves optical spectral lines toward the *red* (longer wavelength) end of the spectrum if the source is moving *away*, and towards the *blue* (shorter wavelength) end if it is moving *towards* us. The 21 centimetre hydrogen line is affected in the same way.

If the telescope is tuned to precisely 21.106 centimetres it detects only hydrogen clouds that are not moving either towards or away from us. If it is re-tuned to 21.105 centimetres, it gives a different view – the clouds that are moving towards us at a speed of 15 kilometres per second (50 000 kph). To see clouds moving *away* at this speed, the receiving wavelength must be changed to 21.107 centimetres. The hydrogen line view of a galaxy showing the relative velocities of the clouds making up the image is best portrayed pictorially by using a colour code. Gas clouds moving away are coloured in successively darker tints of red for increasing velocity, and clouds approaching us in corresponding shades of blue. A rotating galaxy appears multi-tinted, from dark red at the edge that is moving away from us, to dark blue at the other edge where the gases are coming towards us. (Fig. 10.4)

The hydrogen line was predicted during the war by the Dutch astronomer Hendrick van de Hulst and it was first detected in 1951. In the past two decades, radio astronomers have discovered that hydrogen atoms are not the only components of interstellar gas that emit radiation at their own characteristic wavelengths. The others are molecules, groups of atoms chemically bonded together. To date, radio astronomers have discovered more than seventy kinds of molecule out in space, ranging from just two atoms joined together, like carbon monoxide, to a molecule of thirteen atoms called cyano-deca-penta-yne. The range includes ammonia, hydrogen cyanide and ethanol – ordinary alcohol.

Most of these molecules are hidden in the dense clouds of dust and gas where new stars are forming. Here isolated atoms can link together to form molecules; and the obscuring dust protects the fragile molecules from ultraviolet radiation which would otherwise break them up into atoms again. Two of the simple molecules in these clouds are of particular interest. Carbon monoxide (CO) is very widespread, and it is a useful indicator of the extent of these largely invisible clouds. It circumvents the problem that hydrogen, the most abundant element in the clouds, is in the form of hydrogen molecules ($H_2$) here, which do not emit radio waves (unlike single atoms with their characteristic 21 centimetre radiation). There is thus no direct way of knowing how much hydrogen there is in the cloud, nor how it is spread out. Carbon monoxide molecules, however, are scattered throughout the hydrogen. By observing the radiation from carbon monoxide at a wavelength of 2.6 millimetres, radio astronomers can fathom the shape and size of the clouds (Fig. 4.4). The total amount of gas there can also be calculated because the relative strengths of other, weaker carbon monoxide lines indicate that on average there are ten thousand molecules of hydrogen to each carbon monoxide molecule. Until the 1970s, astronomers thought that most of our Galaxy's hydrogen (and other gases) was spread thinly as atoms in the vast spaces between the stars. It has since been found that there is fully as much hydrogen bound up as molecules within the densest of the dark clouds of space, known as *molecular clouds*.

The first molecule found in space, in 1963, was *hydroxyl* (OH), which is a water molecule with one hydrogen atom missing. It radiates at a wavelength of 18 centimetres. Soon after the discovery of hydroxyl radio sources, astronomers found that some clouds were very small and changed in brightness from day to day. There was only one explanation. These small gas clouds are natural *masers* – the radio equivalent of lasers – powered by the infrared radiation from protostars in the dense cloud (Fig. 4.29).

Most molecules radiate at the shortest of radio wavelengths, the millimetre waves and submillimetre waves which border on the far infrared. When astronomers try to pick up these radiations with a conventional radio telescope they run into all the problems of working at short radio wavelengths (the need for a very smooth telescope surface and absorption by water vapour in the Earth's atmosphere). As a result, purpose-built telescopes are now being constructed to work only at these wavelengths, and they will be sited on mountain peaks above the worst of the Earth's atmosphere.

A millimetre wave telescope is intermediate in size between an optical or infrared telescope and a radio telescope. The first was America's 11 metre telescope (Fig. 7.17) sited at Kitt Peak in Arizona – twice the size of the Palomar optical telescope, and about a tenth the diameter of the Effelsberg radio dish. Its successes have prompted a new 'breed' of millimetre wave dishes now under construction – a German 30 metre and a more accurate British 15 metre dish. These telescopes will probe the gas in dense clouds, investigating the events leading to the formation of new stars and planetary systems.

As the first of the 'invisible astronomies', radio astronomy has turned up some of the most exciting objects and places in the Universe – from the dark clouds where new stars are forming to the ultra-compact pulsars left over from a 'star death'; and in the most distant reaches, the colossal explosions of the quasars.

**Fig. 7.17** *The pioneering 11 metre radio telescope at Kitt Peak, Arizona, has a surface accurate enough to observe the shortest radio waves, down to 2 millimetres wavelength. The dome protects it from distortion due to wind pressure and heating by the Sun. At an altitude of 2000 metres, the telescope is above the worst of the water vapour which absorbs millimetre waves.*

# 8 Milky Way System

The luminous band of the Milky Way stretching across the night sky is a broadside view of the galaxy in which we live. Astronomers call it the Milky Way Galaxy, or simply the Galaxy. It is a vast swarm of stars, gas and dust, shaped like a Catherine-wheel with a bulging centre, and measuring some 100 000 light years from edge to edge. Its main constituent is stars: about 200 000 million in total. Our Sun is just one very ordinary star amongst this unimaginable number. Space between the stars is filled with interstellar matter: gases, dust grains, magnetic field and very fast subatomic particles called cosmic rays. There is sufficient gas and dust to build a further 20 000 million stars.

From the outside, our Galaxy would look much like the Andromeda Galaxy (Fig. 10.2); but we see the Galaxy from within, and as a result it is all around us. To 'see' our Galaxy in its entirety, we must look at the whole sky. The photographs and maps on the next few pages are thus views of the entire sky, spread out into an oval frame in much the same way that the surface of the Earth can be represented by a single oval map. Seen edge-on, the flat disc of the Galaxy appears as the narrow band of the Milky Way, and astronomers project the sky onto the maps in such a way that the Milky Way band runs horizontally across the oval. The sky is 'unfolded' such that the left-hand end of the Milky Way on the map in reality is joined up with the right-hand end and the centre of each map is the view directly towards the centre of the Galaxy.

It is impossible to photograph the entire sky at one time, so astronomers at the Lund Observatory, Sweden, have prepared this optical view (**Fig. 8.1**) by redrawing details from a mosaic of photographs onto a single drawing. The original is two metres across, and as well as showing the intricacies of the Milky Way band it shows the positions and brightnesses of all the stars visible to the unaided eye, a total of about 7000 individual images.

The Milky Way band consists of thousands of millions of more distant stars, too faint to be seen individually. In the plane of the Galaxy's disc (along the 'equator' in these projections) there are plenty of distant stars and the sky appears bright. But in other directions, we are looking 'up' or 'down' through the narrow extent of the Galaxy's thickness; there are few distant stars and the sky is dark. Thus

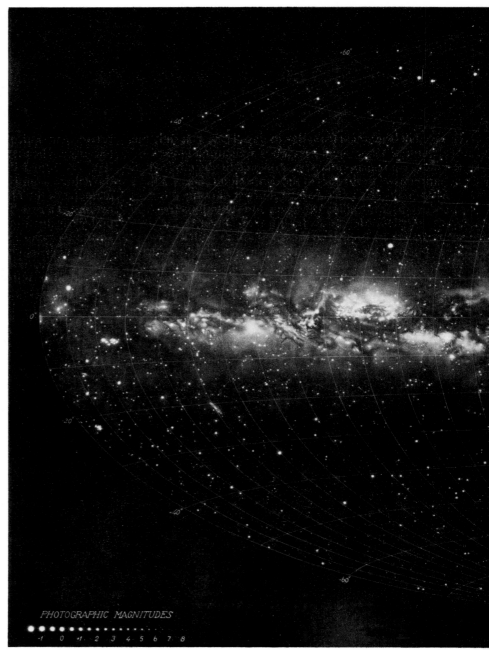

8.1 *Optical, photographic mosaic, Lund Observatory*

we get the appearance of a bright band round the sky.

The dark patches within the Milky Way are dense clouds whose dust particles are blocking off the light from stars beyond. In fact, there are many more stars towards the Galaxy's centre, and the optical view should show a really brilliant region at the centre of this picture, but where there are more stars, there is also more dust. This absorbs much of the extra light, to make the Milky Way rather uniform.

Although this projection distorts the familiar constellations, some patterns can be made out. On the extreme right, just below the Milky Way, is Orion, with the brightest star, Sirius, just above. The

Plough is at the top left, upside down. Vega is the bright star above the Milky Way left of centre, and below is a prominent dust band which appears to split the Milky Way in the constellation Cygnus. Almost mirroring this on the right is the small dark triangle of the Coal Sack. Three small bright 'clouds' below the Milky Way are neighbouring galaxies, the only objects here which are not part of our Galaxy. On the left is the small oval of the Andromeda Galaxy. In the right half lie the two satellite galaxies of the Milky Way, the Small (left) and Large (right) Magellanic Clouds. Although these are separate galaxies, they are attached to ours not only by gravity but by streamers of invisible hydrogen gas.

The objects most prominent at optical wavelengths, however, are not generally the brightest sources detected at other wavelengths. The key diagram (**Fig. 8.2**) shows the locations of the objects which appear prominently on the radio (Fig. 8.3), gamma ray (Fig. 8.4) and X-ray (Fig. 8.6) maps of the sky, as well as a few of the optically bright stars and galaxies. At other wavelengths, the Milky Way band is more striking than in visible light, and sources near this band are generally within our Galaxy (with a few exceptions like the distant radio galaxy Cygnus A), while objects near the top and bottom of the maps are mainly galaxies beyond our own.

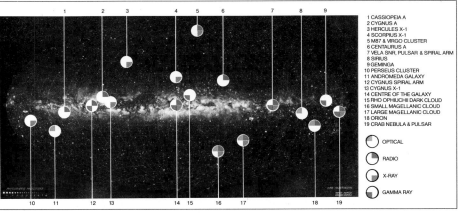

1 CASSIOPEIA A
2 CYGNUS A
3 HERCULES X-1
4 SCORPIUS X-1
5 M87 & VIRGO CLUSTER
6 CENTAURUS A
7 VELA SNR, PULSAR & SPIRAL ARM
8 SIRIUS
9 GEMINGA
10 PERSEUS CLUSTER
11 ANDROMEDA GALAXY
12 CYGNUS SPIRAL ARM
13 CYGNUS X-1
14 CENTRE OF THE GALAXY
15 RHO OPHIUCHII DARK CLOUD
16 SMALL MAGELLANIC CLOUD
17 LARGE MAGELLANIC CLOUD
18 ORION
19 CRAB NEBULA & PULSAR

OPTICAL
RADIO
X-RAY
GAMMA RAY

**8.2** *Diagram, location of bright sources at different wavelengths*

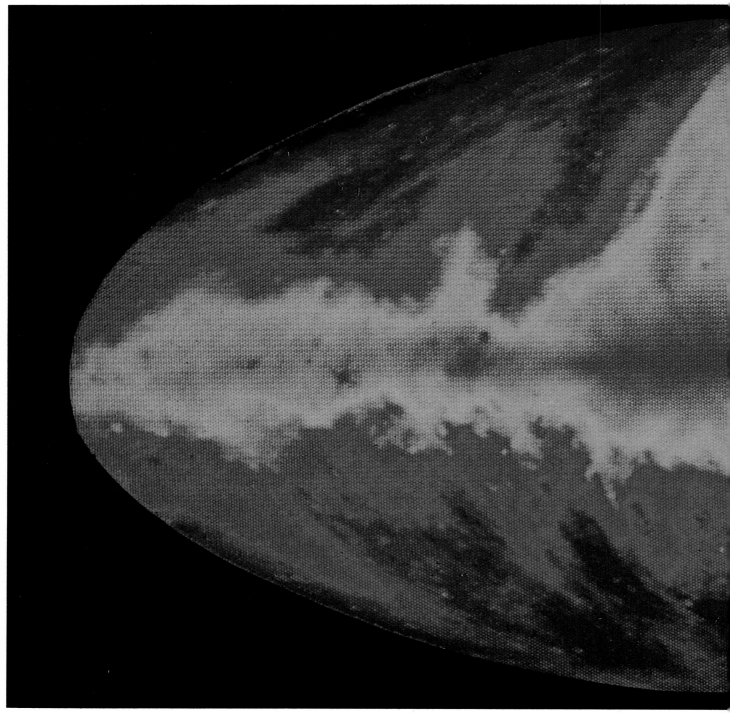

**8.3** *Radio, 73 cm, combination of observations from 100 m Effelsberg, 76 m Jodrell Bank, and 64 m Parkes telescopes*

The radio sky is dominated by the Milky Way, as seen in this whole-sky map (**Fig. 8.3**). It is drawn out in galactic coordinates, in the same way as the optical view (Fig. 8.1) so the two can be compared directly. In this false-colour view, the brightest regions are coded red, and successively fainter regions yellow, green and blue. On the right-hand side away from the Milky Way, black regions show where the radio sky is almost completely dark.

No single radio telescope can see the whole sky (unless it is sited on the Equator), and Glyn Haslam has compiled this radio view from observations made with three of the world's largest radio telescopes: the 76 metre at Jodrell Bank, England; the 100 metre at Effelsberg, West Germany; and the 64 metre at Parkes, in Australia. All the telescope observed at a wavelength of 73 centimetres, and built up different parts of the map by scanning back and forth across the sky, measuring the

radio brightness at each point. After twelve years of observing, Haslam's team spent a further three years reducing the 3 000 000 000 individual measurements with the help of a computer at the Max Planck Institute for Radio Astronomy in Bonn. It ended up with an array of half a million numbers giving the radio brightness at evenly spaced points in the sky, and they used a computer at Bonn's Rheinisches Landesmuseum to convert them into this colour-coded map.

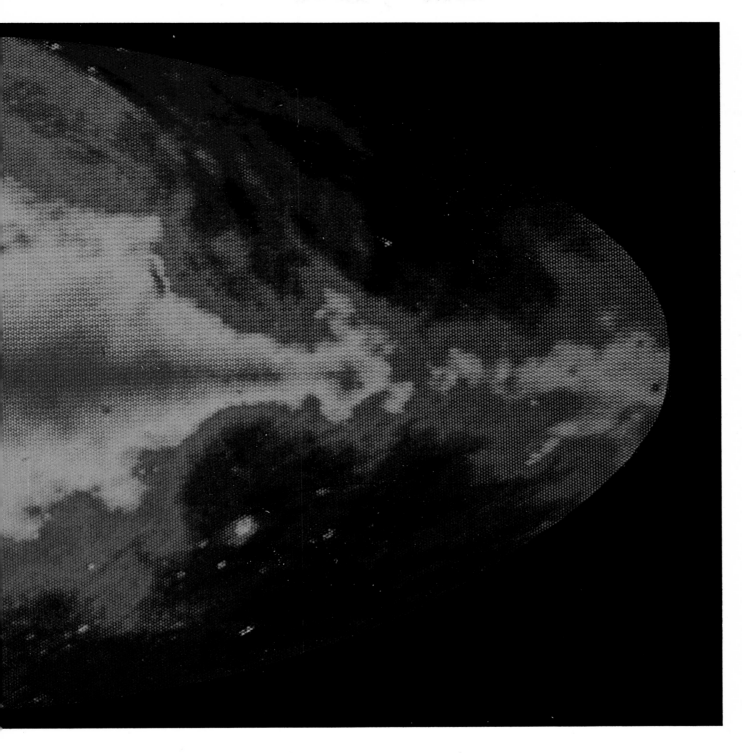

Bright individual sources are identified in Fig. 8.2. They include the galaxies Cygnus A and M87 (both appear very compact here), Centaurus A (elongated) and the Large Magellanic Cloud. Sources within our Galaxy lie along the Milky Way band: Cassiopeia A (with false 'spikes' produced during telescope scans because of its extreme brightness), the Vela region, the Orion region (the yellow 'boomerang' below the Milky Way) and the Crab Nebula.

The radio emission from the Milky Way itself is mainly synchrotron radiation, generated in interstellar space by fast electrons moving through the Galaxy's magnetic field. The radio map is not obscured by dust (unlike the optical view, Fig. 8.1), and the central parts of the Galaxy appear in their true brilliance.

Nearer to us, great filaments marking the magnetic field stretch out of the Galaxy's disc. The most obvious filament, the North Polar Spur, stretches from near the centre of this view upwards, almost to the top of the map. Here it bends over (just below M87) and comes back down (near the position of Centaurus A) to make a complete loop. It is probably the giant shell of an ancient supernova remnant, still slowly expanding from an explosion some 300 000 years ago. The supernova would have been only 500 light years from the Sun, the shell having now swept out 400 light years in every direction, to fill most of the northern sky.

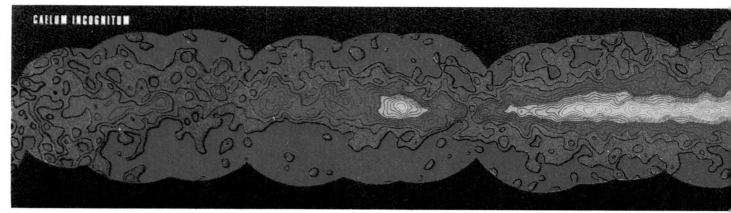

8.4 *Gamma ray, 0.000 000 2–0.000 02 nm, COS-B satellite*

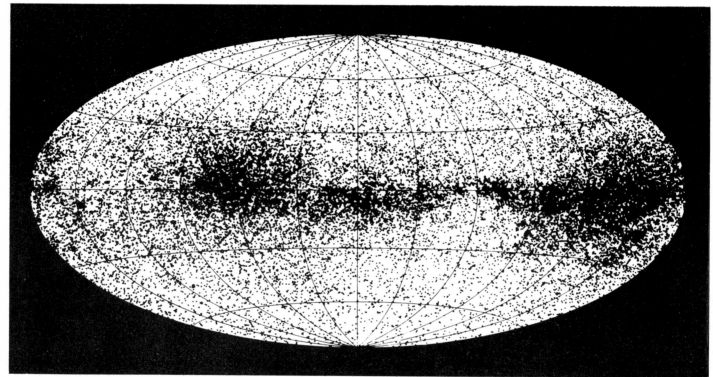

8.5 *Ultraviolet, 135–255 nm, negative print, TD-1A satellite*

Short wavelength views of the sky – from ultraviolet down to gamma rays – reveal successively more energetic objects in our Galaxy.

The European satellite TD-1A compiled the ultraviolet view of the sky (**Fig. 8.5**) from observations in 1972, 1973 and 1974. The satellite could not photograph individual objects, but as it rotated continuously, its telescope surveyed the whole sky and picked out the brightest ultraviolet stars. It located 31 215 stars, about five times the number seen by the human eye, and in this picture each of these stars is marked by a dot. All of them lie within a few thousand light years of the Sun, as the detector was not designed to record the general background glow of distant stars in the Milky Way.

The stars appearing here are the hottest

normal stars, with temperatures of 15 000 K to 50 000 K. Although some are hot white dwarfs (the recently exposed cores of old stars), most are very young, massive stars similar to the stars of Orion's Belt, Sword and Nebula (Figs. 4.2–4.16). According to the TD-1A map, these stars lie mainly in two diametrically opposite directions: Cygnus (middle left) and Orion (right). Young hot stars occur only in the arms of spiral galaxies, so these regions mark the local spiral arm of our Galaxy – probably, in fact, a short spur off a major spiral arm (Fig. 1.9).

The X-ray sky is dominated, not by multitudes of stars, but by a small number of very powerful individual X-ray sources. The first American High Energy Astrophysics Observatory, HEAO-1, scanned the whole sky to produce this X-

ray view (**Fig. 8.6**). The faintest regions of sky are coded green, and successively brighter regions through blue and pink to red. The X-ray detector had a resolution of only a few degrees, so that tiny intense X-ray sources are blurred out to quite large red blobs.

Several dozen powerful sources towards the centre of the Galaxy (Fig. 11.7) are overlapping here, to produce an apparently uniform glow in the middle of the picture. Away from this region, individual X-ray sources can be identified (see the key in Fig. 8.2). Sources beyond our Galaxy include the nearby Large and Small Magellanic Clouds, and the distant Virgo and Perseus clusters of galaxies. The general emission (blue and grey) covering the entire sky is the enigmatic 'X-ray background', possibly due to millions of

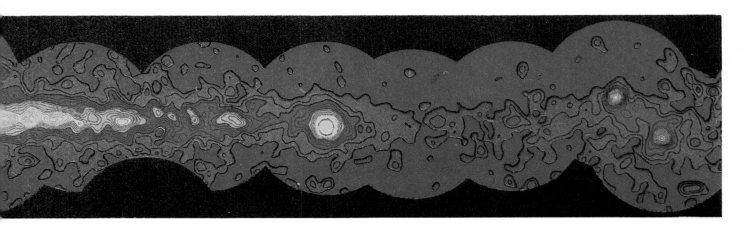

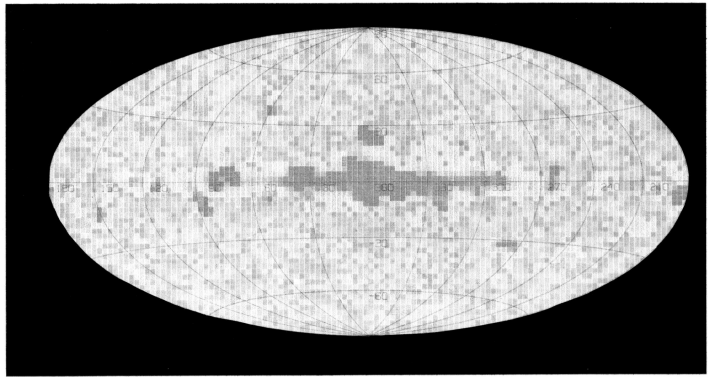

**8.6** *X-ray, 0.02–0.6 nm, HEAO-1 satellite*

very distant quasars.

Two of the brightest sources within the Milky Way are the Crab and Vela supernova remnants, with their pulsars. But most are very close double star systems where a compact star is devouring gases from a normal companion star. In Scorpius X-1 and Hercules X-1, the compact star is a neutron star, while the Cygnus X-1 system probably contains an even more bizarre entity – a black hole ten times heavier than the Sun. It is thought that gases ripped from the companion star shine briefly in X-rays before they disappear into the 'hole'. Whether the companion is a neutron star or a black hole, the evanescent gas streams in an X-ray binary shine 10 000 times more brightly in X-rays than does the Sun at all wavelengths.

The European satellite COS-B mapped the Milky Way at gamma ray wavelengths. The colour-coded map (**Fig. 8.4**) covers only the Milky Way band, with colours running from blue for dim regions, through purple and red to yellow for the brightest gamma ray regions. The COS-B 'telescope' had a resolution of only 2°, so individual small sources are blurred out to this apparent size.

Detailed analysis reveals two dozen individual gamma ray sources. (The identified sources are labelled in Fig. 8.2) The most intense is the Vela Pulsar, flashing regularly thirteen times per second. The second brightest, Geminga, was identified in 1983 as a fairly nearby neutron star, very faint at other wavelengths. Just to the lower left of the bright Crab Pulsar, the map reveals weak gamma rays from the Orion region. The weak gamma ray source

above the galactic centre is another nebula, the rho Ophiuchi dark cloud. The other gamma ray sources are still unidentified but most are probably neutron stars similar to Geminga.

The rest of the gamma rays from the Milky Way form a diffuse 'glow' and are from interstellar matter. They are generated when gas atoms are struck by the high-speed protons which make up the majority of the 'cosmic rays'. Hence the gamma ray map shows up the regions where the interstellar gas is densest. The intense narrow yellow band over the central region of this map reveals the presence of a previously unknown ring of dense gas clouds situated about 10 000 to 15 000 light years out from the Galaxy's centre (one-third to one-half the distance to the Sun).

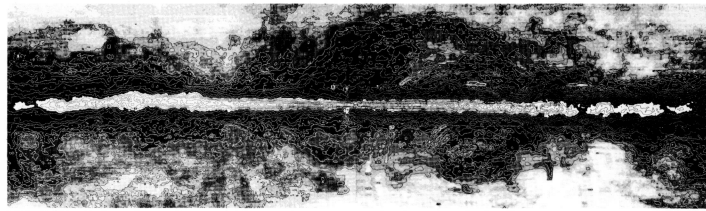

8.7 *Radio, 21 cm hydrogen line, combination of observations from 26 m Hat Creek and 18 and 64 m Parkes telescopes*

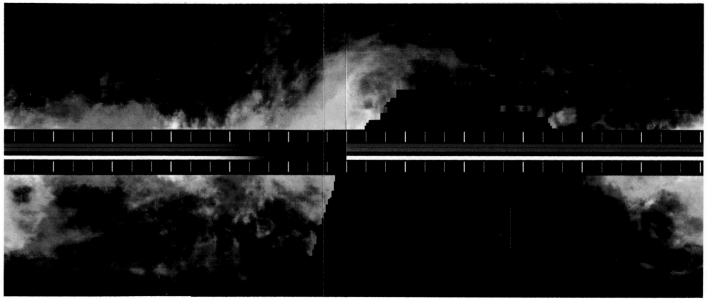

8.8 *Radio, 21 cm hydrogen line, velocity coded, 26 m telescope, Hat Creek Observatory*

Radio telescopes tuned to a wavelength of 21.1 centimetres show us one particular constituent of our Galaxy: hydrogen atoms which make up the bulk of the tenuous, relatively cold interstellar gas. Dutch astronomer J. B. G. M. Bloemen has constructed this view of the Milky Way in emission from cold hydrogen (**Fig. 8.7**) by combining the results from several radio telescopes. The Milky Way runs across the picture, with the Galaxy's centre in the middle. The region of the Milky Way can be compared directly with Figs. 8.1 to 8.6, although the map does not extend far enough vertically to cover the entire sky. The intensity of hydrogen emission is shown by successively darker shades of grey, up to black; then by white contours within the black; and, for the intense Milky Way band, by black contours on a white background.

The interstellar gas is concentrated into a much thinner disc than the Milky Way's stars. While the latter can lie up to several thousand light years from the Galaxy's mid-plane, the cold gas forms a layer along the plane less than a thousand light years thick – in proportion to the Galaxy's 100 000 light year diameter, it is thinner than a gramophone record.

Thus the Milky Way band in Fig. 8.7 is considerably narrower than the visible Milky Way (Fig. 8.1) or the radio view (Fig. 8.3) at wavelengths where the radiation comes from electrons and magnetic field distributed like the stars. The narrow band of hydrogen looks thinnest at the centre of Fig. 8.7, where we are detecting the gas at greatest distances, around the centre of the Galaxy and on the far side of the disc.

This hydrogen gas is moving, and so the precise wavelengths at which we detect the hydrogen emission is altered by the Doppler effect (p. 50). It is possible to 'tune' a radio telescope so that it detects only hydrogen atoms moving at one particular velocity. By making a series of maps at slightly different wavelengths, American astronomer Carl Heiles has produced a set of radio photographs (**Fig. 8.9**) showing gas moving at different speeds. All the thin horizontal strips cover the same part of the Milky Way band, centred roughly on Cassiopeia and stretching from the galactic centre (on the right) to Orion (at the left)–the right-hand edges of the strips correspond to the centre of Fig. 8.7, and they extend past the left-hand margin of Fig. 8.7 to include cold hydrogen seen at the right of Fig. 8.7. The gas visible in the top map is approaching us at 43 kilometres per second; while the hydrogen in the lowest map is receding at 41 kilometres per second. The strip immediately below the centre of Fig. 8.9 shows gas which is almost stationary (receding at only 1 kilometre per second). It reveals local hydrogen around the Sun which covers most of the strip, and is not detected at the approach velocity of 3 kilometres per second in the strip immediatly above.

The Milky Way's appearance changes dramatically between these maps because of the way the Galaxy rotates. The Sun is orbiting the Galaxy's centre, as are all the stars and the interstellar matter, but they do not keep the same distances from one another, because the inner parts of the Galaxy are orbiting more rapidly. From our moving viewpoint, therefore, we see gas catching up with us, when we look in two particular directions, and gas receding from us between.

The top maps reveal the approaching gases in Cassiopeia. These hydrogen clouds must be thousands of light years away from us for the Galaxy's differential rotation to produce such a high speed relative to the Sun. Astronomers regularly calculate the distance of hydrogen clouds in different directions from their speeds.

Moving down the strips, to the progressively lower speeds, we see gas which is closer to us. The central, low-velocity strips are also bright to the right of centre, where we view gas following the same orbit as the Sun, and to the left of centre, in the opposite direction from the galactic centre, a region where gas is moving directly across our line of sight. At the slightly longer wavelengths, in the lower strips, we are seeing, once more, hydrogen clouds at distances of thousands of light years, but this time moving away from us.

Heiles has also mapped larger regions of the sky to investigate the way the interstellar gas is distributed. The false-colour picture (**Fig. 8.8**) from the Hat Creek telescope covers the same region of sky as Fig. 8.7, except for the blank region (to the right of centre), which is never visible from California. The gas along the band of the Milky Way is not shown: it would lie behind the dark horizontal strip. Heiles has colour coded the picture according to each cloud's velocity. The colours range from deep blue for gas approaching us at 20 kilometres per second, through yellow–green for the gas with no Doppler shift, to deep red for cold hydrogen moving away at 20 kilometres per second. This colour coding instantly shows the effect of the Galaxy's rotation.

Both Figs. 8.7 and 8.8 also show faint hydrogen clouds stretching well away from the main concentration along the Milky Way band. The gas is concentrated in long, thin filaments which curve over huge regions of sky. One obvious arc of gas extends upwards to the right of centre, and three other large filaments curve below the Milky Way. The gases in these filaments have a slightly different speed ('colour' in Fig. 8.8) from the rest of the gas at those positions in the Milky Way. The reddish arc at bottom right, for example, is a shell of gases expanding outwards at a speed of 23 kilometres per second. Heiles suggests that the huge expanding shells were 'blown' by a very rare and powerful type of exploding star, a thousand times more powerful than an ordinary supernova.

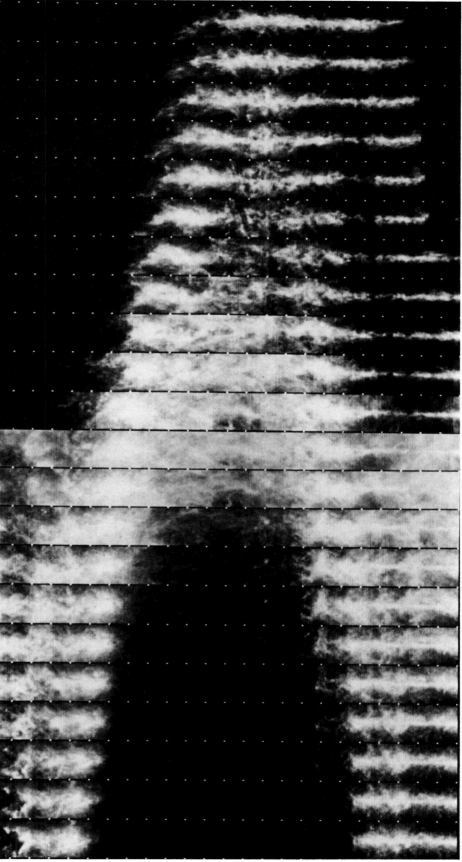

**8.9** Radio, 21 cm hydrogen line, scans at velocity intervals of 4 km/s, ranging from −43 km/s (top) to +41 km/s (bottom), 26 m telescope, Hat Creek Observatory

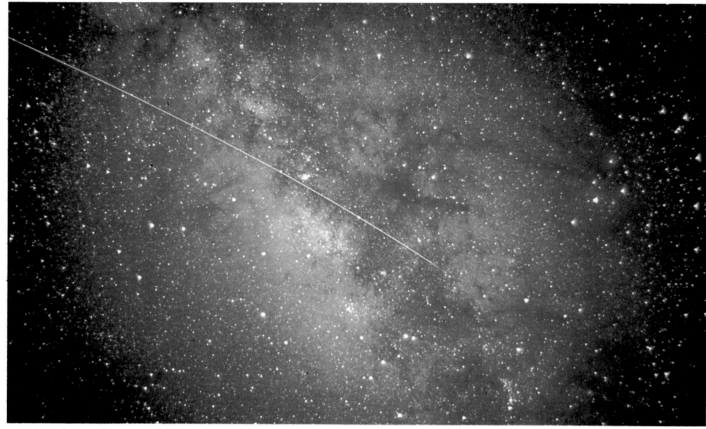

**8.10** *Optical, true colour, with satellite track, 35mm camera with wide-angle lens mounted piggyback on 1 m telescope, US Naval Observatory*

# The centre of the Galaxy

The centre of our Galaxy the Milky Way lies 30 000 light years from us, in the direction of the constellation Sagittarius – well down in the southern part of the sky as seen from Earth. The wide-angle optical photograph (**Fig. 8.10**) covers about 30° by 40° in the constellations Sagittarius and Scorpius. Like all the pictures on this spread and the next, the photograph has North at the top, so the Milky Way band does not run horizontally (as in the previous spreads of this chapter), but stretches diagonally from the top left to the bottom right. The streak is the trail of an artificial satellite which crossed the picture during the exposure. Stars a few hundred light years away make up the winding 'tail' of Scorpius to the right, and the 'teapot' of Sagittarius at the left. The small glowing patch just above the satellite trail is the Lagoon Nebula, a bright nebula of glowing gases some 5000 light years away. Six times farther off, in the background, is the great central bulge of our Galaxy, and cutting right across is a narrower band of dust which dims the stars at the centre of the Galaxy to a million-millionth of the brightness we would otherwise see.

Other radiations can, however, easily penetrate the dust. Previous spreads have shown how the radio, X-ray and gamma ray skies are dominated by the brilliant central regions of the Milky Way, where stars and interstellar matter are closely packed. In this spread and the next, we home in on the galactic centre.

The Einstein Observatory picked out a jumble of X-ray sources at the centre of the Milky Way. The colour-coded view (**Fig. 8.11**) covers an area of about one square degree (the central half-centimetre of Fig. 8.10). The colour coding runs from blue for the dim background, through red, orange, purple, grey, yellow and green to white for the most brilliant X-ray regions.

The intense blob at the very bottom is an X-ray double star system. The glow of X-rays at the galactic centre itself stretches out over 300 light years, appearing as large as the Full Moon. The radiation may come from a pool of very hot gases at the Galaxy's centre, or from very fast electrons there. The jumble also contains a dozen rather faint individual X-ray sources which are difficult to pick out in this picture. These could be weak X-ray double stars, ordinary, very hot, young stars like those in the Orion Nebula (Fig. 4.16) or small supernova remnants, like Cassiopeia A (Fig. 6.26).

But one X-ray source is different. The brilliant 'white' source at the middle of the picture coincides exactly with the position of the centre of the Galaxy (as determined by infrared and radio astronomers) and it shines as brightly in X-rays alone as would a thousand Suns at all wavelengths.

This powerhouse shows up even better at the very short wavelength gamma rays, as a very bright, flickering 'gamma ray star'. Although gamma ray telescopes cannot map it in any detail, the fluctuations of this source tell astronomers that it must be about the size of the Solar System – minute on the scale of the Galaxy. The gamma rays are mostly of one particular wavelength – 0.0023 nanometres – emitted when ordinary electrons annihilate with their 'antimatter' equivalent, positrons. The powerhouse is strong enough to generate not just radiation, but antimatter too.

At the other end of the wavelength scale, early radio astronomers found an intense source which they called Sagittarius A. Modern radio telescopes probe Sagittarius A in detail. The radio view (**Fig. 8.12**) from the Very Large Array corresponds to just the central region of the X-ray picture above, magnified fifteen times. The 'frame' is four arcminutes across (one-eighth of the Moon's apparent diameter). The colour coding shows the background sky in black, dim radio-emitting regions in blue, with the successively brighter parts green, yellow and red.

The radio emission in the right-hand half of this picture comes from a region of hot gas surrounding the galactic centre. Some fifteen light years across, it is known as Sagittarius A West. The denser central

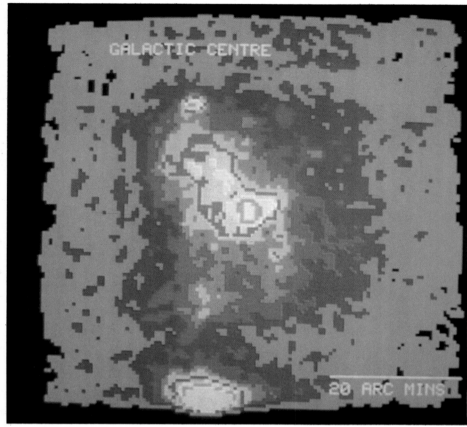

**8.11** *X-ray, 0.3–2.5 nm, Einstein Observatory IPC*

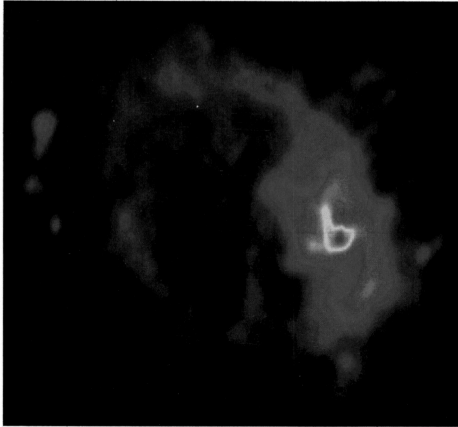

**8.12** *Radio, 6 cm, Very Large Array*

gases make a strange spiral shape, and increase to a peak of brightness (red) which marks the precise position of the galactic centre. Detailed observations by the technique of Very Long Baseline Interferometry have shown that it is only the size of Jupiter's orbit around the Sun. These radio waves apparently come from the same small power-house which produces the X-rays and gamma rays.

Distant quasars and active galaxies like Centaurus A (Chapter 12) have tiny central power-houses which shine at some wavelengths as powerfully as the galaxy itself. At X-ray, gamma ray and radio wavelengths, our Galaxy's core looks like a smaller version of Centaurus A's centre, scaled down to one-hundredth the size, and a hundred-thousandth the power.

The strong power-houses of exploding galaxies and quasars often eject beams of matter millions of light years into space. The radio photograph (Fig. 8.12) shows some signs of ejection from our Galaxy's centre. American astronomer Robert Brown has suggested that the spiral shape of the gas within Sagittarius A West is due to the ejection of two beams of gas, which have swung round like the beams of SS 433 (Fig. 6.36).

In the past, the Galaxy's core may also have ejected high-speed electrons. The large blue loop, extending out 20 light years, is brighter when observed at shorter radio wavelengths, indicating that it is due to synchrotron radiation from fast electrons being whirled along a great curving magnetic field. This loop, Sagittarius A East, could be the remains of an old supernova – but it may, instead, be the trace of a beam of electrons, shot out by our Galaxy's central powerhouse.

The study of the Galaxy's centre is so new, however, that such interpretations are constantly changing. There is no incontrovertible evidence even for the massive black hole which many astronomers now routinely talk about. More non-optical observations are needed to understand the centre of even our own Galaxy.

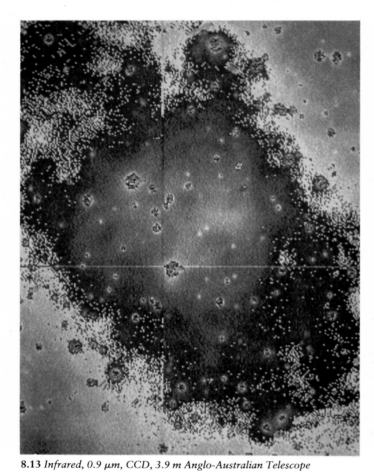

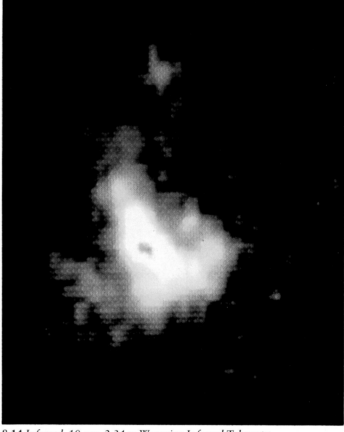

8.13 *Infrared, 0.9 μm, CCD, 3.9 m Anglo-Australian Telescope*

8.14 *Infrared, 10 μm, 2.34 m Wyoming Infrared Telescope*

Infrared radiation can penetrate the interstellar dust and reveal more about our Galaxy's hidden centre. The relatively short-wavelength infrared view (**Fig. 8.13**) encompasses a region 2 by 1½ arcminutes in size, the central half of the area shown in the radio photograph (Fig. 8.12). Colour coding makes the faintest stars yellow, and brighter stars blue, green and red.

This 0.9 micron radiation is much more penetrating than light, and astronomers thought at first that the two faint stars at the centre of Fig. 8.12 were objects at the Galaxy's centre. Recent spectra, however, have shown that they are ordinary stars lying only two-thirds the way to the galactic centre, and dimmed (at optical wavelengths) a million times. Behind them, palls of thicker dust still obscure the actual heart of the Galaxy.

The 10 micron view (**Fig. 8.14**), however, shows clearly the Galaxy's centre. Colour coding runs from red for the least intense regions, through yellow and white to blue at the brightest region, called IRS 1 (infrared source 1). This radiation comes from dust clouds at about room temperature (300 K). The curving band of dust clouds coincides almost exactly with the innermost part of the hot gas spiral seen at radio wavelengths (Fig. 8.12). As in nebulae, ultraviolet radiation is heating the gas and warming the interspersed dust: the

radiation here probably comes from a hundred recently born hot stars – or possibly from the central powerhouse.

Observations like Fig. 8.14 do not show how the dust is moving, but the hot gas mixed with it produces various spectral lines at particular infrared wavelengths, and the Doppler shift in these spectral lines reveals the gas clouds' motions. One of the easiest to study from Earth is a line from neon atoms, at a wavelength of 12.8 microns. Its Doppler shift shows that the gas and dust clouds in the left-hand part of the curve are moving away from us at a speed of 150 kilometres per second, and the right-hand part is approaching us at a similar speed. The precise centre of the Galaxy lies just to the right of the blue peak IRS 1. The 10 micron picture thus reveals a ring of gas and dust about a light year out from the galactic centre, circling the centre at high speed.

The ring's speed is held in check by the gravitational pull of matter within it, which must be equivalent to six million Suns. But shorter wavelength infrared observations (like Fig. 8.15) indicate that the stars within the ring amount to only three million solar masses. Many astronomers believe that the 'missing' three million solar masses are in the form of a massive black hole at the Galaxy's very centre. An accretion disc of gas around this hole

would then be the powerhouse that we detect as the very compact radio source, and which produces the galactic centre's ultraviolet, X-rays and gamma rays.

The colour composite (**Fig. 8.15**) at three wavelengths in the near infrared resolves details as fine as an optical photograph would achieve (1 arcsecond). The picture shows a region only 10 arcseconds (half a light year) across. It covers the central part of Fig. 8.14, enlarged about ten times. The dust clouds are seen here as 'reddish' regions. The bright orange object (on the left) is IRS 1, and the bright red blob is visible on Fig. 8.14 just to the top right of IRS 1: it may be a rather cold dust cloud. The brightest object, the white source at the top (IRS 7), is a red supergiant star, shining as brightly as 100 000 Suns. Most of the other bright objects are red giants.

But the central trio of 'blue' infrared sources (IRS 16) is rather unusual. Spectra of its components suggest that the two outer objects are hot gas nebulae, without any dust. The central object, however, seems to be the dense cluster of stars we expect to find at the galactic centre, and it probably coincides with the compact radio source. This central cluster is only one-twentieth of a light year across, and its small size is more evidence for a central massive black hole, holding the stars close together by its gravity.

**8.15** Infrared, 1.6 (blue), 2.2 (green) and 3.8 (red) μm, 3.9 m Anglo-Australian Telescope

# The Large Magellanic Cloud

Our Galaxy's nearest neighbour in space is a nondescript galaxy called the Large Magellanic Cloud. Although it is relatively small and dim, the Large Magellanic Cloud is so close to us on the cosmic scale that it can be seen easily with the naked eye. This 'cloud' and its diminutive companion in the southern skies, the Small Magellanic Cloud, are named after the Portuguese navigator Ferdinand Magellan who first reported them in 1521 after his epic circumnavigation of the Earth.

The Large Magellanic Cloud lies 170 000 light years from us, and its companion further off at 200 000 light years – roughly twice the diameter of the Milky Way Galaxy. Even so, both Clouds lie at less than one-tenth the distance of the nearest major galaxy, the great Andromeda Galaxy (Fig. 10.2). The Magellanic Clouds are presently heading away from our Galaxy, having passed closest to it about 500 million years ago. They are probably satellites of the Milky Way, orbiting it every 6000 million years just as the Earth orbits the Sun – making the Clouds a part of the 'Milky Way system'.

The Large Magellanic Cloud consists of some 10 000 million stars (one-tenth as many as the Milky Way) and enough gas to make one new star for every ten in the galaxy at present. It is a microcosm of our own Galaxy, and in many ways, the Large Cloud is easier to study because we can look at it from the outside and see immediately the locations of stars and gas clouds relative to one another.

The photograph of the Large Magellanic Cloud (**Fig. 8.16**) is a composite of three black and white pictures taken through blue, green and red filters on the large Schmidt telescope at the European Southern Observatory in Chile, and recombined to produce a sharp, true-colour photograph. It covers a region of sky 4° (eight Moon breadths) across, the central one-third of the galaxy, which is 35 000 light years in total extent. The Large Magellanic Cloud has traditionally been classified as an 'irregular galaxy', but its central stars clearly concentrate into a broad 'bar' running across this picture, so it is probably a smaller relative of the barred spiral galaxies (see also Fig. 8.20).

Most of the stars in the bar are yellow, orange and red – old stars which have existed since the Large Magellanic Cloud was born soon after the Big Bang. But groups of white and bluish-white stars show regions where stars have been born more recently. The youngest of all are still lighting the gases they were formed from – nebulae glowing red in the light from hydrogen atoms. The largest gas cloud (on the left) is the massive Tarantula Nebula (Figs. 8.21–8.24).

**8.16** *Optical, true colour, 1 m Schmidt telescope, European Southern Observatory*

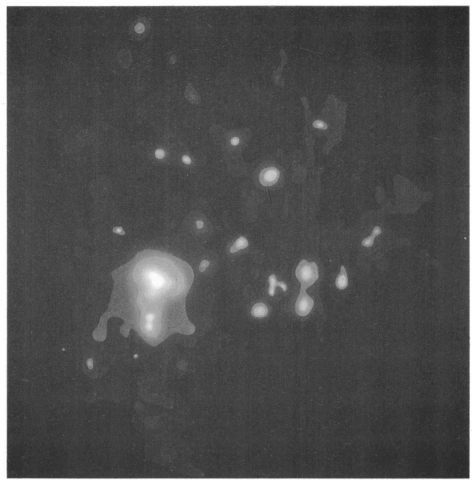

8.17 *Radio, 11 cm, 64 m Parkes Telescope*

Radio observations of the Large Magellanic Cloud show up the hot gas in the galaxy, without the background glow of stars. The radio contour map (**Fig. 8.17**) covers the same area as the previous colour photograph (Fig 8.16), with the background sky red, and orange, yellow and white showing more intense radio emission. The strong radio sources coincide with the bright gas clouds visible in the optical photograph. Observations at different radio wavelengths can, however, reveal unambiguously whether each glowing cloud is a nebula surrounding young stars, or a supernova remnant producing radio waves by synchrotron emission.

The structure of the nebulae and supernova remnants shows up best on optical photographs taken through a red filter 'tuned' to the wavelength emitted by hydrogen gas (656 nanometres). British astronomer John Meaburn has constructed an unusually large filter and placed it in front of the UK Schmidt Telescope in Australia to obtain this spectacular picture (**Fig. 8.18**). The photograph is 6° across (a region fifty per cent larger than Figs. 8.16 and 8.17). The stars of the galaxy's central bar are dimly visible, but outshining everything else is the Tarantula Nebula (on the left). The picture is sprinkled with other gas clouds, loops and filaments, showing that the entire galaxy is a vigorous region of star birth, and star death.

The irregular clumps of gas here are nebulae, shining in the ultraviolet radiation from new stars born within them. (To indicate the scale, the Orion Nebula would appear less than 1 millimetre across here.) The rings of gas are supernova remnants, blown out by the type II supernova explosions of massive stars at the end of their short lives. Some of these remnants are over 1000 light years across. They are easily seen and studied in the Large Magellanic Cloud, but such huge remnants in our own Galaxy fill large regions of the sky, and only recently have radio observations revealed their existence in the Milky Way (Figs. 8.3 and 8.7)

The supernova remnants are not the largest gas loops in the Large Magellanic Cloud, however. The hydrogen light photograph (Fig. 8.18) reveals several, even larger, very faint 'supershells'. The largest – 4000 light years across – lies at the top left of the figure. Fig. 8.19 shows it to a larger scale, but printed by the technique of unsharp masking to bring out the faintest filaments. The supershells are apparently caused by a runaway combination of star birth and star death. At the centre of the shell, a large cluster of new stars has formed. The heaviest soon exploded as supernovae, and the expanding remnants compressed the surrounding interstellar gas. The compression made a ring of new star-forming nebulae, a few dozen light

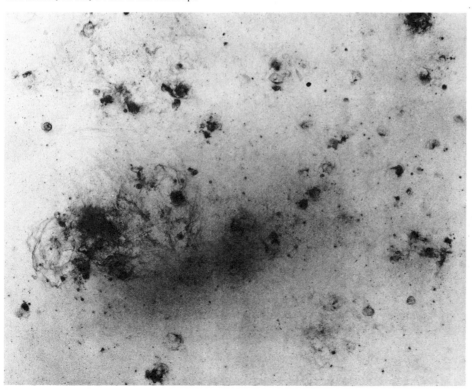

8.18 *Optical, 656 nm hydrogen line, negative print, 1.2 m UK Schmidt Telescope*

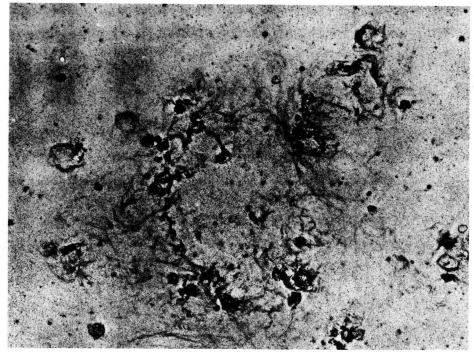

**8.19** *Optical, 565 nm hydrogen line, unsharp-masked negative print, 1.2 m UK Schmidt Telescope*

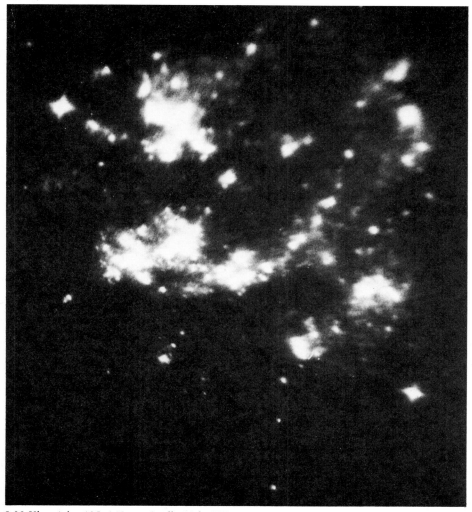

**8.20** *Ultraviolet, 125–160 nm, Apollo 16 far UV camera*

years across. The heavy stars here in turn exploded, to produce a ring of compression further out. The cycle repeated again and again, so that after a few million years the ring of nebulae has become an enormous supershell.

Ultraviolet telescopes can show directly the very hot and massive O and B type stars which have only recently been born, and will soon die as supernovae. Ultraviolet is absorbed in the Earth's atmosphere, and this photograph (**Fig. 8.20**) was taken by the Apollo 16 astronauts during their mission to the Moon in 1972. It covers 8° of sky, a region one-third larger than Fig. 8.18. The central bar of old red stars is invisible: it would lie at the centre of the photograph, and cover one-third of the picture's width.

The ultraviolet view reveals how the young stars generally form well away from the galaxy's centre – and follow curves reminiscent of the arms of a spiral galaxy. Many of these clusters of young stars coincide with the nebulae seen in hydrogen light; for example, the bright patches at the centre, stretching in a curve towards the top right. Dust in the nebulae strongly obscures ultraviolet radiation from stars within them, however. So at ultraviolet wavelengths, the great Tarantula Nebula is surprisingly inconspicuous – no brighter than one of the central nebulae. The two brilliant regions to the left of the centre contain young stars which have already cleared away the surrounding gas and dust – these clusters are visible on the optical colour photograph (Fig. 8.16) as sprinklings of bluish stars, one immediately below and the other to the right of the Tarantula. The most luminous region seen in the ultraviolet (top left) is the large cluster of young stars which has blown out the supershell seen in hydrogen light (Fig. 8.19).

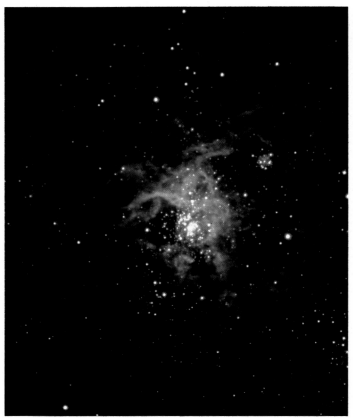

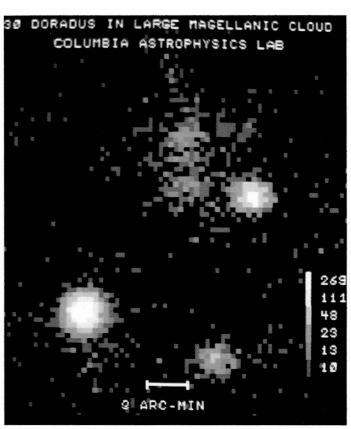

8.21 *Optical, true colour, 0.9 m telescope, Cerro Tololo Inter-American Observatory*

8.22 *X-ray, 0.3–2.5 nm, Einstein Observatory IPC*

The Large Magellanic Cloud contains one of the largest and most powerful nebulae known. Officially designated 30 Doradus or NGC 2070, the gas cloud's filamentary 'legs' have earned it the nickname 'Tarantula Nebula'. The true-colour picture (**Fig. 8.21**) stretches 15 arcminutes (700 light years) from top to bottom, covering almost exactly the main nebulosity seen in Fig. 8.16. (Note, however, that the very faint extensions seen in Fig. 8.18 stretch to a total size of 5000 light years!) Fig. 8.21 is a short-exposure photograph, revealing the cluster of young stars at the centre of this gas cloud.

The stars total a mass of half a million Suns, and they have been formed from a huge cloud of gas and dust originally containing over a million solar masses. Stars still being born in dark clouds around the Tarantula, and the heat from the collapsing fragments of cloud produces long wavelength infrared radiation. The pioneering infrared satellite IRAS detected the Tarantula as a powerful source (**Fig. 8.23**). Here IRAS has scanned bands in the upper part of the Large Magellanic Cloud, nearly parallel to the central 'bar' of stars: the scan covers roughly the length of the bar left to right (4°). The Tarantula is the peak on the left. Dust in the remaining clouds shows as smaller peaks covering the whole left-hand half of the scan.

The Tarantula is a most active object.

The pressure of ultraviolet radiation from the hot central stars and explosions of short-lived massive stars as supernovae are pushing gas outwards in huge bubbles. But the stars' gravity is also pulling external gas inwards. This gas is forced to travel around the edges of the bubbles, and hence forms overlapping shells of gas – which appear as the Tarantula's 'legs'.

The X-ray picture (**Fig. 8.22**) gives us another view of this activity. It covers an area twice as large as the neighbouring optical photograph (Fig. 8.21), and the Tarantula fills the top right-hand quadrant. (The bright source to the lower left is LMC X-1, a close double star system where the gas streams from one star are falling towards the other, which is probably a black hole.) The young stars in the Tarantula have coronas (outer atmospheres) of gas so hot that they emit X-rays, forming the diffuse glow of X-rays above the centre. The bright source to the right of the centre is a supernova remnant – the multi-million degree gases from the explosion of a young star at the edge of the cluster.

Detailed studies show that the powerhouse of the Tarantula is the brilliant object at the centre, known as R136. For many years, astronomers believed that it was a very compact cluster of about a hundred stars, each about fifty times heavier than the Sun (O type stars). But in

1979, J.V. Feitzinger took exceptionally sharp photographs of R136 at the European Southern Observatory, Chile. The colour-coded picture (**Fig. 8.24**) is enlarged a hundred times relative to Fig. 8.21. The green contours enclose the regions of lowest intensity, orange more intense parts, while the blue contours show the peaks of intensity. R136 clearly consists of three parts. There are two fainter companion stars to the left, but almost all its light comes from the object on the right, R136a. Even here, however, the stars' images appear slightly blurred, and John Meaburn later used the technique of speckle interferometry (which 'freezes' atmospheric blurring with many short exposures) to show that R136a is less than one-hundredth of a light year across – only ten times the size of the Solar System. A cluster of a hundred heavy stars, plus the many lightweight stars which would have formed with them, could not fit into such a small region.

Ultraviolet spectra from the IUE satellite are also at odds with the spectra expected from a cluster of O stars. The conclusion seems inescapable: R136a is a single star. And what a star! It has a surface temperature of 60 000 K, among the highest known for a star. With a mass of 2500 Suns and a luminosity equal to 100 million Suns, it is by far the heaviest and brightest star known of in the Universe.

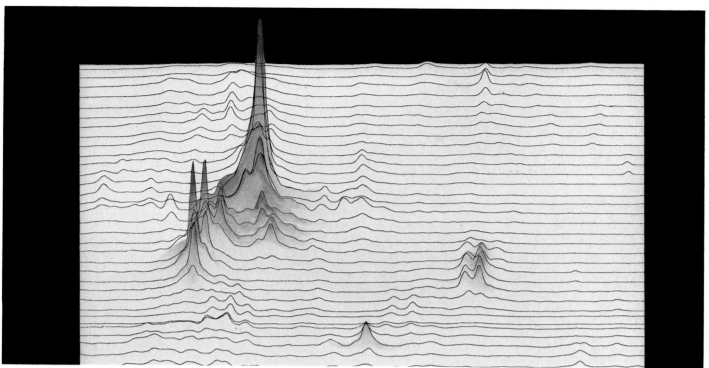

**8.23** *Infrared, 100 μm, Infrared Astronomical Satellite*

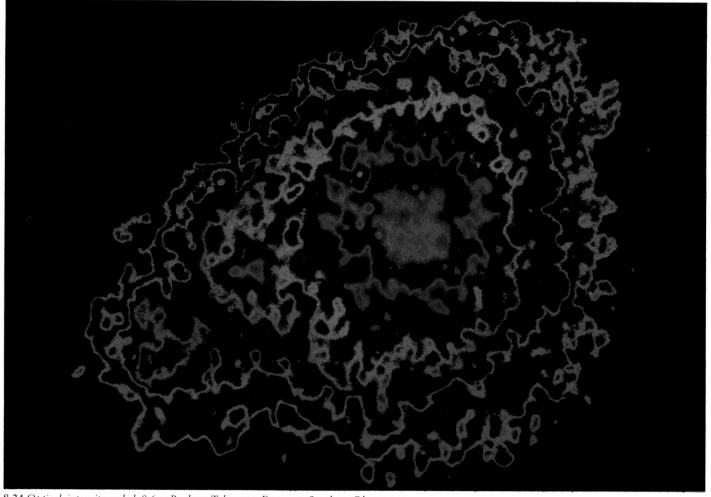

**8.24** *Optical, intensity coded, 0.6 m Bochum Telescope, European Southern Observatory*

# 9 Ultraviolet Astronomy

Everyone knows that ultraviolet radiation – in small quantities – is beneficial. It gives us a suntan, and builds up vitamin D within our bodies. But anyone who has fallen asleep under the midday Sun has felt the other effect of ultraviolet radiation: too much exposure kills the living cells of our skin leaving the raw blisters of sunburn. In fact, only a small fraction of the Sun's ultraviolet power penetrates the Earth's atmosphere to sea-level, and this is the less dangerous, longer wavelength ultraviolet. If the full flood of shorter wavelength ultraviolet poured down whenever the Sun shone it would scarcely be safe to venture out-of-doors in the daytime.

Our shield against this dangerous radiation is a diffuse layer of invisible ozone gas, high in the tenuous upper regions of the Earth's atmosphere. Ozone is an unusual form of oxygen. The ordinary oxygen molecules we breathe consist of two atoms joined together, while the ozone molecule is a triplet of oxygen atoms. Ozone gas is transparent to ordinary visible light, but it absorbs ultraviolet radiation very strongly. As a result, the Sun's visible light flows to Earth virtually unhindered, but the ultraviolet is stopped by the ozone screen about thirty kilometres above our heads.

The ozone layer has been essential for the existence of life on Earth. Early life evolved in the protection of the oceans, and animal life could only emerge onto dry land

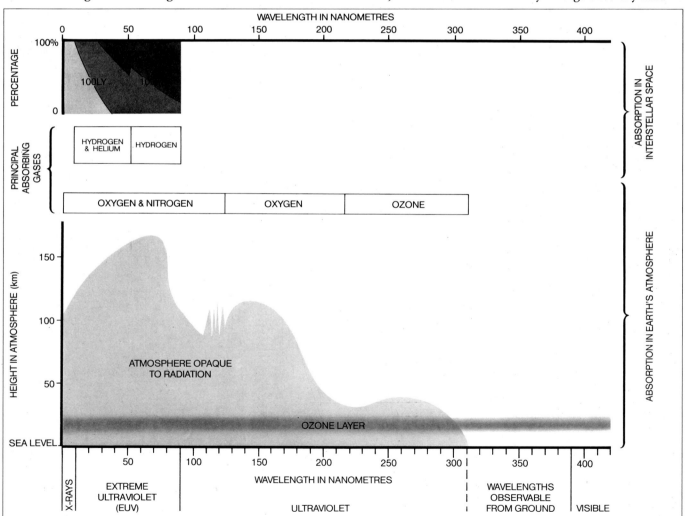

**Fig. 9.1** *Ultraviolet radiation is readily absorbed by gases, both in space and in the Earth's atmosphere. Hydrogen and helium, the most abundant gases in space, absorb radiation of less than 90 nm and 50 nm respectively, thus dimming stars at these short wavelengths. Gases in the Earth's atmosphere absorb shorter ultraviolet rays at very high altitudes, while a lower layer of ozone blocks off ultraviolet with wavelengths up to 310 nm. Only the longest – and the least interesting – ultraviolet wavelengths can be observed from the ground; most discoveries in ultraviolet astronomy have come from shorter wavelength detectors flown above the atmosphere on rockets or satellites.*

**Fig. 9.2** *The last of three Orbiting Astrophysical Observatories undergoing prelaunch tests at Kennedy Space Center, Florida. Renamed Copernicus after launch, the satellite obtained the first high-quality ultraviolet spectra with the largest reflecting telescope yet launched. Copernicus lasted a record nine years before it was switched off in 1981.*

because the ozone layer shielded it from the lethal solar ultraviolet. But astronomers see the ozone layer as a mixed blessing. As well as blocking solar radiation, the ozone layer absorbs all other ultraviolet rays coming from space. An astronomer trying to see the Universe in the ultraviolet is as badly off as an optical astronomer faced with a completely overcast sky; in fact, he is in a worse situation, because the 'ozone clouds' never clear away. The ozone layer forms a continuous permanent opaque cover to the sky.

Strictly speaking, the ultraviolet region of the spectrum (Fig. 9.1) consists of wavelengths shorter than 390 nanometres – the wavelength of violet light, the shortest radiation visible to the human eye. The ozone layer is, however, transparent to the longer ultraviolet radiation, down to a wavelength of 310 nanometres. So the longer ultraviolet (310 to 390 nanometres wavelength) can penetrate down to sea-level. (It is this radiation which gives us a suntan.) The sky in the longer ultraviolet can be studied in much the same way as the visible light from the Universe, using ordinary optical telescopes with the same kind of photographic plates or photoelectric devices.

From the astronomer's point of view, in fact, it does not make sense to draw the distinction between visible light and ultraviolet at 390 nanometres. The boundary is traditionally drawn here because it marks the boundary between radiation which is visible and invisible to the human eye. Since astronomers no longer use their eyes as radiation detectors, to them it is more logical to draw the distinction at the wavelength where the ozone layer limits observations, at 310 nanometres. *Ultraviolet astronomy*, therefore, is the study of wavelengths shorter than 310 nanometres, and it involves a whole new technology because ultraviolet telescopes must be raised above the ozone layer.

Astronomers observing the long wavelength infrared radiation can overcome the problem of atmospheric absorption by putting their telescopes on high mountains (Chapter 5). Ultraviolet astronomers have a much worse problem, since the ozone layer lying way up in the stratosphere is three times higher than the summit of Mount Everest. The first unmanned ultraviolet telescopes were flown on high-altitude balloons, like the weather balloons which meteorologists used to investigate the stratosphere. Later telescopes have been launched on brief rocket flights which lifted them above the ozone layer for a few minutes at a time. The best answer of all is an ultraviolet satellite. From an orbit above the Earth's atmosphere, a satellite can survey the ultraviolet sky for years on end.

The first ultraviolet satellites, for example the second Orbiting Astronomical Observatory (OAO-2) and the European TD-1A, carried out general surveys of the ultraviolet sky. At first sight, there seems to be little difference between the sky in the ultraviolet and the familiar night sky. It is scattered with thousands of stars with a concentration in the band of the Milky Way (Fig. 8.5). But a closer look shows that the familiar constellation patterns have disappeared.

Stars whose temperatures are a few thousand degrees above absolute zero (degrees K) produce most of their radiation in the form of visible light. A star that is hotter than 10 000 K shines most brightly at ultraviolet wavelengths. The stars of the ultraviolet sky are fiercely burning suns with temperatures in the range from 10 000 to 100 000 K. As a result, ultraviolet detectors show a very different view of the constellation Orion for example. The prominent red star Betelgeuse is cool on this scale of temperature; with a temperature of 'only' 3600 K it emits very little short wavelength radiation and is invisible in the ultraviolet. The other optically bright star, Rigel, is much hotter: this 11 500 K star is somewhat brighter in the ultraviolet than in the optical. But even Rigel is overshadowed by the brilliance of Orion's belt, three of the brightest ultraviolet stars in the sky in a closely set triplet. Each of these stars is some five times hotter than the Sun, with a temperature around 30 000 K.

Ultraviolet telescopes thus 'see' the hottest stars of all, and because of this they naturally pick out the youngest star groups in the sky. Stars are developing all the time in a galaxy like our Milky Way. They are formed – invisible to all but infrared detectors – in dark dusty nebulae; but eventually the newly born group of stars blows away the dust and shines undimmed as a new star cluster. The cluster contains stars of all masses, from heavyweights some fifty times heavier than the Sun to diminutive stars of only one-twentieth the Sun's mass. The lightweight stars are dim yellow or red stars, with long lifetimes of several

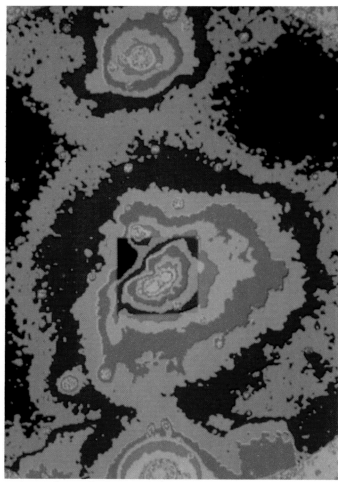

**Figs. 9.3–9.4** *Ralph Bohlin and colleagues launched a 0.3 metre telescope by rocket to observe the Orion Nebula at 247–277 nanometres. The image was detected and intensified with a microchannel plate (Fig. 9.5) and recorded on ordinary film (Fig. 9.3).* **Fig. 9.4** *is the same image, colour coded for intensity. The dimmest regions (invisible in Fig. 9.3) are green and brown, with brighter parts yellow, dark blue, light blue and purple; details of the brilliant centre are shown in a repeat of the colours.*

thousand million years or more. During their lives, they can drift away from their birthplaces and spread out around the galaxy. But the heavy stars are profligate with their fuel. They 'burn' much more fiercely, at such high temperatures that they shine most brightly at ultraviolet wavelengths, and they devour their fuel so quickly that they reach the ends of their lives in only a few million years. This is only the blink of an eyelid in the astronomical timescale, so short that the star has no time to travel away from its birthplace before it reaches the end of its natural span and destroys itself in a supernova explosion.

Ultraviolet stars are thus comparatively young stars. They are found only in the youthful clusters of stars which lie close to regions of star birth. The constellation Orion marks the nearest such region in our own Galaxy, and as a result it is bejewelled with ultraviolet stars.

The distribution is more obvious when we turn to galaxies beyond our own. Spiral galaxies are large, disc-shaped conurbations consisting of thousands of millions of stars. New stars are born in the curved 'spiral arms', where gravity clumps the interstellar gas into nebulae. The lightweight stars diffuse away from the arms, to fill the regions in between. Optical telescopes pick up the yellow or reddish light from these stars, and show the whole disc

of the galaxy as a gently glowing spiral, shining most brightly at the centre where the old, red stars are concentrated. Ultraviolet telescopes, on the other hand, are blind to these old stars. They see only the youngest, hottest stars (Fig. 10.27). Because the galaxy's arms are star birthplaces, ultraviolet detectors emphasize the spiral arms and show little of the galaxy's overall structure and inner regions.

Although ultraviolet instruments must be designed to operate automatically from a balloon, rocket or satellite carrier, the actual telescopes and detectors can be very similar to optical instruments. Ordinary reflecting telescopes with an aluminium reflecting surface will also focus ultraviolet light. The image can be recorded with an electronic detector, or even on ordinary film – provided the film contains a minimum amount of the gelatin used to hold the sensitive silver salts in place, because gelatin absorbs ultraviolet radiation shorter than 220 nanometres.

Photographic film can show up fine details in an image, but it is not very sensitive to radiation falling on it. Just as optical astronomers have turned to image intensifiers, so have their ultraviolet colleagues. One very simple kind of ultraviolet image intensifier is the *microchannel plate* (Fig. 9.5). It consists of thousands of small, very thin, glass tubes stuck together side by side, to form a block that may be a few centimetres across and only a millimetre thick. The tiny tubes form a myriad of parallel channels, each a millimetre long and only one-fortieth of a millimetre in diameter. For ultraviolet astronomy, the plate is housed within a glass envelope which is evacuated to prevent air molecules interfering with its operation. The front window of the envelope supports a sensitive metal electrode, and the rear window a phosphor screen which produces light when electrons fall on it. A voltage of around a thousand volts is applied across the plate's one millimetre thickness, and a photographic plate is placed at the back to record the final image.

This microchannel plate assembly is fixed at the telescope's focus so that the ultraviolet image falls onto the front electrode. When an individual packet of ultraviolet radiation (a *photon*) hits the electrode, it dislodges an electron into the tube directly beneath. Once inside this microchannel, the electron is accelerated down the tube by the electric voltage. It does not travel far, however, before it hits the side of the narrow channel. The collision dislodges a few more electrons from the glass wall. These too are accelerated down the tube, and collide with the walls. By the time the first electron reaches the bottom of its microchannel, it is accompanied by an avalanche of around 100 000 electrons knocked out by successive collisions. These electrons plough into the phosphor, and produce a bright spot of light – just as the electrons in a television tube make the phosphor of the screen glow. The spot's position corresponds exactly to the location of the ultraviolet photon in the original image, but it is much brighter. The microchannel plate thus acts as a simple and robust image intensifier, and the image in the phosphor screen is permanently recorded on the photographic emulsion behind.

Photographic plates are a very convenient way of recording the intensified image when the whole telescope assembly is destined to return to Earth, and balloon and rocket ultraviolet telescopes are designed to float down intact on parachutes. Even on the odd occasion when the parachutes fail, the plates can usually be recovered intact from the mangled remains of the telescope. But it is not easy to return photographic plates from satellites in orbit. So far, an ultraviolet satellite which takes pictures of the sky has not been launched, but detectors are currently being designed for use on future missions. These must scan the image electronically, so it can be radioed back to Earth like a television picture. Microchannel plates can be adapted for electronic 'reading' by replacing the phosphor and photographic plates with an array of small electrodes (anodes), one at the base of each channel to catch its avalanche of electrons. These multi-anode microchannel plate arrays (MAMAs) will be used on purpose-built ultraviolet satellites later in the 1980s. The first imaging ultraviolet telescope, however, will be the Space Telescope.

Although this 2.4 metre reflector is usually described as an optical telescope (Chapter 3), the Space Telescope and its electronic light detectors can also observe in the ultraviolet, right down to a wavelength of only 115 nanometres.

The ultraviolet satellites launched to date have either been survey instruments, or they have been designed to study the spectrum of the ultraviolet radiation from stars and galaxies. The third successful Orbiting Astronomical Observatory, named Copernicus (Fig. 9.2), carried a 0.8 metre ultraviolet telescope into space in 1972. During its nine year life, Copernicus took the first detailed look at ultraviolet spectra over the wide range from 95 to 300 nanometres. It was followed in 1978 by the joint American–European International Ultraviolet Explorer, IUE (Figs. 9.6–9.9). Although IUE's 45 centimetre telescope is somewhat smaller than that of Copernicus, its more modern ultraviolet detectors mean that it can observe much fainter stars.

Astronomers were keen to use their first ultraviolet space observations to study stars' spectra, rather than to produce images, because the ultraviolet spectra contain clues to some of the commonest elements in the Universe. Carbon and nitrogen, for example, do not produce very informative spectral lines at the wavelengths of visible light, but their strong ultraviolet spectral lines show readily the abundance of these elements. Astronomers

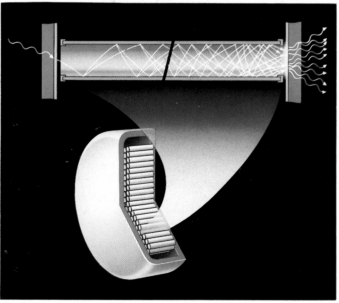

Fig. 9.5 *A microchannel plate consists of thousands of narrow glass tubes side-by-side, in an evacuated glass envelope. Ultraviolet falling on a metal electrode (left) dislodges an electron. A high voltage draws it down the nearest tube, and collisions with the wall multiply the electrons. They hit a phosphor layer (right) to produce a bright spot of light.*

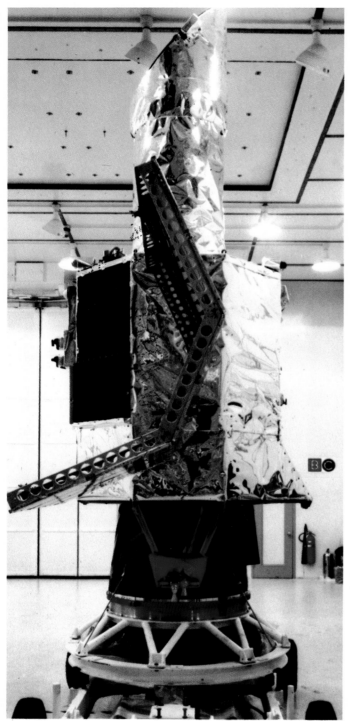

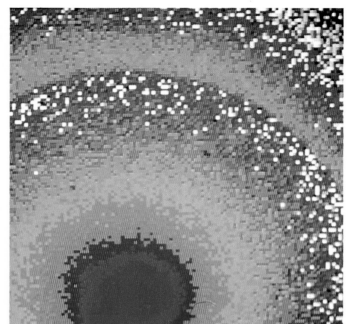

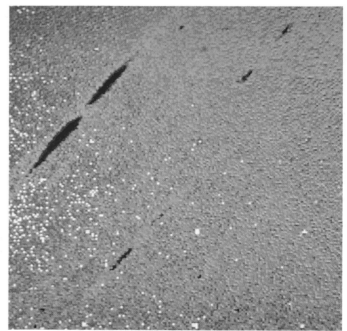

**Figs. 9.6–9.9** *The International Ultraviolet Explorer (IUE) satellite (Fig. 9.6), launched in 1978, has a 0.45 metre telescope for obtaining ultraviolet spectra. British and European astronomers operate IUE from Spain, while American astronomers control it from the Goddard Space Flight Center, Maryland (Fig. 9.9). As well as receiving the ultraviolet data, astronomers can 'see' what IUE is looking at, through an optical TV camera. Both the optical view and the ultraviolet spectra are colour coded for intensity, from white and blue for the faintest parts, through green to red for the brightest regions. Sirius (Fig. 9.7) is so brilliant that the optical view has an overexposed centre and false rings. The ultraviolet spectrum of Capella (Fig. 9.8) has wavelengths of 115–190 nanometres spread along successive diagonals. The brightest emission lines (red) come from hot gases – hydrogen, silicon, helium and carbon – in this giant star's atmosphere. The broad hydrogen line (upper left) is split by absorption from interstellar hydrogen.*

using IUE have discovered, for example, that a nova explosion in Cygnus in 1978 produced a lot of nitrogen, while the supernova that created the famous Crab Nebula threw out relatively little carbon. These are important clues both to the formation of new elements, and to the mechanisms behind star explosions. In addition, the radiation from any distant star has to pass through the interstellar gas on its way to us. This gas between the stars is invisible, and optical astronomy can tell us little about it. But atoms in the gas absorb ultraviolet radiation of specific wavelengths. Observations from Copernicus and IUE show that the gas is far from being uniformly spread: supernova explosions have blasted out 'holes' where the gas is even more tenuous than it is generally in interstellar space, so that the structure of the invisible gas in our Galaxy resembles a Swiss cheese.

In the early days of ultraviolet astronomy it was thought that the interstellar gas would produce a natural limit to the observations of the sky at the short wavelengths of the *extreme ultraviolet* (EUV), radiation with wavelengths from 91 nanometres down to the beginning of the X-ray region of 10 nanometres (Fig. 9.1). The vast majority of the interstellar atoms are hydrogen, and hydrogen atoms absorb very strongly radiation which has a wavelength of 91 nanometres or less. Every photon of radiation carries a certain amount of energy with it, which depends directly on its wavelength: the shorter the wavelength, the higher the energy. The photons of ultraviolet radiation of 91 nanometre wavelength have just enough energy to break up a hydrogen atom, knocking apart the electron and the proton which constitute the atom. Radiation of this wavelength should, therefore, not be able to travel far through space. Ultraviolet photons of slightly shorter wavelength have higher energy, so they too are absorbed by hydrogen in space. These higher energy photons, however, are less likely to hit a hydrogen atom, and as we move towards the X-ray wavelengths, we find that photons can once again travel freely through space.

It has turned out that this interstellar 'fog' at the EUV wavelengths is not quite as dense as astronomers had feared, and it is sufficiently clumpy that we should be able to see right through it in some directions. Telescopes to explore the skies at EUV wavelengths are in the planning stages. The Americans have designed an EUV Explorer Satellite, while the British will fly an EUV telescope piggyback on the German Rosat X-ray satellite.

Extreme ultraviolet detectors see even hotter objects in the sky – gas with temperatures between 100 000 and 1 000 000 K. No ordinary stars are this hot. EUV telescopes should, however, pick out the very tenuous and extremely hot outer atmospheres – coronae – which surround some stars, and which have already been observed at longer wavelengths by the IUE satellite. They should also see some much more bizarre objects, including the extremely hot cores of old stars which have suddenly become exposed as they have blown away their outer, cooler layers of gas. And they will see the hot streams of gas which are already familiar from X-ray astronomy: gas flowing from one star to a close companion, or – on a much larger scale – the gas streams that are tumbling into a massive black hole within a quasar (Chapter 11). What should be most interesting, however, is the discovery of new kinds of object which astronomers have not predicted.

The EUV radiation from one particular star has already been investigated. The Sun is sufficiently close that no interstellar gas lies in the way to absorb its radiation and its proximity also means that we can make out fine details of its surface and atmosphere. Since the Sun seems to be a fairly typical star, it is hoped that a detailed investigation of it can tell us about the structure and lifestyle of stars in general.

Visible light shows us the Sun's 'surface', the *photosphere*, which has a temperature of 5800 K and is a relatively smooth, unruffled ball of light. The Sun's outer atmosphere, the *corona*, is completely different: it is a complex, turbulent, ever-changing tangle of gas streams. Oddly enough, the corona is very much hotter than the photosphere. With a temperature of a few million degrees, the corona 'shines out' in X-rays. The key to the corona's structure and temperature must lie in the layers between it and the photosphere below.

The layer of gas immediately above the photosphere is the *chromosphere*, visible during eclipses of the Sun as a thin ring of reddish light – light from hydrogen atoms. The chromosphere is only a few thousand kilometres thick; it is at roughly the same temperature as the photosphere, but the gases here are much less dense. One big difference is that the tenuous chromospheric gases are marshalled by the Sun's magnetic fields into large-scale patterns, the *chromospheric network*. Above the chromosphere lies an extremely thin, and still enigmatic layer of gas. In this transition region, the temperature rises from about 6000 K at its lower boundary to 1 000 000 K at the top – in a vertical distance of only a thousand kilometres, less than one-thousandth of the Sun's diameter. The gas in the transition region emits radiation at ultraviolet and EUV wavelengths. Astronomers have been particularly keen to investigate our local star at these wavelengths, to unravel the secrets of its atmosphere.

The first detailed pictures of the Sun with ultraviolet and EUV detectors came with the unmanned Orbiting Solar Observatory satellites of the 1960s and early 1970s; and the exploration has continued more recently with the sophisticated Solar Maximum Mission satellite. But the most spectacular results have come from the American manned space station Skylab. Launched in 1973, the 91 tonne Skylab was the most massive object ever put into Earth orbit. Three three-man crews inhabited the space

**Fig. 9.10** *The solar observatory – the Apollo Telescope Mount (ATM) – of the American Skylab space station is the separate structure at the far end, with four solar panels to provide its own power. The various telescopes looked out through the circular face plate.*

station for a total of five and a half months, and amongst their many tasks in orbit was a continual surveillance of the Sun with Skylab's own solar observatory, the Apollo Telescope Mount, or ATM (Fig. 9.10).

The ATM carried eight telescopes which watched the Sun at wavelengths ranging from visible light down through ultraviolet and EUV to X-rays: from 700 nanometres down to just 0.2 nanometres. Skylab was so huge that these telescopes were not the miniaturised versions needed for most spacecraft; they were full-scale instruments which any solar observatory on the ground would be glad to possess – if it were not for the Earth's atmosphere! In total, the telescopes alone weighed almost a tonne. The manned crews carried results down to Earth,

so the ATM could afford the luxury of fine-grain film to record its images of the Sun. Skylab was launched with film stowed in a radiation-proof lead locker (a locker so massive that American space officials were worried about the damage it might cause when Skylab eventually re-entered the Earth's atmosphere in 1979), and each crew brought the exposed film back on its return to Earth. The three crews took a total of 150 000 pictures of the Sun at different wavelengths.

In the ultraviolet and EUV, our star appears as a blotchy, tempestuous ball of fire (Fig. 9.11). Views at different wavelengths within this region reveal different layers of its chromosphere, transition region and lower corona. We see tiny, brilliant points of 'light' scintillating

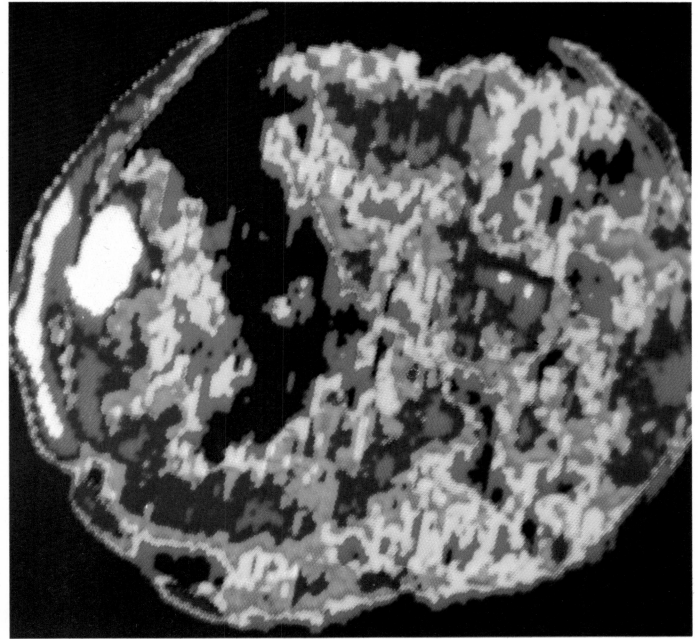

Fig. 9.11 *Astronauts manned the ATM control desk almost continuously while aboard Skylab, and a monitor on board showed them the appearance of the Sun at extreme ultraviolet wavelengths (17–55 nm) in colour coded form. This radiation comes from the transition region between the Sun's chromosphere and corona.*

in the Sun's atmosphere; they may be a more accurate measure of the Sun's magnetic activity than the well-known sunspots which blemish the visible photosphere. The longer wavelengths show tall prominences of cooler gas stretching lazily upwards, or exploding violently into the corona (Fig. 2.10). The corona itself does not exist as a uniform atmosphere, as astronomers had thought before Skylab took its EUV pictures. The hot coronal gas exists only in large individual blobs where it is confined by magnetic fields stretching up from below. In between, there are dark *coronal holes* through which gas streams unhindered into space.

Our views of the Sun in the ultraviolet are superb, mainly because the Sun is by far the brightest object in the sky at these wavelengths, just as it is in the visible. As with most of the new astronomies, ultraviolet observations started with the Sun and they revealed many different aspects of our nearest star. The years since the last Skylab astronauts splashed down in 1974 have, however, seen great advances in the rest of ultraviolet astronomy. Spectra from IUE, for example, are giving us clues to the coronae of other stars – which tie in well with the Sun's coronal hole structure spotted by the Skylab telescopes. With the further unmanned ultraviolet satellites now planned, and especially with the advent of the Space Telescope, ultraviolet astronomy promises to catch up quickly with the other, longer established 'invisible astronomies'.

# 10 Normal Galaxies

Beyond the Milky Way lie an estimated 100 thousand million other galaxies. Although many had been catalogued as 'nebulae' in the past, it is only during this century that their true nature was realised. Much of the credit for this work must go to Edwin Hubble who, in the 1920s and 1930s, systematically photographed hundreds of galaxies and classified them into different types according to their appearance.

Hubble's classification has stood the test of time. Although the original scheme has been extended, refined and subdivided, astronomers still classify galaxies into three basic types – spiral, irregular and elliptical – because a galaxy's appearance really does reflect its internal conditions. (Some spiral and elliptical galaxies also have small, very active cores; they are the 'active galaxies' covered in chapter 12.)

Spiral galaxies resemble our Milky Way in both size and form. All have a central bulge made of older stars, surrounded by a disc of young stars, and dust and gas in which the spiral arms are embedded. Some spirals have bar-shaped rather than circular bulges ('barred spirals'), and there are individual variations in the smoothness of the arms and the degree to which they are coiled around the bulge. Each spiral galaxy has its own individuality, but all are obviously part of the same family.

Irregular galaxies, like the Small Magellanic Cloud, do not share this coherence. Although they resemble the spirals in having a similar make-up of gas, some dust, and young stars, they are smaller and have no particular shape.

Elliptical galaxies are quite unlike the other two types. They have already used up virtually all their dust and gas in star formation. As a result, we see them today as smooth, featureless balls of old, red stars almost devoid of interstellar matter. What they lack in individuality they compensate for in size, however, and many elliptical galaxies are among the biggest and most massive in the Universe. But they can also range down to the other extreme. Some dwarf ellipticals contain less than a million stars and are scarcely visible unless they lie close to us. These are probably the commonest kind of galaxy in the Universe.

Galaxies can exist singly, but it is more common to find them in pairs, groups or clusters, held together by their mutual gravitational attraction. All the galaxies described in this chapter belong to small groups of a few dozen members, most of

them dwarfs. At the other extreme are the giant clusters of galaxies, in which hundreds or thousands of galaxies – mainly ellipticals – cover volumes of space up to 50 million light years in extent.

**Fig. 10.1** is an optical photograph of a typical spiral galaxy, NGC 253. Its designation means that it is the 253rd object in the New General Catalogue of Nebulae and Clusters of Stars, published in 1888 by Johan Dreyer. Other bright, nearby galaxies are prefixed by an 'M', which indicates that they were included in the catalogue published by comet-hunter Charles Messier in 1784 listing objects liable to be confused with comets!

NGC 253 belongs to the Sculptor group of galaxies which, at just under 10 million light years away, is the nearest small cluster to our Local Group and very similar to it in size. With a diameter of 40 000 light years, NGC 253 is the group's largest member. Although less than half the size of the Milky Way, NGC 253 is, nevertheless, four times brighter and, with a mass of 200 thousand million Suns, just as massive. It must look very similar to our Galaxy, but because we see it from only 17° above its rim, we miss a lot of its details. However, Fig. 10.1 clearly shows the smooth yellow glow from the central bulge, contrasting strongly with the dusty blue spiral arms. Bands of dust are particularly obvious on the North (top) side. This is the edge of the galaxy closest to us, and the disc's thick dust lanes can be seen silhouetted against more distant star-clouds.

Optical photographs tell only part of the story, however. In the infrared, NGC 253 is the third brightest galaxy in the sky. This is because – like the galaxy M82 – it is unusually rich in dust, which absorbs light from stars and re-radiates it at longer wavelengths. At radio wavelengths, it has a central source which is a hundred times more powerful than its counterpart in our Galaxy, and a thousand times more powerful than that in the Andromeda Galaxy. X-rays tell of a recent burst of star formation near the centre, an event given additional support by radio and infrared observations. Yet little of this activity is visible at optical wavelengths.

The same is true of all galaxies. Appearances are not always as they seem. Today, multi-wavelength observations are 'peeling the layers off galaxies' as never before, to reveal that even the most ordinary have some extraordinary features.

**10.1** NGC 253, *true colour, 3.9 m Anglo-Australian Telescope*

10.2 *Optical, blue light, 0.9 m telescope, Lick Observatory*

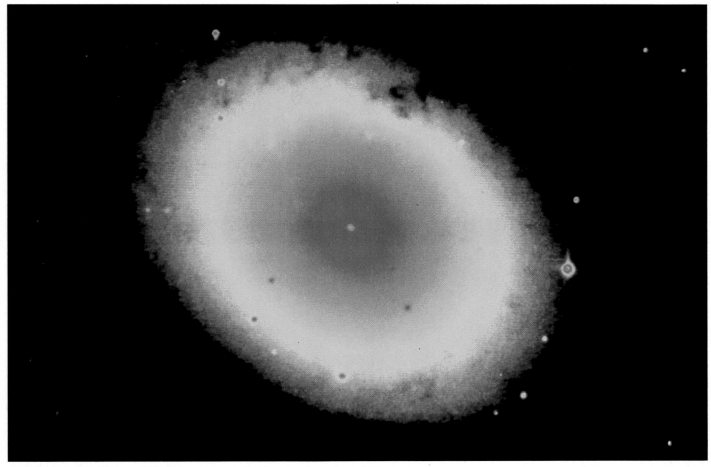

10.3 *Optical, 520–720 nm, CCD, 0.6 m telescope, Smithsonian Astrophysical Observatory*

# Andromeda Galaxy (M31)

At a distance of 2.2 million light years, the Andromeda Galaxy (**Fig. 10.2**) is the closest major galaxy to our Milky Way. Like the nearby spiral in Triangulum, M33 (Figs. 10.15–10.20), it is a member of our Local Group of galaxies, a flattened cluster of about three dozen mostly 'dwarf' star systems, which span a region of space nearly five million light years across. The Andromeda Galaxy (M31) and the Milky Way are the largest and most important members of this small group.

Although M31's true nature as a galaxy was not realised until this century, its existence has been known since at least AD 964, when the Persian astronomer Al-Sufi recorded it as a 'little cloud'. It is dimly visible to the naked eye as a small oval patch on November evenings. However, our eyes pick out only the bright central bulge, and the faintest outer spiral arms extend to about 4° by 1.25° (nearly twenty times the Moon's apparent area); Fig. 10.2 shows the central 2° by ½°, regions out to a radius of 40 000 light years. In fact, M31 is even bigger than the Milky Way. Its full diameter is nearly 150 000 light years, making it one of the largest spiral galaxies known, and it contains some 300 000 million stars.

In many ways, however, the Andromeda Galaxy is similar to our own Galaxy: Fig. 10.2 shows a view of M31 often described as 'how the Milky Way would look from outside'. The two spiral galaxies resemble each other closely in form, and the similarity is reinforced by M31 having two orbiting companion galaxies, NGC 205 (top) and M32 (bottom) in the place of our Magellanic Clouds. Unlike the Clouds, with their active star formation, Andromeda's companions are much smaller, elliptical galaxies made up of old, red stars.

The Andromeda Galaxy would be far more spectacular if it were not presented to us at such a steep angle. We see it inclined at only 13° to the line of sight – virtually edge-on – and so we miss many of the intricacies of its structure. But the bright, young, blue stars and glowing gas clouds which, as in our Galaxy, dominate the spiral arms, outline the spiral pattern clearly in this blue-light optical picture.

The optical picture also reveals that the western (right) side of M31 is closer to us than the other side. Dark lanes of dust belonging to the innermost spiral arms show up in silhouette against the bright nucleus on the nearer side only, although the distant eastern side actually has a brighter rim. The whole disc is distinctly warped as a result of the gravitational interaction of the two satellite galaxies.

Because the galaxy's central bulge is so bright, it is featureless and burnt out on a photograph designed to capture detail in the fainter arms. But when it is specially photographed with a CCD camera, which has a far wider response to intensity variations, the bulge shows structure. The CCD picture (**Fig. 10.3**) is colour coded for intensity, with blue for the faintest levels, through yellow, orange, red and finally white. The outer parts of the bulge are noticeably dimmed by dust, and the ragged details of the innermost dust lanes are now much more obvious – even though this photograph was taken using a smaller telescope with poorer resolution. Inside the bulge the dust disappears. The light becomes dominated by even more densely packed, old, red and yellow stars, whose crowding shows up in the smoothly increasing luminosity towards the centre. Right at the heart of the Andromeda Galaxy is a brilliant point of light, a tightly packed star cluster similar to that found by infrared astronomers at the centre of our Milky Way (Fig. 8.14).

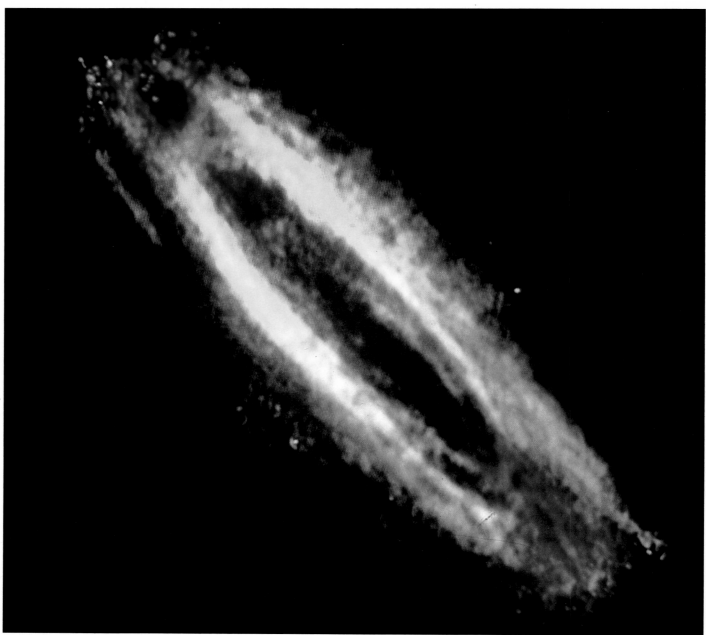

10.4 *Radio, 21 cm hydrogen line, velocity coded, Westerbork Synthesis Radio Telescope*

Like the Milky Way, M31 is a galaxy whose spiral arms are rich in cold hydrogen gas. The gas emits its characteristic signal at a wavelength of 21.1 centimetres, enabling its distribution and its movement to be mapped by radio telescopes. Observations at this wavelength reveal a great deal about the dynamics of the galaxy.

Mapping the Andromeda Galaxy at 21 centimetres is not entirely straightforward. M31 has an intrinsic velocity of its own within the Local Group, which is superimposed on all the motions of gas within it. In fact, M31 is one of the rare galaxies which are approaching us, rather than rushing away; it lies so nearby that its motion (with respect to the Milky Way) is

not affected by the expansion of the Universe. The net velocity of approach of M31 is 310 kilometres per second – which, despite its apparently high value, means that the two galaxies will move closer together by only 0.000000004 per cent in the next million years!

Fig. 10.4 shows a radio map of the hydrogen in M31, colour coded for velocity, and corrected for the intrinsic motion of the galaxy itself. The brightness of the emission reveals the amount of gas present, while the colour shows its speed towards or away from us: green and blue regions are approaching while yellow and red regions are receding.

Although Fig. 10.4 shows the same extent of the galaxy as the optical

photograph (Fig. 10.2), there is a striking reversal of M31's appearance. The optically dominant central bulge is dark at these wavelengths, while the formerly inconspicuous rim of the galaxy shines brilliantly at 21 centimetres. The glowing regions here reveal the galaxy's cold hydrogen reserve. Towards the central bulge, the zone dominated by old, red and yellow stars, all the hydrogen has already been used up, or is in the form of hydrogen molecules ($H_2$). But in the disc there is still a considerable number of hydrogen atoms, which show up particularly where the gas has been compressed by the passage of a spiral arm. For instance, strong concentrations here correspond to the dust lanes on the right and above the central

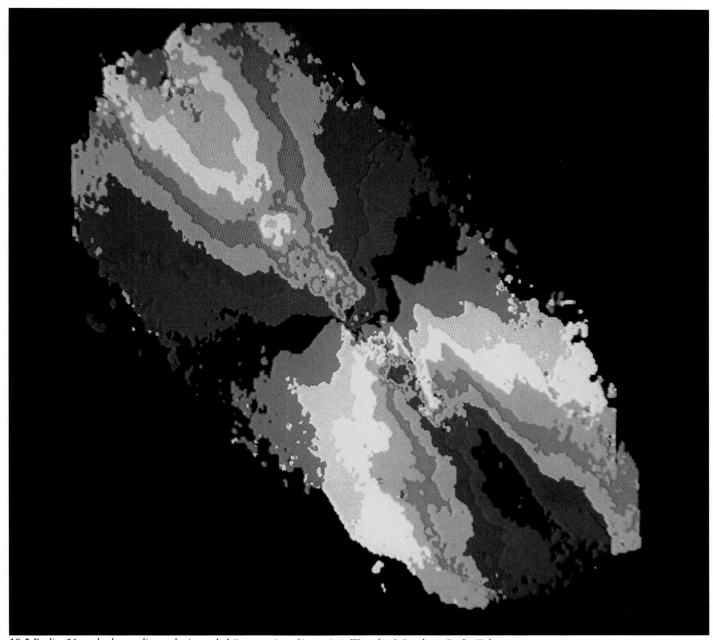

**10.5** *Radio, 21 cm hydrogen line, velocity coded (irrespective of intensity), Westebork Synthesis Radio Telescope*

bulge; and gaping 'holes' occur in the outermost regions where there are spaces between the spiral arms.

The velocity colour coding shows that M31 is in a state of systematic motion. The lower portions are approaching, while the upper parts are receding, revealing that the whole galaxy is rotating in space with its spiral arms trailing behind it. This is shown in more detail in **Fig. 10.5**, which codes the velocity of hydrogen gas right across the galaxy, irrespective of its amount (and hence its intensity at 21 centimetres). The black contour shows gas across the minor axis, which is effectively at rest. The other colours are at velocity intervals of approximately 25 kilometres per second: towards the lower right, red, orange,

yellow, green, blue, and purple regions show increasing velocities of approach; while to the top left the opposite sequence of colours indicate increasing speeds of recession.

Two intriguing features emerge from this 'scarab beetle' plot. First, the contours are not smooth parabolas pointing towards the galaxy's centre – which they would be, if the Andromeda Galaxy were rotating smoothly. There are 'bumps' of anomalous velocity throughout, such as the yellow and purple patches just above and below the centre. These correspond to the positions of the spiral arms, and they may be a result of the 'density wave' (which creates the arms) disturbing the speed of the gas as it passes through.

Second, the rotation curve here enables us to find the mass of M31, because the rotation speed at a given radius is dictated by the amount of matter lying within. The rotation speed should increase with radius near the centre, but after a maximum it should then fall off gradually with distance. Fig. 10.5 shows that the outermost parts of M31 have a definite tendency to maintain a high rotation rate, implying that there must be some unaccounted mass whose gravitational field holds the high-speed outer arms in check. Astronomers suspect that this mass may lie hidden in a huge, dark, almost starless halo which surrounds the Andromeda Galaxy, and contains almost ten times the mass of the visible galaxy.

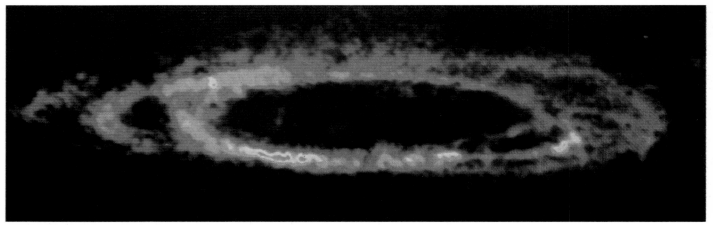

10.6 *Radio, 21 cm hydrogen line, Cambridge Half-Mile Telescope*

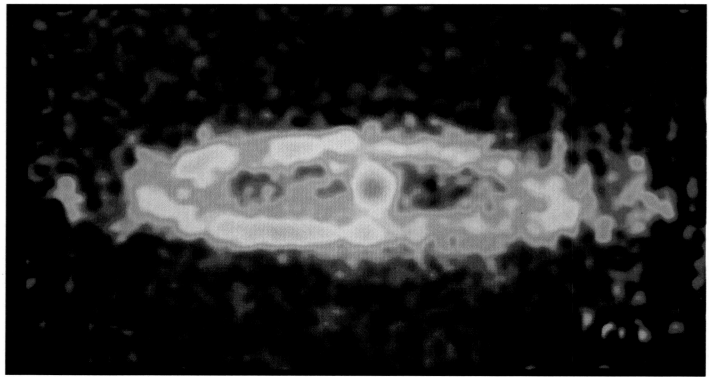

10.7 *Radio, 11 cm, 100 m Effelsberg Telescope*

These radio and ultraviolet wavelength pictures of the Andromeda Galaxy reveal in detail the 'young' active regions in the galaxy's disc, where stars are still in the process of forming. The radio maps (Figs. 10.6 and 10.7) show the distribution of various forms of interstellar matter here. **Fig. 10.6** is a 21 centimetre picture of M31, covering a slightly larger area (3½° by 1½°) than Fig. 10.4. It is also rotated anticlockwise by 45° relative to Fig. 10.4, and colour coded for intensity alone instead of velocity. The photograph reveals the concentrations of cold hydrogen gas in the arms of M31 and shows an extra spiral arm (on the left), extending beyond the arms seen in the optical photograph (Fig. 10.2). The extremely narrow, intense bands, coded light green and pink, are

dense ribbons of gas, following the spiral arm pattern. Some of these hydrogen arms are 30 000 light years long, but only 100 light years wide.

Fig. 10.7 is a radio map of M31 at the shorter wavelength of 11 centimetres, shown on the same scale. As with Fig. 10.6, the map is rotated and colour coded for intensity, with the red regions being the brightest and blue the faintest. Like Fig. 10.6 the map shows a bright 'ring' of superimposed spiral arms, which lies some 35 000 light years from the centre of the galaxy; although at 11 centimetres wavelength, the structure is much patchier than the arms of hydrogen which shine at 21 centimetres. Most of the 11 centimetre radiation is synchrotron emission from fast-moving cosmic ray electrons spiralling

around large-scale magnetic fields in M31's arms. The magnetic field pervades the whole galaxy, and the synchrotron radiation is brightest in the ring of spiral arms because this is where the high-speed electrons are generated. They are accelerated by supernovae which explode without ever leaving the spiral arm of their birth, and by flare stars and X-ray sources.

Synchrotron emission is also responsible for the apparently intense source right at the centre of the galaxy in Fig. 10.7 As we have seen in Fig. 10.3, this is a region of very crowded stars where the density of cosmic rays and magnetic fields is correspondingly high, and where the synchrotron emission would be expected to peak. However, the radiation is by no means as strong as it looks on the map: the apparent intensity is a result of the emission being so concentrated. M31's central source contributes about 1 per cent of the galaxy's total synchrotron emission, and is only one-twentieth as bright as our own galactic centre source, Sagittarius A (Fig. 8.12).

Although most of the 11 centimetre radiation from the 'ring' is synchrotron emission, some particularly intense patches represent radio emission from very hot gas surrounding newly formed stars. These hot gas clouds are also visible as bright nebulae on the optical photograph (Fig. 10.2).

M31's regions of young stars also dominate the view at ultraviolet wavelengths. Young, extremely hot O and B stars produce a great deal of their energy in this part of the spectrum, as demonstrated in two images obtained by a 0.3 metre ultraviolet imaging telescope, carried into space above our absorbing atmosphere aboard a sounding rocket. **Figs. 10.9** and **10.10** show views at short ultraviolet and long ultraviolet wavelengths respectively, with an optical photograph (**Fig. 10.8**) for comparison. All three have been rotated like Figs. 10.6 and 10.7. Both ultraviolet images reveal the 'ring of activity' in the region of the Andromeda Galaxy's spiral arms. Tight knots of emission spotlight the sites – corresponding to bright nebulae in the optical picture – where star formation is currently taking place and where young O and B stars are present. The images clearly show that M31 is an 'evolved' galaxy, with a dominant population of middle-aged stars (invisible in the ultraviolet) and only a little new star formation, narrowly restricted to its outermost regions. The peculiar smudges in Fig. 10.9 are completely unconnected with M31: the parachute returning the ultraviolet telescope to Earth at the end of the mission failed, and these 'pressure-induced marks' are the result of the subsequent crash!

M31's central regions emit ultraviolet radiation too, and in common with the centres of other galaxies there is a great deal more than would be expected from a population of old stars. One possibility is that the stars are indeed old, but very deficient in elements heavier than hydrogen and helium. The effect of this as yet unexplained deficiency would be to increase the ultraviolet flux. Alternatively, the stars responsible for the ultraviolet brightness, particularly at longer ultraviolet wavelengths (Fig. 10.10), may actually be young, but with a distribution which mimics that of an old population of stars. The ellipsoidal central bulge in Fig. 10.10 certainly follows the distribution of old stars seen in Fig. 10.3, and it is difficult to see how young stars could distribute themselves so smoothly. At the moment, the excess central flux in M31 and many other galaxies is still a mystery.

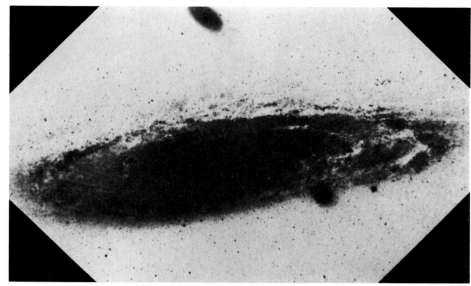

10.8 *Optical, negative print, 1.2 m Palomar Schmidt Telescope*

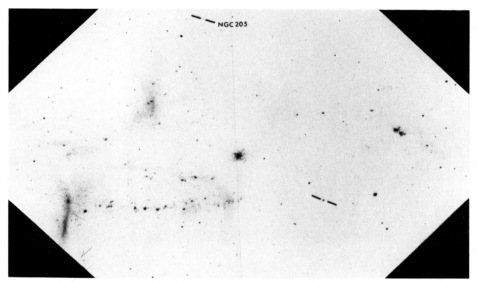

10.9 *Ultraviolet, negative print 125–175 nm, 0.3 m rocket-borne telescope*

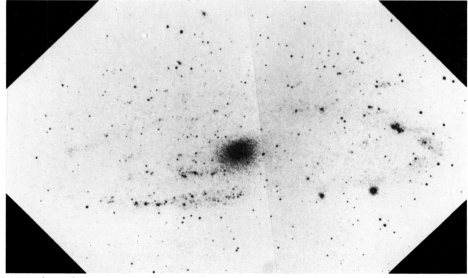

10.10 *Ultraviolet, negative print 175–275 nm, 0.3 m rocket-borne telescope*

10.11 *X-ray sources detected by Einstein Observatory overlaid on optical photo by Harvard College Observatory*

10.12 *Optical, true colour, 1.2 m Palomar Schmidt Telescope*

The X-ray luminosity of an average galaxy is only a thousandth or a ten-thousandth of its light output. This may seem very small, but while the optical luminosity is spread amongst thousands of millions of stars, the X-ray output is concentrated into just a few dozen exceptionally powerful sources, and the orbiting Einstein Observatory pinpointed eighty-eight X-ray sources in the nearby Andromeda Galaxy (**Fig. 10.13**). **Fig. 10.11** shows their positions superimposed on an optical photograph of the galaxy.

The outer parts of the galaxy, bluish in the optical colour photograph (**Fig. 10.12**), are young, active regions. The X-ray sources here follow the spiral arm closely – like the curve of the faint arm at the top left. Most are probably close binary systems consisting of a young, blue supergiant and a neutron star, in which the latter's strong gravity pulls streams of hot gas into an X-ray emitting disc around itself. Other sources in the arm regions are

the remnants of supernovae.

The X-ray sources in the central bulge – made up mainly of middle-aged stars which give it a yellowish colour in Fig. 10.12 – are also binaries. They consist of a neutron star with an old red giant companion. Although more luminous, their output varies erratically as streams of gas 'feed through' at different rates. One of the most striking differences between the Andromeda Galaxy and our own is the large number of X-ray sources in M31's globular clusters, the dense balls containing thousands of old, red stars which surround the galaxy itself. Twenty of Andromeda's globular clusters contain X-ray sources, and even the average ones are one and a half times brighter than the brightest in orbit about the Milky Way. Fig. 10.13 shows this clearly: the bright spot at the top right, for example, is an exceptionally powerful globular cluster source.

Fig. 10.13 also highlights the concentration of bright sources towards the

centre of M31. Within 2 arcminutes (1300 light years) of the centre there are at least nineteen sources – more than six times the number in the corresponding region of the Milky Way. However, this may simply reflect M31's far greater number of stars. Striking differences are seen in **Fig. 10.14**, which shows the innermost 5 arcminutes of M31 (rotated anticlockwise by 45°) in two exposures made seven months apart. Many of the sources have varied markedly in brightness, particularly the source which coincides with the precise centre of the galaxy (1 arcminute left of the brightest source in the left-hand exposure). This is almost a hundred times more powerful than the source at the centre of our own Galaxy (Fig. 8.11), and it varies by at least a factor of ten. While the other variable sources in Fig. 10.14 are probably normal binary systems, the nature of the 'core' source is still uncertain.

GREAT NEBULA IN ANDROMEDA (M 31)

30 ARC-MIN:

**10.13** *X-ray, 0.3–2.5 nm, mosaic, Einstein Observatory IPC*

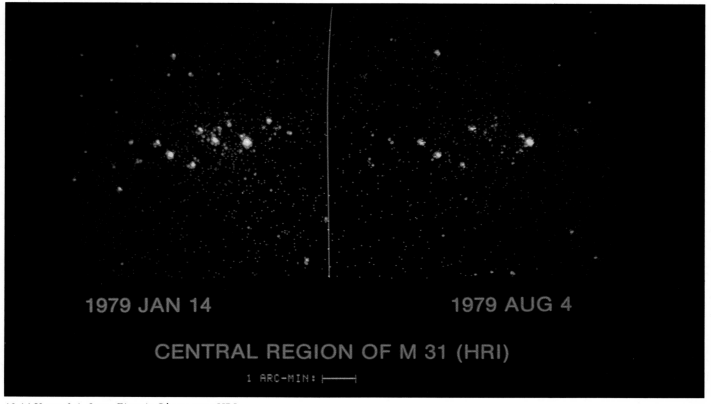

1979 JAN 14        1979 AUG 4

CENTRAL REGION OF M 31 (HRI)

1 ARC-MIN:

**10.14** *X-ray, 0.4–8 nm, Einstein Observatory HRI*

**10.15** *Optical, enhanced true colour, 1.2 m Palomar Schmidt Telescope*

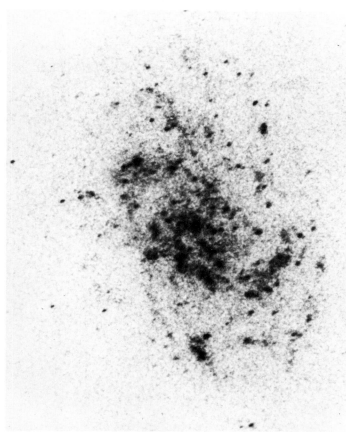

**10.16** *Ultraviolet, 125–175 nm, negative print, 0.3 m rocket-borne telescope*

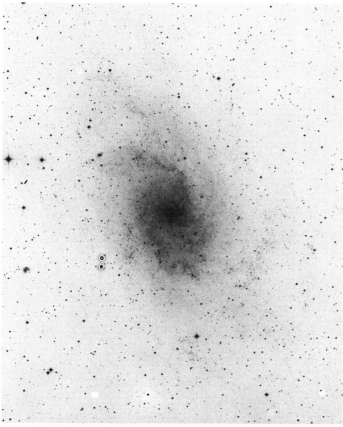

**10.17** *Infrared, 0.71–0.88 μm, negative print, 1.2 m Palomar Schmidt Telescope*

# M33

M33 lies close to the Andromeda Galaxy (M31) in the sky, in the neighbouring constellation of Triangulum, and it is another major member of our Local Group of galaxies. At 2.4 million light years, slightly further away than M31, it is sometimes claimed to be the most remote object visible to the unaided eye. However, M33 is much fainter (with an apparent magnitude of 6.5) and not visible unless sky conditions are exceptionally good.

The optical view of M33 in **Fig. 10.15** is enlarged to twice the scale of the similar view in M31 in Fig. 10.2. Even allowing for M33's slightly greater distance, it is obvious that the Triangulum galaxy is much smaller than its giant neighbour. With a diameter of 40 000 light years and a mass of 15 000 million Suns, it is much more similar in size to the Large Magellanic Cloud (Figs. 8.16–8.20), and like the Large Magellanic Cloud, M33 is vigorously forming new stars, which show up in Fig. 10.15 as ragged, blue spiral arms. In this computer-enhanced optical picture, photographs taken through blue, green and red filters have been combined with their respective colours printed as strongly as possible, to emphasise the slight colour differences in M33. Star formation at work can be seen in several giant clouds of glowing hydrogen – vivid pink in Fig. 10.15 due to emission from the hydrogen atoms at 656 nanometres – especially those strung out along the spiral arm which unwinds from just right of centre and extends upwards to the left. Most prominent of all is NGC 604, at the end of the arm. In size, mass and appearance it is very similar to the Large Magellanic Cloud's enormous Tarantula Nebula (Figs. 8.21–8.24). M33 is a more 'open' spiral than M31 or the Milky Way, with wide, far-flung arms and a proportionally smaller centre bulge.

Mixed in with the young stars is a good deal of dark interstellar dust, which shows up as reddish-brown – its actual colour – against the underlying yellow glow from the older stars in the galaxy. The latter's smooth distribution contrasts strongly with the patchy spread-out pattern made by the young stars.

**Fig. 10.16** shows an ultraviolet view of M33. In contrast to similar pictures of M31 (Figs. 10.9 and 10.10), where the emission is restricted to an outer 'ring', there is ultraviolet emission from all over the galaxy. In fact, the ultraviolet view is little different from the view at optical wavelengths. Both are dominated by the radiation from hot young O and B stars, which have been recently formed throughout the whole galaxy.

Fig. 10.16 also reveals the true extent of M33, highlighting clumps of O and B stars at great distances from the centre. The emission is strongest in regions where stars have already formed (below centre). It is, however, weak in areas where stars are still forming, such as in the nebula-studded spiral arm above centre, because the gas absorbs ultraviolet radiation from the embedded stars. NGC 604, like the Tarantula Nebula, is hence not very prominent at ultraviolet wavelengths.

In complete contrast, **Fig. 10.17**'s 'photographic infrared' view shows a less extensive, much smoother distribution of stars. (The two spots to the left of M33 are photographic defects.) The main body of the galaxy at this wavelength is half the size of the ultraviolet image. But the older, red and yellow stars which emit this radiation are the types which make up by far the greater proportion of the galaxy's mass; the young, blue, spiral arm stars amount to comparatively little mass. The old stars, and hence the galaxy's mass, are concentrated into the prominent central bulge. Silhouetted against it are the patchy dust lanes, while the surrounding disc shows only smooth, spiral arms in contrast to the narrow, patchy zones of recent star formation so dominant in Fig. 10.15.

16 000 light years. Fig. 10.18 shows this enhancement around the edge of the galaxy, and also the concentration of emission at the top left. This is a region where there has been little star formation so far (compare especially with Fig. 10.16, which shows the distribution of young stars), but it is evidently earmarked for star birth in the not-too-distant future.

Star formation is also responsible for M33's appearance at a wavelength of 11 centimetres (**Fig. 10.19**). If our eyes were sensitive to these radio waves, we would classify M33 as a completely different kind of object from the Andromeda Galaxy (see Fig. 10.7). In Fig. 10.19, the colour-coded map reveals strong emission from M33's smooth, glowing disc, in which the strength of emission decreases from red through yellow to green. (The intense red blob to the right of M33 is a background source.)

Emission at this wavelength is dominated by radiation from hot gas and by synchrotron radiation from fast-moving cosmic ray electrons. In M33, the radio waves come (on average) 40 per cent from hot gas, and 60 per cent from electrons. This, coupled with Fig. 10.19's prominent spiral arms, reveals that stars are actively forming throughout M33, not just in its outer regions. Star formation is even continuing close to the central bulge, where the fraction of hot gas rises to 70 per cent. But the most obvious region of star formation is NGC 604 (the red blob left of centre), which is several times larger than any comparable region in the Milky Way or the Andromeda Galaxy.

Although M33 is less than a tenth as massive and only a tenth as luminous as M31, its total radio output is very similar. M33 has brighter, more numerous clouds of glowing hydrogen, and its higher supernova rate leads to the production of more cosmic ray electrons. Once again, the underlying cause is the high rate of recent star formation.

At least two of the glowing hydrogen clouds associated with star birth show up on the Einstein Observatory X-ray image of M33 (**Fig. 10.20**). There are nine sources in the galaxy, but the central source is ten times brighter than any of the others. It is a remarkably strong source for such a small galaxy, at least a thousand times more powerful than the X-ray source at the centre of the Milky Way (Fig. 8.11). The Einstein Observatory saw it halve in brightness in just a few months. Its nature is a mystery, for in its strength it resembles the source at the centres of 'active' galaxies. However, the light from M33's innermost core shows no signs of activity, and there is no detectable radio source here.

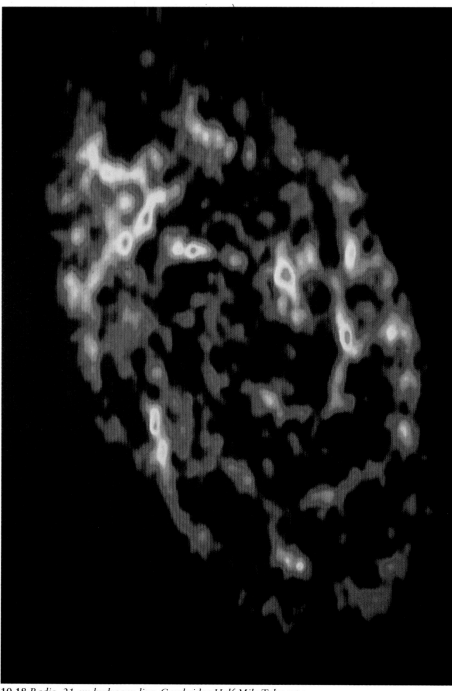

10.18 *Radio, 21 cm hydrogen line, Cambridge Half-Mile Telescope*

M33's present burst of star formation shows up at nearly all wavelengths. **Fig. 10.18** reveals that the galaxy is also rich in reserves to continue the process. It is a map of M33's distribution of cold hydrogen, colour coded to show intensity (and hence the amount of interstellar gas). The coding runs from dark green and blue for weak emission, through pink, yellow and finally bright green for the most intense peaks.

The spiral structure stands out very clearly. Although the arms on this map are slightly 'fuzzy' (not fully resolved by the radio telescope), two arms can be traced out to 800 light years from the centre, beyond which four arms are faintly visible. The arms are about 1000 light years wide and extend for 12 000 light years before their structure becomes jumbled.

Unlike the Andromeda Galaxy (Fig. 10.6), M33's cold hydrogen is spread throughout the disc and not restricted to an outer 'ring'. There is, however, a definite increase outwards from the centre; 6000 light years out, the thickness of the hydrogen layer is about 800 light years, but this increases to 2000 light years thickness at a radius of

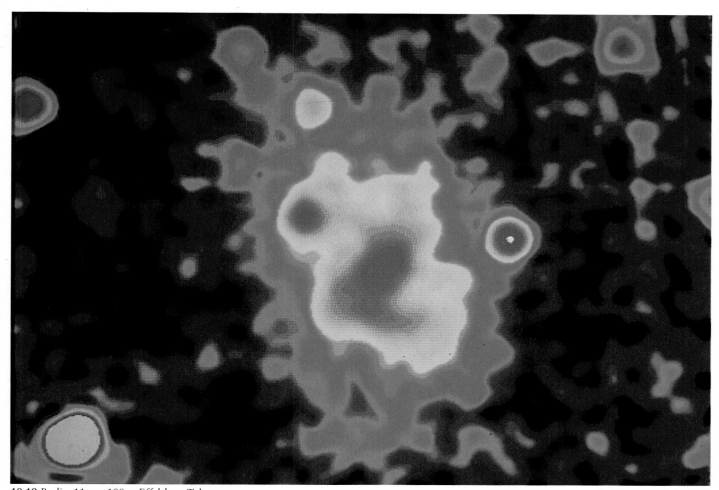

**10.19** *Radio, 11 cm, 100 m Effelsberg Telescope*

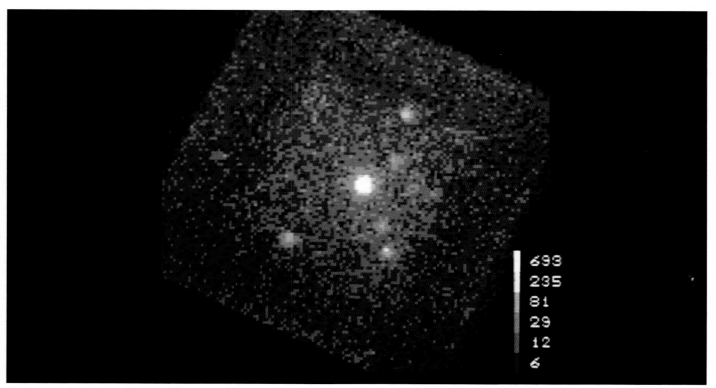

693
235
81
29
12
6

**10.20** *X-ray, 0.3–2.5 nm, Einstein Observatory IPC*

**10.21** *Optical, true colour, 1 m telescope, US Naval Observatory*

# M51

'Spiral convolutions; . . . with successive increase of optical power, the structure has become more complicated . . . .' So wrote the Third Earl of Rosse about this galaxy in 1845. With his monster telescope at Birr, Eire, he was the first astronomer to discover spiral structure in a 'nebula'.

M51's nickname – the 'Whirlpool Galaxy' – is apt. Although well outside our Local Group, at a distance of 20 million light years, M51 presents its beautiful spiral structure face-on to us. **Fig. 10.21** shows the Whirlpool's appearance at visible wavelengths: a symmetrical, wide-open spiral with a narrow bridge of matter connecting up to its peculiar spiral companion, NGC 5195.

Fig. 10.21 tells little of the true extent of the Whirlpool. The galaxy's arms extend much further into space, stretching to a total diameter of 65 000 light years. With a mass of 50 000 million Suns, M51 is both smaller and less massive than our Galaxy, yet it is four times brighter overall. Even the small NGC 5195 is three-quarters as bright as the Milky Way, and the centres of both galaxies are particularly brilliant.

Although M51's visible appearance is dominated by the young, blue stars and glowing nebulae in its spiral arms, the 'photographic infrared' image (**Fig. 10.22**) – which shows the distribution of older red stars – differs surprisingly little from the optical image. As in the case of M33 (Fig. 10.17), the galaxy's extent is smaller. But its spiral structure, although slightly broadened, is still powerfully defined. This points to the presence of a long-lasting, confining mechanism which can maintain even relatively old spiral arms. Also noticeable are the very bright centres of both galaxies. That of the companion is particularly prominent, and it is clearly not circular. NGC 5195 is actually a barred rather than an ordinary spiral, and the bar shows up very clearly in this image.

The Whirlpool Galaxy's unusual brightness stems largely from recent star formation, with the consequent dominance of bright, hot O and B stars. The ultraviolet image (**Fig. 10.23**) shows their distribution throughout the galaxy, in a colour-coded intensity map. The red (most intense) regions reveal the locations of young, ultraviolet-emitting stars, while yellow, green and finally blue areas have successively fewer newly formed stars. NGC 5195 has no new stars itself; but the Whirlpool's major region of new star formation lies just opposite it, and corresponds with the brightest section on the outer optical spiral arm. Fig. 10.23 reveals new stars on the other side of the Whirlpool galaxy's centre, and in a ring all around the centre itself. There is a bright ultraviolet source marking the galaxy's innermost core.

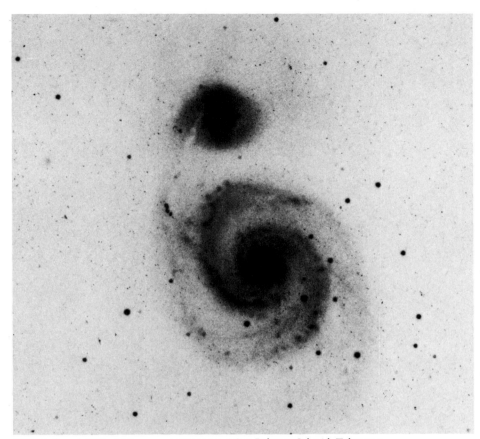

10.22 *Infrared, 0.71–0.88 μm, negative print, 1.2 m Palomar Schmidt Telescope*

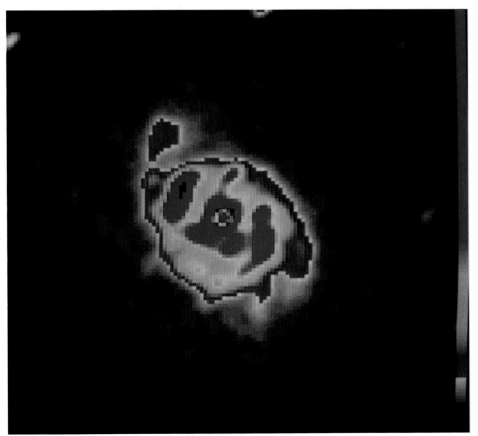

**10.23** *Ultraviolet, 175–275 nm, 0.3 m rocket-borne telescope*

M51's spiral arms are the most perfect 'textbook' example possessed by any nearby galaxy. The arms may result from the tidal pull of NGC 5195, because the gravitational effects of a close companion can help to generate spiral 'density' waves' in a galaxy. These zones of compression propagate around the galaxy as a spiral-shaped 'standing wave' with its own speed, which is slower than the rotation of the stars and gas at the same distance from the centre. When gas moves into the wave, it is slowed down and compressed to form into new stars. When stars enter the wave, they are slowed by the pull of stars already there and stay longer in the wave than outside it. The net result is a prominent pattern of spiral arms studded with young stars and glowing nebulae.

Although companionless spirals have weak density waves, those with companions should have the sharpest, best-defined arms, because of the greater amount of compression. **Fig. 10.24**, a colour-coded radio view of M51, reveals that this galaxy does indeed have sharp, narrow arms.

The intensity of the emission decreases from red, through yellow and green to blue, and the arms stand out as extremely narrow regions of constant brightness. This wavelength reveals emission from fast-moving electrons in magnetic fields, and here the strong density wave is compressing the interstellar field into a prominent spiral pattern. There is also a bright radio source at the Whirlpool's centre – one of the strongest in any normal galaxy – and a much less powerful one in NGC 5195. **Fig. 10.25** shows a bas-relief optical view of the two galaxies, produced by the same computer-subtraction technique described in Fig. 12.11, reveals not only the complexity of the arms but also the faintest outer extensions of the system. There is a great deal of low-luminosity matter surrounding NGC 5195, which appears to have been disrupted by its interaction with the larger galaxy.

These faint extensions figure prominently on the six spectacular computer-graphics images of **Fig. 10.26**. This three-dimensional representation of the system was obtained by summing sixteen million 'picture elements' (pixels) from several optical photographs and representing the images as a landscape of 'mountains and valleys' (the surrounding spikes are stars), with the peaks corresponding to the regions of highest luminosity. In fact, M51's centre is so bright that its 'peak' has been artificially truncated. The rotated images highlight the complex structure in the faint 'foothills'. The two central images reveal the extent of disturbed material around NGC 5195. It covers an area as large as the Whirlpool Galaxy itself.

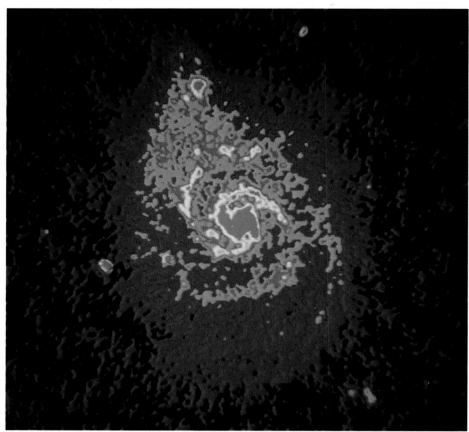

**10.24** *Radio, 21 cm continuum, Very Large Array*

**10.25** *Optical, bas-relief representation, 1.2 m Palomar Schmidt Telescope*

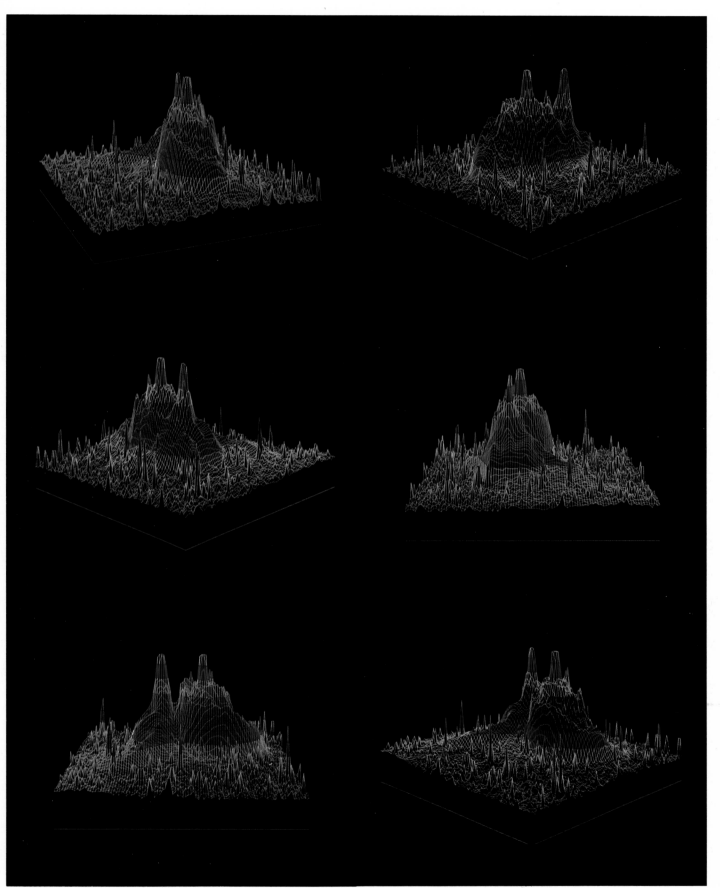

**10.26** *Optical, computer graphic representation of intensity, six orientations, derived from 1.2 m Palomar Schmidt Telescope photos*

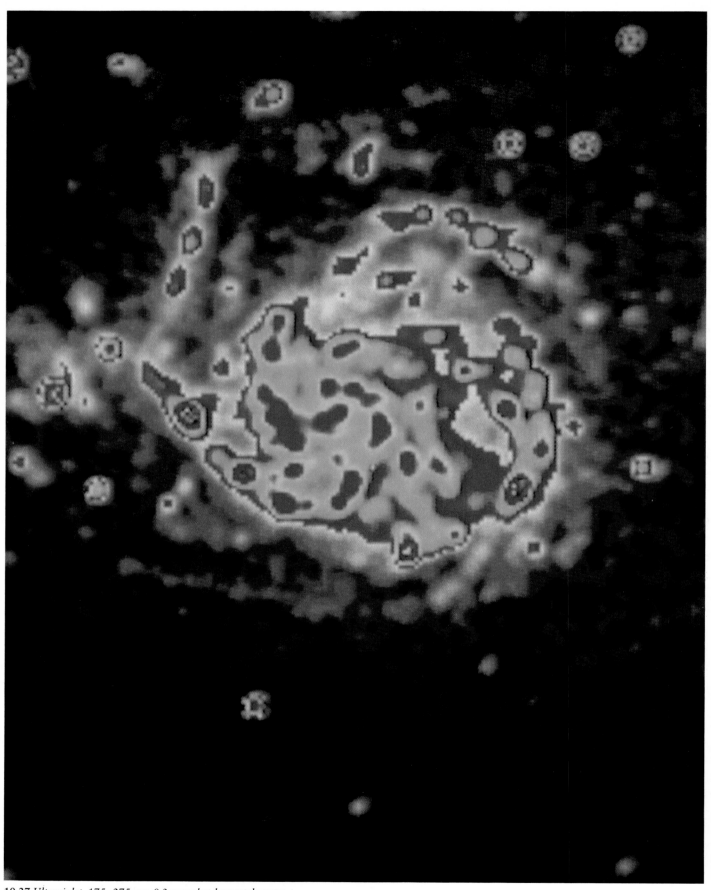

**10.27** *Ultraviolet, 175–275 nm, 0.3 m rocket-borne telescope*

**10.28** *Optical, 3.8 m telescope, Kitt Peak National Observatory*

# M101

Spanning nearly half a degree of sky in the constellation of Ursa Major, M101 (**Fig. 10.28**) is one of the largest, brightest spiral galaxies known. At its distance of 20 million light years, this angular size corresponds to a diameter of 200 000 light years – twice that of our Milky Way, which would cover only the inner bright region of the galaxy seen here.

M101, the 'Pinwheel Galaxy', has a mass of three thousand million Suns and belongs to a loose group of galaxies which also includes M51 (Fig. 10.21). Its extremely

narrow, far-flung spiral arms are dominated by huge clouds of glowing hydrogen, some of which are 3000 light years across. A number of them in the outer arms in Fig. 10.28 are so bright that they have been allocated their own NGC numbers: from left to right the most prominent are NGC 5462, 5461, 5455, and 5447.

Because M101 is so rich in young hot stars, it is very bright at ultraviolet wavelengths (**Fig. 10.27**). The brightest regions are coded red in this map, with the intensity decreasing through yellow, green and blue to grey. There is a prominent

band of young stars to the left of centre, and other concentrations along the inner arms. This ultraviolet image also reveals the true extent of the galaxy, well beyond the limits visible on optical photographs. The emission from these faint 'ultraviolet arms' does not, however, come directly from young stars, which would be easily visible at optical wavelengths. Instead, it is radiation from other parts of the galaxy which has been scattered by dust. The dust, with cold hydrogen gas (Fig. 10.31), forms invisible extensive reserves in the outer arms.

At a radio wavelength of 2.8 centimetres,

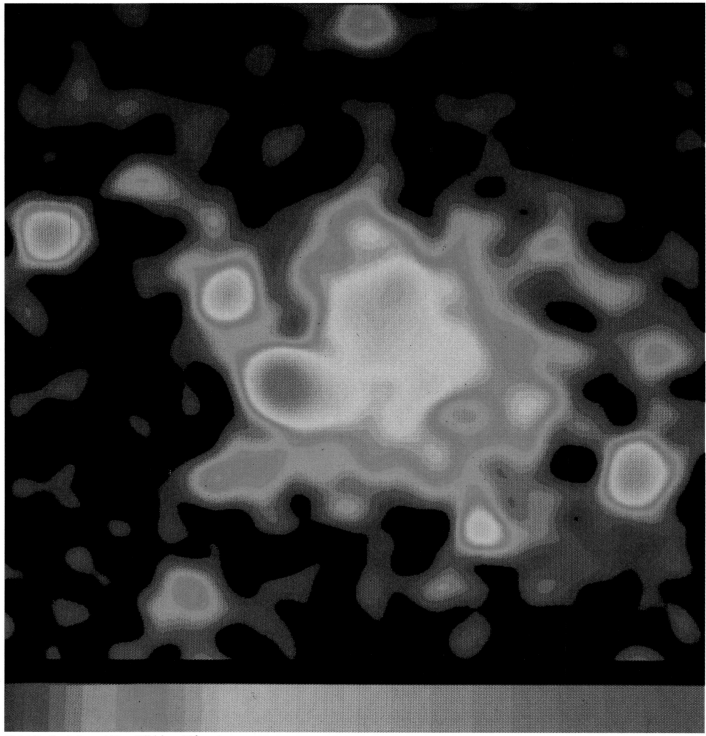

**10.29** *Radio, 2.8 cm, 100 m Effelsberg Telescope*

M101's appearance is dominated by strong emission from its gigantic, hot hydrogen clouds (**Fig. 10.29**). This colour-coded map shows the highest intensities as red, decreasing through orange, yellow, green and finally blue. The spiral arms can just be picked out, but both they and M101's central source are extremely weak at this wavelength when compared to the nebulae. The brightest source on this map is

NGC 5461; to its left are NGC 5462 and NGC 5471 (out of the picture in Fig. 10.28) while NGC 5455 and NGC 5447 lie to the right.

With an average diameter of 3000 light years, these nebulae are ten times larger than their counterparts in the Milky Way, and give out twenty times more energy at radio wavelengths. They are three times the size of giant hydrogen clouds in other

galaxies, such as the Tarantula Nebula (Figs. 8.21–8.24) in the Large Magellanic Cloud, and NGC 604 in M33 (Figs. 10.15–10.20). One cloud, NGC 5471, contains 1000 O stars within a region 600 light years across, and they heat up a surrounding cloud of gas with a mass equivalent to ten million Suns. At radio wavelengths, NGC 5471 produces ten thousand times the power of the Orion

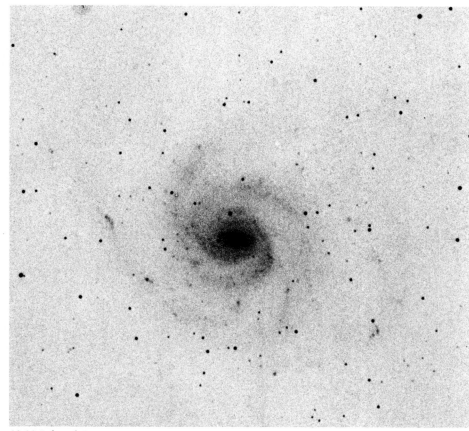

10.30 *Infrared, 0.71–0.88 μm, negative print, 1.2 m Palomar Schmidt Telescope*

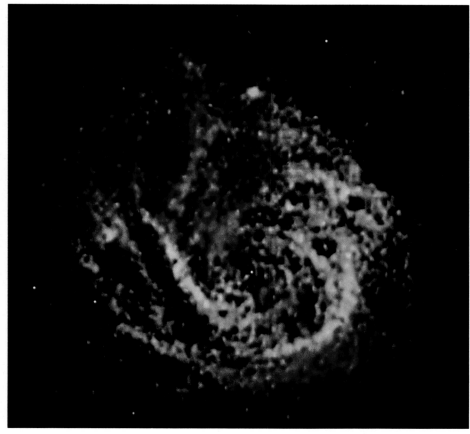

10.31 *Radio, 21 cm hydrogen line, velocity coded, Westerbork Synthesis Radio Telescope*

Nebula. It is still not certain whether these regions are 'supergiant' versions of the Orion Nebula scaled up hundreds of times, or whether they are vast clusters of hundreds or thousands of individual nebulae.

At 'photographic infrared' wavelengths, however, the cloud complexes all but disappear, and M101 looks a lot smaller (**Fig. 10.30**). This image, which shows the distribution of older, red and yellow stars in the galaxy, reveals how little star formation has occurred in the outer disc until now. By contrast, the central bulge is very prominent, for it is made up of older stars. The old spiral arms are much less tightly confined than those of M51 (Fig. 10.22), possibly because M101 does not have a companion galaxy to help maintain them by gravitational perturbations.

Like M31, M101 is a galaxy where the rotation shows up vividly in velocity-coded maps at the 21 centimetre wavelength of radiation from cold hydrogen (**Fig. 10.31**). Green and blue arms show successively higher velocities of approach, while orange and red arms are receding. M101's southern (lower) side is clearly rotating towards us, while the northern (top) side is being carried away. The map also reveals that M101 is actually tipped up with respect to us along a line joining the most intense red, and not face-on as it looks at first sight.

Fig. 10.31 covers an area similar to the ultraviolet map in Fig. 10.27 – a region of space 400 000 light years across which extends to nearly twice the diameter of the optical image. Future generations of stars will be born in this vast, extended disc, and as the site of star formation continues to migrate from the centre, M101's already enormous visible extent will grow still further.

10.32 *Optical, 1.2 m Palomar Schmidt Telescope*

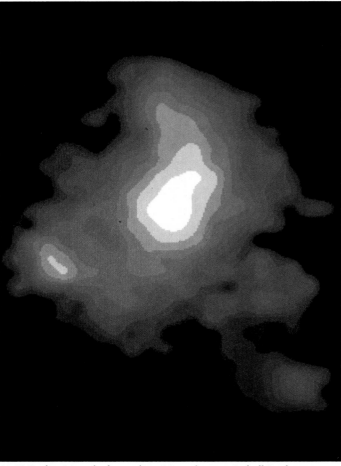

10.33 *Radio, 21 cm hydrogen line, 76 m telescope, Jodrell Bank*

## M81/82 Group

At a distance of 10.5 million light years, the M81 group of galaxies is one of the nearest small clusters to our Local Group. It covers a smaller region of space (2 million by 1 million light years) but has a similar number of members (thirty-two have so far been identified) and, like the Local Group, is dominated by just a few bright galaxies.

**Fig. 10.32** is an optical photograph of the central region of the cluster. (The whole cluster is four times larger, covering an area of Ursa Major more than 350 times that of the Full Moon's disc.) M81 (centre) dominates the picture. With a diameter of 100 000 light years, the same as the Milky Way, and a mass of more than 200 thousand million Suns, this spiral galaxy is by far the biggest in the group. Above it in Fig. 10.32 is the peculiar galaxy M82. The other two galaxies are NGC 3077 (left), a small irregular, and NGC 2976 (bottom right), a dwarf spiral.

All the galaxies are strongly interacting with one another. Unlike the interaction between the Whirlpool Galaxy and NGC 5195 (Fig. 10.21), this is not obvious at all in visible light. But at radio wavelengths, the picture changes

dramatically. **Fig. 10.33** is a map of the distribution of hydrogen gas throughout the cluster, colour coded for intensity (and hence amount of gas). White regions are richest in cold gas, while the amounts decrease through grey, orange, red, crimson and finally, purple. The plot covers the same area as Fig. 10.32, and shows very clearly that the galaxies are a close-knit group embedded in a huge cloud of material. As well as the gas associated with the individual galaxies, there is a great deal filling the space between the galaxies. In fact, over a quarter of the hydrogen in the group is estimated to be in this form, a mass equivalent to 1400 million Suns.

This intracluster gas probably originated from tidal interaction between the members of the M81 group, like similar gas found in the M51 and M101 groups. Gas has been (and is being) pulled out of the galaxies in streamers, and is slowly diffusing out into the cluster. The main source of gas is M81 itself, as Fig. 10.33 reveals; but NGC 2976 and particularly NGC 3077 are sources too. Although a bridge of strong emission connects M81 with M82, the latter – the second most massive galaxy in the group – is not as strong a source as the smaller NGC 3077.

The motions of the intracluster gas are very complicated, and streams close to one another sometimes have completely different velocity patterns. However, the behaviour of the streams, and in particular that of the bridge between M81 and M82, is consistent with their being the result of intergalactic tides. The distribution of the gas is almost as complicated as its velocity patterns. There are several marked concentrations, the most prominent being to the right and just below M81. This is the 'South-West Cloud': a huge, intergalactic gas cloud which appears to be linked to both M81 and to NGC 2976 (below). The cloud might belong to our own Galaxy and just happen to lie superimposed on the M81 group, but this is unlikely.

If the South-West Cloud does belong to the M81 group, it is very substantial. It has a mass of hydrogen equivalent to 320 million Suns: exactly twice that associated with the spiral galaxy NGC 2976 nearby. But it is just a gas cloud, not a galaxy; even the most sensitive optical photographs do not reveal any stars or nebulae. Its estimated total mass, including molecular hydrogen, is 1600 million Suns, which is larger than many dwarf galaxies.

Whether or not the South-West Cloud

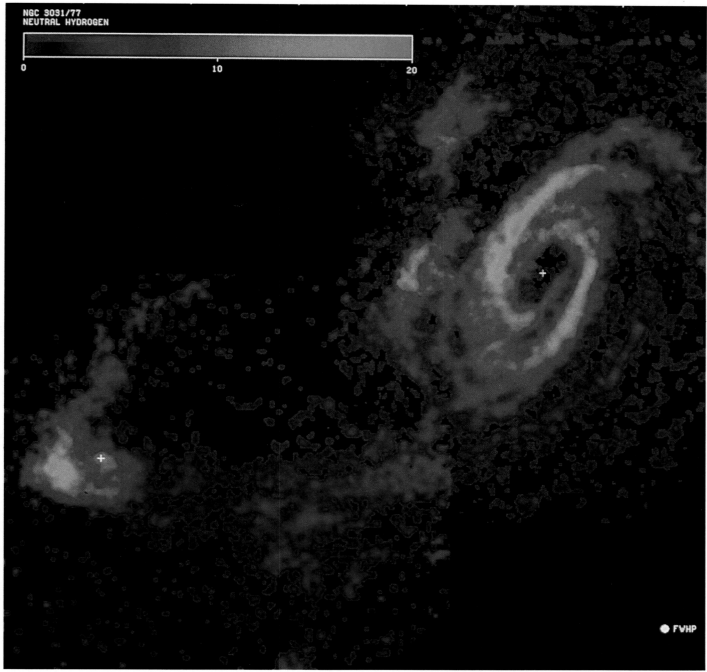

NGC 3031/77
NEUTRAL HYDROGEN

0          10          20

● FWHP

10.34 *Radio, 21 cm hydrogen line, Westerbork Synthesis Radio Telescope*

will eventually form into a galaxy depends upon its stability. It is just on the borderline between permanence and impermanence. The currently estimated mass and extent are almost sufficient to make it stable, but searches for low-luminosity material which would bind it more securely are planned. Some astronomers, however, believe that it is just a temporary cloud of gas which M81 has stripped away from NGC 2976, and that it will expand and lose its identity within 300 million years.

M81 is also undergoing a strong interaction with NGC 3077, as **Fig. 10.34** reveals. This colour-coded intensity map shows two complicated bridges of hydrogen extending from M81 (right) to NGC 3077 (left). Yellow regions are densest concentrations of gas, and the amounts decrease through red and blue. As well as revealing the degree of the tidal interaction between these two galaxies, the map shows the galaxies' enormous extent when viewed at this wavelength. It also highlights the differences between the two. M81 is a typical spiral whose gas is strongly concentrated into arms, where it is currently forming into young stars. In the central regions, however, many generations of star formation have exhausted the gas

and the result is an obvious depletion of hydrogen right in the middle.

NGC 3077, on the other hand, is an irregular galaxy. This type of galaxy seems relatively inefficient at converting gas to stars and consequently has a much higher proportion of gas even now. Although NGC 3077 has a mass of only 3300 million Suns, one-third of this is gas. Fig. 10.34 shows clearly that the smaller galaxy is gas-rich, even in the central regions. However, sustained interactions with M81 will eventually drive this gas out of the NGC 3077 and into the intracluster medium.

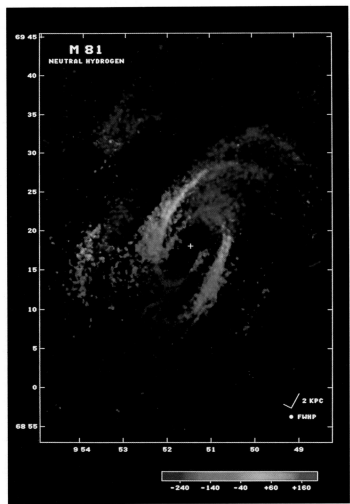

10.35 *Radio, 21 cm hydrogen line, velocity coded, Westerbork Synthesis Radio Telescope*

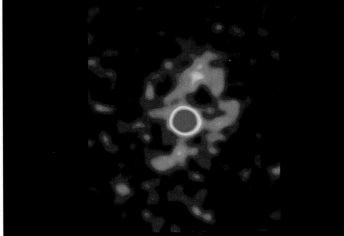

10.36 *Radio, 2.8 cm, 100 m Effelsberg Telescope*

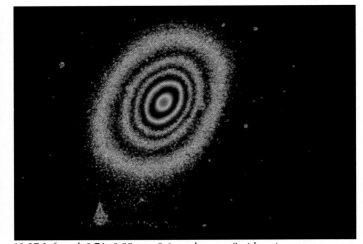

10.37 *Infrared, 0.71–0.88 μm, 0.6 m telescope, Smithsonian Astrophysical Observatory*

## M81

M81, the dominant member of its small group, is a beautiful spiral galaxy similar to our Milky Way, and it is inclined to us at an angle of 32°. Its rotation is vividly revealed in **Fig. 10.35**. This cold hydrogen picture is colour coded for velocity: red regions have the highest velocities of recession, blue the highest velocities of approach, while yellow and green lie in between. The spiral arms at the bottom are therefore coming towards us, while the upper portion is turning away. Since the right-hand side of the galaxy lies closest to us, M81 is clearly rotating in an anticlockwise direction with its arms trailing as it spins.

In Fig. 10.35, the brightness of the image represents the intensity of the radio emission, showing the extent of the hydrogen in M81. It reaches well beyond the optical disc, pulled out as a result of the tidal interaction with nearby NGC 3077 (compare with Fig. 10.34). However, there is a marked lack of gas close to the centre, and most is located in the outermost

regions. In this, M81 resembles M31 (Fig. 10.6), whose gas-rich star-forming regions are also restricted to an outer ring. The underlying cause may be the same in both cases. Repeated close passage by a companion galaxy through the body of a gas-rich spiral can scour the centre of gas, and gravitationally 'splash' the site of star formation into the outermost parts of the disc. In the case of M31, M32 is probably the culprit; here, NGC 3077 may be responsible for M81's 'ring'.

However, the centre of M81 is not completely dead, as **Fig. 10.36** shows. This colour-coded intensity map at a radio wavelength of 2.8 centimetres reveals the location of hot gas and fast-moving charged particles throughout the galaxy. Although the arms show up faintly (blue), the emission at this wavelength is dominated by strong radiation (red) from the central regions. M81's central source is probably similar to that of the Milky Way (Fig. 8.12), but it is twice as powerful, and like the central source in our Galaxy, it may be slightly variable. Neither source, however, is as powerful as its counterpart

in M51 (Fig. 10.24).

Like the Milky Way, M81 shows a large increase in the concentration of stars towards its centre. **Fig. 10.37** is a 'photographic infrared' scan of the central regions only – for comparison, the isolated dot at the bottom left is the star just below the centre in Fig. 10.38. Each contour represents an increase in brightness of sixteen times, showing a smooth build-up in the numbers of the older, cooler stars which populate this region. There is a dense star cluster right at the centre, two thousand times brighter than the outer regions of the central bulge.

The sharp increase in concentration towards the centre is also obvious in **Fig. 10.38**, an optical photograph processed through a range of filters to emphasise the natural colours, and so highlight the distribution of different types of star. The smooth bulge of old reddish stars in the centre is circled by slender arms of young blue stars, and the 'ring' of recent star formation shows clearly.

10.38 *Optical, enhanced true colour, 5 m Hale Telescope*

10.39 *Optical, 5 m Hale Telescope*

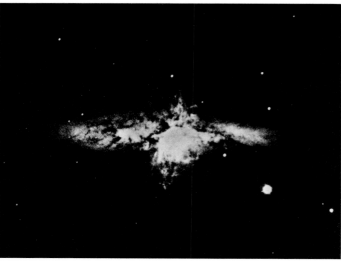

10.40 *Optical, 656 nm hydrogen line, 5 m Hale Telescope*

10.41 *Optical, polarised (perpendicular to major axis of galaxy), 5 m Hale Telescope*

10.42 *Optical, polarised (parallel to major axis of galaxy), 5 m Hale Telescope*

# M82

Until recently, M81's prominent companion – the peculiar galaxy M82 – was believed to be the nearest exploding galaxy to the Milky Way. The reasons are not difficult to see. **Fig. 10.39**, a straightforward optical view – which, like Fig. 10.40, has been rotated to a level position – shows an irregular spindle of a galaxy flecked with chaotic patches of dust, with evident signs of a disturbance at its centre. There are none of the easily resolvable regions of star formation or young clusters which we see in neighbouring M81, for M82's stars are dimmed some fifty times by a veil of dust. When a weak central radio source was discovered by Cambridge radio astronomers in the 1960s, it was very tempting to draw together all these phenomena as evidence for violent activity – albeit on a smaller scale than in quasars and radio galaxies (Chapter 12).

In the red light emitted by hydrogen at a wavelength of 656 nanometres (**Fig. 10.40**), M82 appears even more disturbed. From its wrecked-looking centre, tangled filaments of glowing matter

stretch outwards to distances of 10 000 light years above and below the galaxy – equal to half the diameter of M82 itself. These filaments seem to be in violent outward motion. South of the galaxy (the bottom of Fig. 10.40), their light shows a Doppler shift towards the blue, indicating that they are approaching us at 100 kilometres per second, while at the North they show a redshift corresponding to a recession speed of 100 kilometres per second. These velocities are as seen along our line of sight; allowing for the inclination of M82 (we see it at about 10° from an edge-on position), the actual speeds are 1000 kilometres per second.

The final piece of evidence supporting an explosion was the fact that the filaments are strongly polarised. **Figs. 10.41** and **10.42** show the appearance of M82 through polarising filters arranged at right angles: the filaments are far more prominent in the right-hand photograph which transmits only light waves vibrating back and forth parallel to the galaxy. One of the most efficient ways in which polarised light is produced is by the synchrotron process. This produces very directional radiation from fast-moving

electrons spiralling around strong magnetic fields, and it is encountered only in extremely energetic sources – such as the Crab Nebula (Figs. 6.11 and 6.12).

However, there is more to M82 than meets the eye. More recent observations at wavelengths other than light reveal none of the activity normally associated with active galaxies. There are no jets, radio-emitting lobes or powerful central sources. In particular, a careful series of measurements of the filaments has shown that their hydrogen light is also polarised. Atoms of hydrogen have no free electrons, and so the polarisation cannot result from the synchrotron process. Instead, it is almost certainly produced by light from elsewhere (the galaxy's centre) being scattered off dust particles in the filaments – the same mechanism which polarises light from the daytime sky.

The current picture of M82 is of a small edge-on spiral galaxy which has recently ploughed into a huge cloud of gas and dust. The cloud was produced in a tidal interaction with M81, and it is clearly visible on the radio view of the whole cluster in Fig. 10.33. If it is assumed that the cloud is slowly drifting from M81 towards and past M82 at a speed of 100 kilometres per second, then the filaments' red and blueshifts can be explained without the need for the occurrence of an explosion. On the southern (lower) edge of the galaxy, the 'upward-moving' dust in the cloud reflects light from the 'approaching' galaxy's disc: consequently, that light is perceived by us as blueshifted. North of the galaxy, the receding dust redshifts the light it reflects. Even the spectacular filaments themselves are not the aftermath of a violent event. They are almost certainly an artefact of patchy illumination of the surrounding cloud by pockets of young stars in the dust-laden galaxy's disc.

So M82 has turned out to be not quite as exotic as its tangled appearance first suggested. It should be classified as a 'normal', if slightly disturbed galaxy. M82 is basically a spiral galaxy, with a mass of 6000 million Suns, and inclined so that its North (top) side lies closer to us. There is evidence for rotation, even though disturbed at the galaxy's centre, as seen in **Figs. 10.43** and **10.44**. These colour-coded maps, made at a wavelength of 2.6 millimetres, show the distribution of dense molecular clouds containing carbon monoxide molecules in the central 2000 light years of the galaxy. The 'star diagram' at the top left of Fig. 10.43 indicates the orientation and extent of the galaxy, while below it are 'cuts' showing the distribution of molecular clouds at different velocities in the disc, with the densest concentrations marked in red. At the top right is a velocity map (shown in more detail in Fig. 10.44), while the lower frames plot the distribution of clouds of all velocities throughout the galaxy. Fig. 10.44 is a combined intensity/velocity map of the carbon monoxide distribution in which the colours signify a velocity of approach (increasing blue) or recession (increasing red). The disturbances at the centre have produced a marked difference between the gas clouds in the North (top) which are approaching, and those in the South (bottom) which are receding; but further out the galaxy's rotation becomes apparent, with higher blueshifts toward the right and redshifts to the left.

But nowhere is M82's spiral structure more spectacularly revealed than in a colour-enhanced optical photograph (**Fig. 10.45** overleaf). This deep computer-processed view, taken through a series of colour filters, reveals the natural colours of each region of the galaxy and highlights low-contrast details. The bright red streak at the top right is an artefact. The spiral arms are particularly obvious on the left-hand side — a region which shows only faintly in short-exposure conventional photographs such as Figs. 10.39 and 10.40. But the dazzling yellow-white centre and the splash of red filaments — reflecting hydrogen light — dominate its appearance.

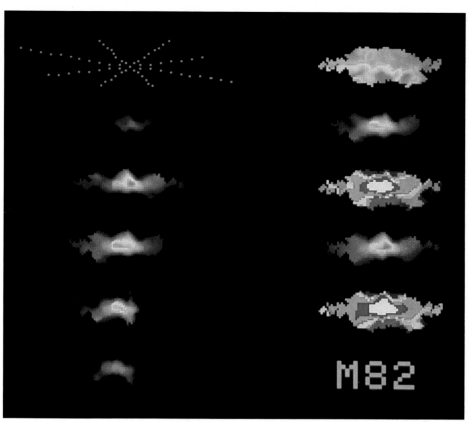

10.43 *Radio, 2.6 mm carbon monoxide line, 13 m telescope, Five College Radio Astronomy Observatory*

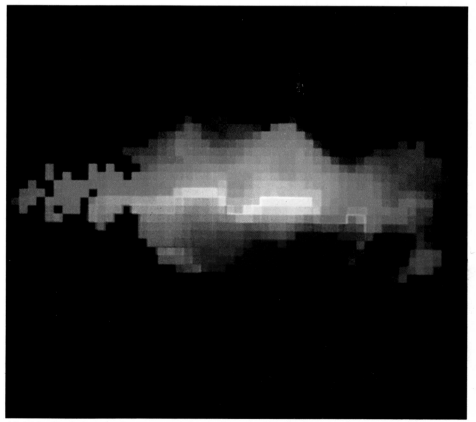

10.44 *Radio, 2.6 mm carbon monoxide line, velocity coded, 13 m telescope, Five College Radio Astronomy Observatory*

**10.45** *Optical, enhanced true colour, 5 m Hale Telescope*

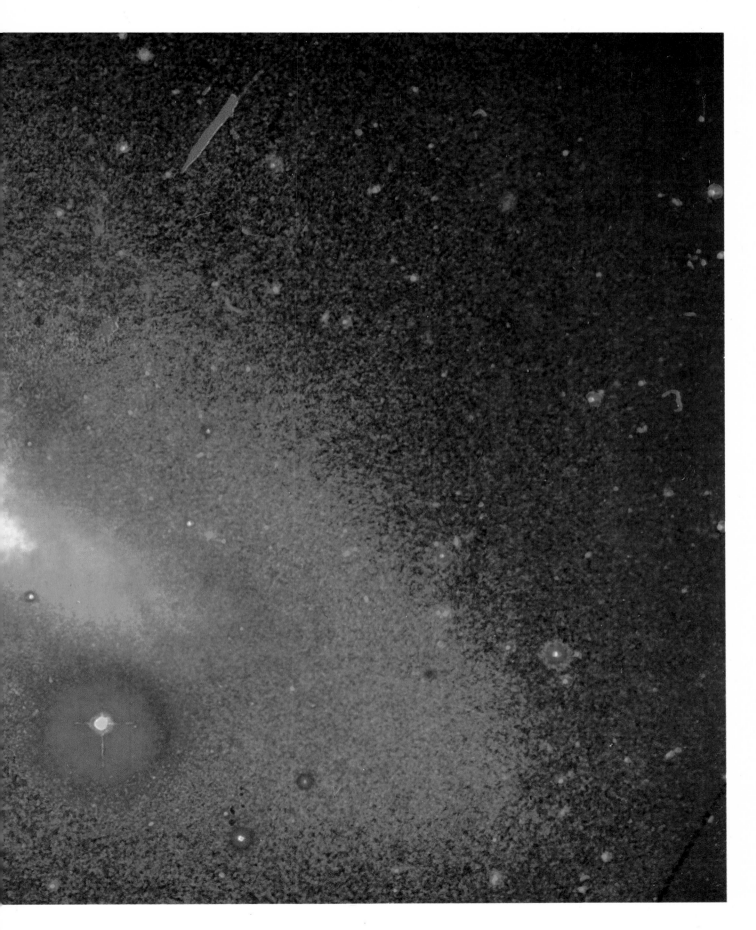

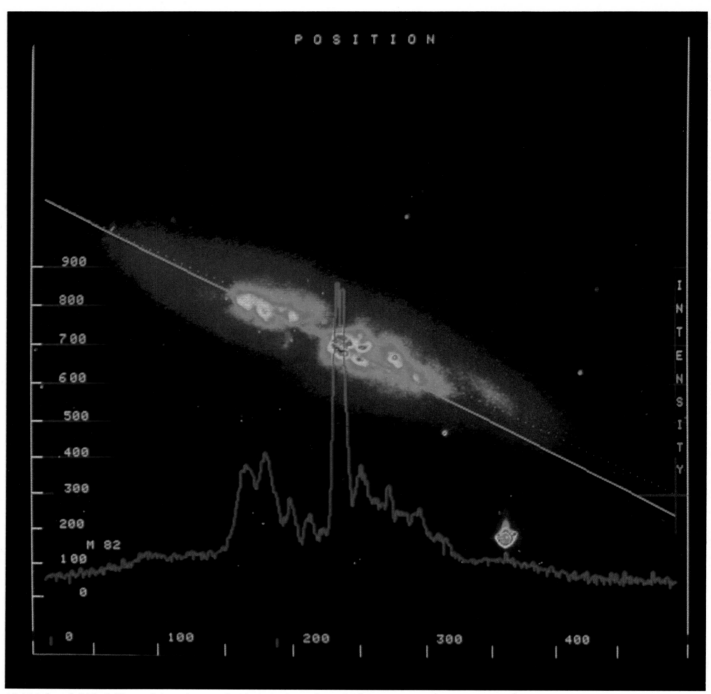

**10.46** *Infrared, 0.71–0.88 μm, 0.6 m telescope, Smithsonian Astrophysical Observatory*

Although M82 is not by official definition an 'active galaxy', it is certainly not inactive. Observations at wavelengths which penetrate the obscuring dust reveal that its central regions are undergoing an intense burst of star formation. **Fig. 10.46**, a 'photographic infrared' view, pinpoints the sites of this new activity. Here, the deep blue contour delineates the outer extent of the galaxy as seen in Fig. 10.39. The intensity of emission, colour coded from blue (faint) through green and yellow to red (bright), and represented on the superimposed graph below, reveals centres of star formation. Activity is concentrated in the central 2000 light years of the galaxy (the graph's peak) where heated clouds of dust cocoon thousands of millions of young stars.

Mapped at a radio wavelength of 3.7 centimetres, the innermost 2000 light years of M82 are shown colour coded in **Fig. 10.47**. This region corresponds to the innermost red zone in Fig. 10.46, and the contours coloured blue, green, yellow and red enclose regions of successively higher radio brightness. One of the strongest pieces of evidence against the idea that M82 is exploding is the fact that the most intense (red) source at radio wavelengths does not coincide with the galaxy's centre, as in, for example, M87 (Fig. 12.25). The centre itself is marked by a weak source at the top left. M82's most powerful radio source is instead another indication of its recent epidemic of star formation, for it is a giant gas cloud heated by young stars inside. All over the central region, there are other radio sources of this kind, and also some which generate radio waves by synchrotron emission. These are probably the remains of exploded, short-lived stars.

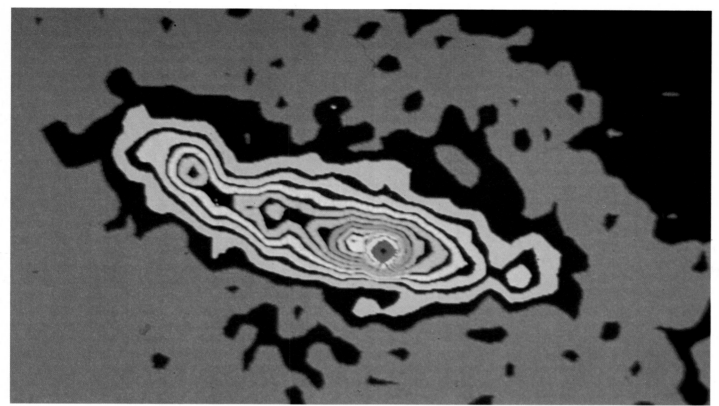

10.47 *Radio, 3.7 cm, National Radio Astronomy Observatory interferometer*

At X-ray wavelengths, the story is the same. **Fig. 10.48** shows a colour-coded intensity map made by the Einstein Observatory, in which the strength of emission increases from pale blue to white according to the colour scale on the lower right. The picture is to roughly twice the scale of Fig. 10.46, and here, too, a peak of emission is centred on the star-forming regions. The emission also extends southwards into the filaments facing M81. A great deal of this appears to be diffuse emission, not concentrated in point sources. One interpretation is that these X-rays are produced by the large number of supernova remnants that should exist in regions of intense starbirth and stardeath.

Most astronomers now believe that the present burst of star formation which we see in the centre of M82 was triggered by its gravitational interaction with M81. Judging from the lifetimes of the stars which have formed, the event happened about 100 million years ago. The interaction must have dumped clouds of matter onto M82's quiet centre. Rapid infall would have precipitated vigorous star formation, generating turbulence, powerful radiation and palls of tangled dust grains — the state in which we see this disturbed galaxy today. M82 may not be an exploding galaxy, but it is unique in our neighbourhood.

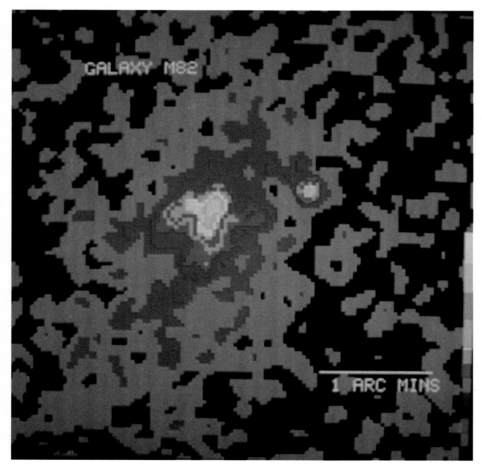

10.48 *X-ray, 0.4–8 nm, Einstein Observatory HRI*

# 11 X-ray & Gamma ray Astronomy

Just before midnight on 18 June 1962, a small Aerobee rocket blasted off from New Mexico under the light of a Moon just one day past full. The rocket soared on an arc which took it 225 kilometres high, well above the atmosphere, before it fell back to Earth. During its few minutes in space, an unusual payload, a trio of Geiger counters, was exposed to the sky.

The flight was supposed to detect X-rays coming from the Moon; but the Geiger counters – the first successful 'X-ray telescopes' – did not 'see' the Moon at all. Nor did they see the bright stars in the sky; nor the band of the Milky Way stretching across the summer heavens. Instead they saw a dim glow from the entire sky, and even more unexpected, they found a brilliant source of X-rays in the constellation Scorpius – in a position where there is no bright star or radio source.

This flight marked the beginning of X-ray astronomy. The Sun had already been observed in X-rays, but now, for the first time, astronomers could observe the depths of the Universe in the shortest wavelengths of radiation – the X-rays and gamma rays (Fig. 11.1). There is no hard and fast dividing line between the shortest ultraviolet rays and the longest X-rays. In fact, the extreme ultraviolet (EUV) waves described in Chapter 9 are something of a no-man's-land between ultraviolet and X-ray astronomy. Astronomers are now working their way into this region from the shorter wavelengths where X-ray astronomy began: they tend to call this disputed part of the spectrum 'X-ray–ultraviolet' (XUV) radiation. For convenience, the boundary between the EUV or XUV and X-rays proper is usually taken as being at 10 nanometres – a wavelength about a hundred times the size of an atom. Going down through the X-ray wavelengths, there is again no particular boundary between the shortest X-rays and the longest gamma rays, but astronomers tend to make the distinction at a wavelength of 0.01 nanometres (about one-tenth of an atom's diameter). X-rays are radiation

with wavelengths between 0.01 and 10 nanometres; the rest – right down to the shortest detectable wavelengths – are gamma rays.

We usually think of X-rays and gamma rays as extremely penetrating radiation; hospital X-ray machines produce radiation which can penetrate our soft tissues as if they were not there, and gamma rays are used in industry to scan metal sheets in search of cracks and faulty welds. But both kinds of radiation are eventually stopped as they travel through matter. The Earth's atmosphere contains quite enough gas to absorb X-rays and gamma rays from space well before they reach the ground (Fig. 1.2). The longest wavelength X-rays are stopped high in the outermost layers of the Earth's atmosphere, at heights of around 100 kilometres, almost half as high as the lowest satellite orbits. The shorter wavelengths are more penetrating. X-rays similar to those produced in hospital X-ray machines (about 0.02 nanometres wavelength) get down to an altitude of about 40 kilometres; but even the shortest of the gamma rays so far detected – with wavelengths of about 0.000 000 001 nanometres – can penetrate only as far down as 10 kilometres altitude, about the height of Mount Everest and only a quarter of the depth of the atmosphere.

So X-ray and gamma ray astronomers must fly their telescopes above the bulk of the atmosphere, on rockets or satellites (although high-altitude balloon flights will suffice for some gamma ray observations). As a result, X-ray astronomy did not begin at all until after the Second World War, which brought the first high-altitude rockets. The German V2 rocket was built as a military weapon, but after the war, captured V2s became a more valuable – and much more successful – weapon, in the scientist's struggle to understand the Universe. For the first time, meteorologists could study the Earth's upper atmosphere; and astronomers could detect radiations which are absorbed by the atmosphere.

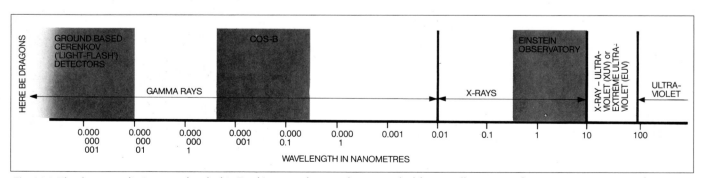

Fig. 11.1 *The shortest radiations are absorbed in Earth's atmosphere, so they are studied from satellites (except for extreme gamma rays, whose absorption causes detectable flashes of light). X-rays are separated from ultraviolet by the EUV (or XUV) wavelengths, which are absorbed in space by hydrogen and helium atoms (Fig. 9.1). The first X-ray images came from the Einstein Observatory. The shortest X-rays emitted by hot gas are about 0.01 nanometres; gamma rays are still shorter wavelengths, produced by high-speed subatomic particles. They cannot be focused, and current gamma ray views are crude, the best being from the COS-B satellite.*

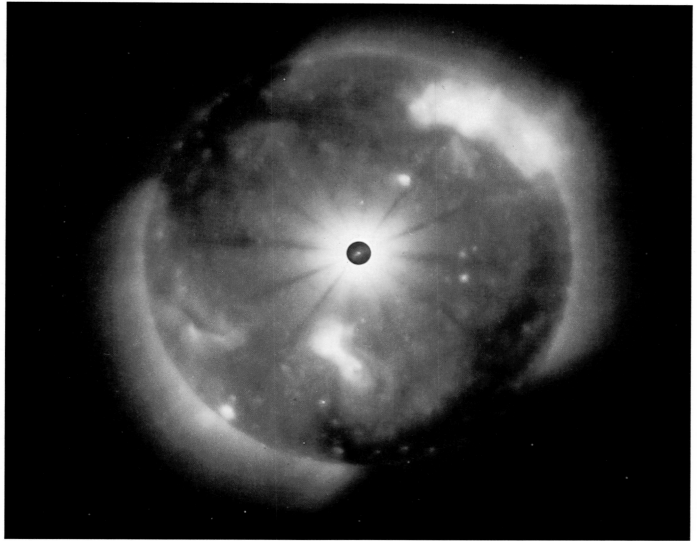

**Fig. 11.2** *X-rays from the Sun, the first detected X-ray source, come from regions of hot gas in its atmosphere, the corona. This Skylab picture was taken during a flare (centre) on 6 August, 1973. The flare's core is dimmed by a small dense filter: in X-rays it shines a thousand times brighter than the rest of the Sun.*

Optical observation had already convinced solar astronomers that the Sun's outer atmosphere, the corona, is extremely hot – at a temperature of one to two million degrees K. Such hot gas should emit X-rays. Geiger counters flown on a V2 in 1948 did indeed detect the predicted X-rays from the solar corona. The subsequent brief rocket flights in the 1950s were followed by satellite observatories which could watch the Sun fairly continuously, the most successful being the American series of Orbiting Solar Observatories. By far the best views of the X-ray Sun, however, have come from the manned Skylab solar observatory (described in Chapter 9). To X-ray detectors the Sun appears quite different. The round disc of the photosphere which is so blindingly bright to human eyes is not hot enough to emit X-rays. It appears as a black globe, surrounded by strands of the million-degree gases which make up the patchy corona (Fig. 11.2).

Despite the early successes of solar X-ray astronomy, most astronomers were pessimistic about detecting X-rays from more distant stars – and most did not even consider that this marked the beginning of a new branch of

astronomy, any more than visual observations of the Sun alone could be considered as 'optical astronomy'. The early rocket-borne Geiger counters could detect the Sun only because it is so nearby. If the Sun lay as far away as the nearest stars, it would appear so dim in X-rays that no detector of the 1950s or 1960s could have picked up its radiation, and astronomers at the time knew of nothing other than stars which could produce radiation of these wavelengths. A few bold pioneers, however, believed that unpredicted astronomical objects might exist which are dim at optical wavelengths but bright in X-rays. The decisive rocket flight of June 1962 was funded to investigate whether the Moon produced X-rays as a result of the impact of high-speed solar wind particles from the Sun. But the scientists involved were hoping for higher dividends: the discovery of some new, unknown powerful X-ray source far beyond the Solar System; and that is exactly what they found, in the brilliant X-ray source now called Scorpius X-1 (the brightest X-ray object in the constellation Scorpius).

This rocket flight was followed by others, which

uncovered a couple of dozen X-ray sources in the sky, and measured their approximate positions. But a detailed look at the X-ray sky had to await orbiting X-ray satellites, which could examine the sky at leisure. The first X-ray satellite, Uhuru (Fig. 11.3), was launched from Kenya on 12 December 1970, the anniversary of the country's independence day, and *Uhuru* is Swahili for 'freedom'. The American Uhuru and its British, American, Dutch (Fig. 11.4) and Japanese successors have located hundreds of X-ray sources in the sky. As with Scorpius X-1, such sources do not usually coincide with a bright star or galaxy or powerful radio source. But because satellites can measure their positions fairly precisely, X-ray sources have been matched up with faint optical or radio objects. As a result, astronomers have been able to discover just what kinds of object in the Universe can produce large amounts of X-rays while remaining comparatively dim at other wavelengths.

Since X-rays are an extremely energetic form of radiation, natural X-ray sources must be excessively powerful or violent types of object. Many are very hot

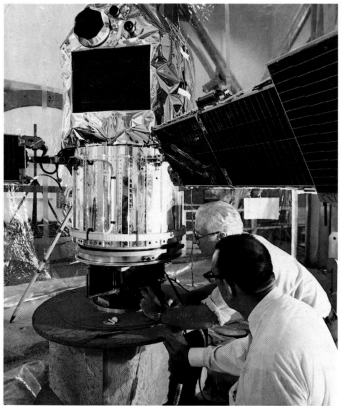

**Fig. 11.3** *The first X-ray astronomy satellite, Uhuru, scanned the sky with a proportional counter behind a collimator (large black rectangle), as it rotated. The optical star- and Sun-sensors (round apertures above) revealed the location of scans on the sky.*

clouds of gas. Just as gas at a temperature of a few thousand degrees (like the Sun's photosphere) emits the comparatively long wavelengths of visible light, while gas of from 10 000 to around 1 000 000 K produces ultraviolet and EUV radiation, so gas that is still hotter shines out in X-rays. The corona of the Sun is one example. Thanks to X-ray astronomy, we now know of many places in the Universe where gas is raised to multi-million degree temperatures. Where a compact star (a neutron star or black hole) is in orbit around another normal star, gases from the latter can be drawn towards the small star by its gravity. The gas swirls around the compact star in a disc where the individual gas streams spiral inwards as friction robs them of rotational energy. The friction also heats up the gas until its temperature is 100 million degrees or more. Scorpius X-1 is a double star system like this, where the gas shines ten thousand times brighter in X-rays than does the Sun at all wavelengths. Another type of X-ray source is the supernova remnant. When a star blows itself apart in a supernova explosion, the expanding gases can also reach multi-million degree temperatures. Much farther out in space, we find that clusters of galaxies are embedded in enormous tenuous pools of gas millions of light years across, at a temperature of some 100 million degrees.

X-rays can also come from regions filled with very energetic subatomic particles like electrons, which are either moving through a magnetic field or interacting with longer wavelength radiation. The hot gas streams falling from one star onto another compact star can generate fast electrons which add to the supply of X-rays. In the centres of quasars, gas streams are spiralling into massive black holes; the electrons convert some of their energy into X-rays before they disappear into the black hole.

The early satellites picked up the X-rays from space with a kind of detector which was a slightly more refined version of the simple Geiger counter, the *proportional counter*. It not only detects the presence of an X-ray, but measures its wavelength too. The proportional counter (and other types of X-ray and gamma ray detector) makes use of the fact that electromagnetic radiation is not just a series of waves, but also comes in distinct packets of energy, called *photons*. The shorter the wavelength, the higher the energy of each individual photon. So each X-ray photon is much more energetic than a photon of ultraviolet or visible light; and within the X-ray region of the spectrum the photon energy increases as we look at shorter wavelengths. (X-ray astronomers, in fact, generally refer to X-rays by energy rather than by wavelength.)

The proportional counter is a 'chamber' filled with gas – often argon – and containing two wire grids or electrodes, which are kept at a high voltage relative to each another. When an X-ray photon enters the chamber, it collides with several atoms of argon and knocks electrons out of them.

**Fig. 11.4** *The Astronomical Netherlands Satellite (ANS) is here mated with its Scout launcher in August 1974. One of several small X-ray satellites launched in the mid-1970s, ANS made the important discovery of X-ray bursters (Fig. 11.5).*

The shorter the wavelength of the X-ray the more electrons it produces. An X-ray photon of two nanometres wavelength would typically knock out some fifteen electrons initially, while for a photon with a wavelength of only one nanometre the number would be around thirty. These negatively charged electrons are repelled from the negatively charged grid, and attracted to the positively charged grid. As they speed through the gas, they knock out additional electrons from the gas atoms, so that an initial bunch of 30 electrons can grow to about 300 000 by the time it hits the positive electrode. This avalanche of electrons constitutes a small electric current, which is picked up and amplified by sensitive electronics attached

to the electrode. Whenever an X-ray photon hits the detector, there is a quick burst of electric current, just like the burst which is heard as a 'click' in the uranium prospector's Geiger counter. But in the proportional counter, the strength of the current depends on the number of electrons liberated by the photon, and so it gives a measure of the photon's energy – and hence its wavelength.

The early satellites did not have telescopes to provide a focused view of X-ray sources. Their detectors were open to the sky, but set behind a *collimator*, an array of metal bars or slats, often arranged like a honeycomb, which blinkered their view to a small region of the sky around 1°

in size – about twice the apparent size of the Moon. They could survey the sky by sweeping systematically across it, determining the directions from which intense X-radiation emanated. They could also look at any individual source, with the collimator cutting out interference from sources lying in other positions in the sky.

By pointing at a particular object, astronomers could study its X-ray spectrum, counting the number of photons of each wavelength picked up by the proportional counter. They could also study how its X-ray brightness varied with time. Most of the brightest X-ray sources in the sky are double star systems, and these are continuously flickering in brightness as the individual gas streams swirl and tumble. If the compact star of a double star system is a neutron star its magnetic field channels the gas down to its poles, so the star's radiation causes apparent X-ray pulses (an X-ray pulsar). Gas may build up on the surface until it explodes as an *X-ray burster* (Fig. 11.5).

Despite these discoveries about X-ray sources, astronomers were handicapped in not having a telescope which could show their structure – particularly when it came to the extended gas clouds like the supernova remnants and the gas pools around clusters of galaxies. Unfortunately, it is very difficult to focus X-rays. No substance can act as a lens to bend X-rays to a focus without absorbing them on the way. Curved bowl-shaped mirrors, of the kind used in optical reflectors and in radio telescopes, are of little use either. An X-ray hitting such a mirror is absorbed rather than reflected.

X-rays are only reflected if they hit a metal surface at a very shallow angle, just grazing it. German physicist Hans Wolter pioneered the design of such *grazing-incidence* X-ray mirrors in the 1950s, when he tried to build an X-ray microscope. His microscope came to naught, but the mirror designs were taken up by Riccardo Giacconi, a pioneer X-ray astronomer involved in the discovery of Scorpius X-1, and they have become the basis of today's X-ray telescopes. The Wolter Type I design for a focusing X-ray mirror looks nothing like a conventional mirror. It is the polished inside of a slightly tapering metal cylinder. The front half of the cylinder is polished to the shape of a paraboloid; the back half to a rather more steeply tapering hyperboloid. An X-ray entering the front end of the cylinder, near to the one edge, strikes the tapering inside surface at a grazing angle, and is reflected so that it travels towards the central axis. After this first reflection, the X-ray hits the more steeply tapering hyperboloidal rear half of the cylinder. Again it is reflected, and travels at a steeper angle towards the central axis of the cylinder. It eventually crosses the axis at a focal point well beyond the cylinder itself. The combination of these two reflections means that the Wolter Type I mirror forms a perfect image of the X-ray source at the focal point.

Such a grazing-incidence mirror must be used to form an image of an X-ray source, but Wolter telescopes have a major disadvantage: they only collect a small amount of the radiation falling on them. Only X-rays falling near the outside of the cylinder's cross-section actually strike its first section, and these comprise only a fraction of the total radiation falling on the front end of the telescope. In fact, an opaque stop must be mounted in the centre of the tube to block off the majority of the X-rays, which would otherwise travel straight down and 'wash out' the image in the focal plane. The answer, in practice, is to mount a second telescope inside the tube of the first to intercept and focus radiation falling nearer the centre. This smaller, 'nested' telescope is constructed so that it focuses X-rays to exactly the same focal point as the first, and so adds its X-ray image in exact register to the image formed by the larger cylinder.

Giacconi's first X-ray telescope formed part of the solar observatory – the Apollo Telescope Mount – on the Skylab space station (Fig. 9.10). The focused X-ray images were recorded on film which was returned to Earth by the Skylab astronauts, and the exquisite views of the Sun's corona proved that an X-ray telescope could produce pictures with detail as fine as that from optical telescopes.

In order to look at the faint, distant sources beyond the Solar System, X-ray telescopes must be much larger and

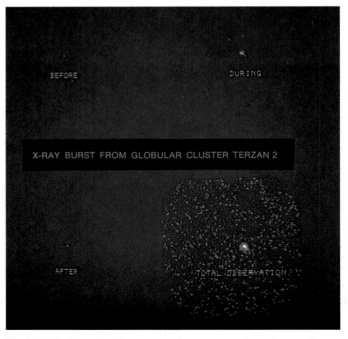

**Fig. 11.5** *The Einstein Observatory witnesses an outburst from an X-ray burster (the nuclear explosion of gas that has fallen onto a neutron star). The burster lies in the Terzan 2 cluster of stars, 2000 light years away. The infalling gas normally produces a faint X-ray source (top left); suddenly the explosion produces a burst of X-rays a thousand times stronger (top right), but after 20 seconds it has faded (bottom left).*

heavier than the simple proportional counters mounted on the early generation of X-ray satellites. The chance to fly one came with the series of three huge High Energy Astrophysics Observatories (HEAOs) in the late 1970s. These satellites (Figs. 11.6 and 11.10) each weighed some 3 tonnes, and were 5.8 metres long -- a HEAO satellite standing on end would reach as high as a house. The second HEAO was built around an imaging X-ray telescope. This epoch-making satellite was renamed the Einstein Observatory, to honour Albert Einstein whose centenary was in 1979.

The Einstein Observatory's telescope (Fig. 11.8) consisted of four Wolter Type I mirrors nested inside one another, with the outermost mirror 58 centimetres across. The concentric mirrors (Fig. 11.11) all brought the X-rays to the same focal point, near the far end of the satellite, 3.4 metres beyond their mid-point. Even though the four mirrors could focus many more X-rays than a single Wolter mirror, the whole nested assembly only intercepted a few per cent of the X-rays falling within the rim of the outermost mirror – the price that X-ray astronomers must pay for getting a focused image at all.

To make the best use of the Einstein Observatory's X-ray telescope, Giacconi's team provided it with four different kinds of X-ray detector. They were mounted on a

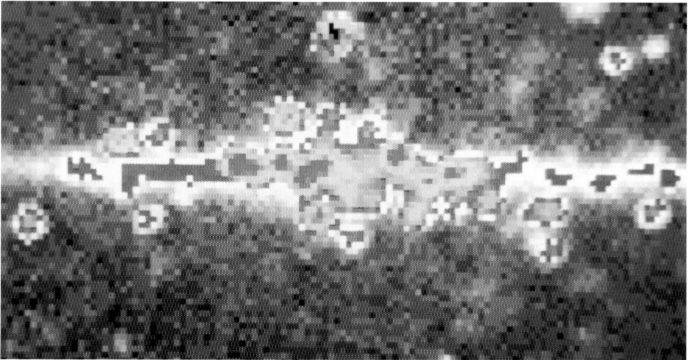

Fig. 11.7 *HEAO-1 revealed many bright sources towards our Galaxy's centre (in the middle here): the region mapped extends one-third of the way around the sky (left to right) along the band of the Milky Way (compare with Figs. 8.6 and 8.2). Colour coding for intensity portrays the dimmest regions blue, and the brightest red and black. The brightest source (above centre) is Scorpius X-1.*

turntable which rotated to bring any one of them into the position of the focused X-ray image. The Einstein Observatory was thus as flexible as a ground-based optical observatory, where astronomers regularly change the light-detectors or spectroscopes according to the kind of observation they want to make. Two of Einstein's detectors were spectrometers, to investigate the amount of X-radiation of different wavelengths being emitted by an X-ray source. The other two 'photographed' the appearance of the source as 'seen' through the telescope. Because the Einstein Observatory was unmanned, it was not possible to follow Skylab's precedent and photograph directly onto film which could be returned to Earth by astronauts. Instead, instruments had to be devised that could measure the X-ray images electronically, so that they could be radioed back to Earth.

The Imaging Proportional Counter (IPC) worked in essentially the same way as the proportional counters used in early satellites, except that it measured not only the energy (wavelength) of the incoming X-ray photon, but also the position at which it hit the detector, thus building up a picture of the source. The IPC could see a region of sky 1° (60 arcminutes) square, and it could measure the positions of incoming photons to an accuracy of about one-fiftieth of its field of view. The pictures were thus blurred to a resolution of about an arcminute, but they provided useful information on the wavelengths of X-ray photons falling at each point. In fact, the Einstein IPC was very similar to the human eye in its resolution of detail and in its ability to see different wavelengths – 'colours' of radiation – albeit at wavelengths a thousand times shorter.

The IPC had a simple way of measuring the position of incoming photons – or, rather, the bursts of electrons they produced. Beneath its wide front plastic window were three grids of wire, one below the other. The middle grid was positively charged, the upper and lower grids were

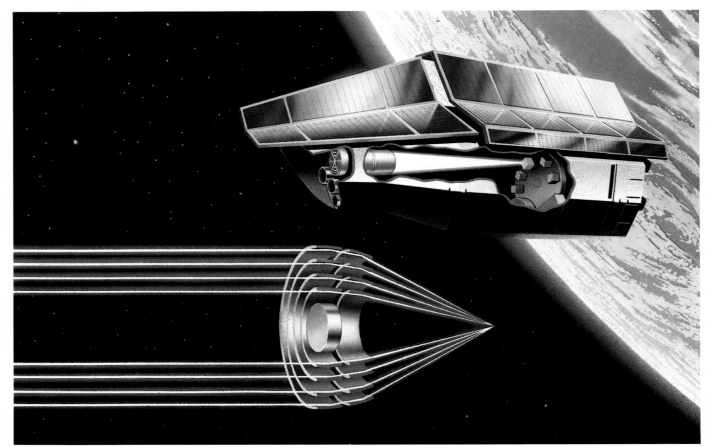

**Fig. 11.8** *Cutaway view of the Einstein Observatory shows the mirror assembly (blue cylinder on the left) focusing X-rays onto a detector (orange cube). The inset (shortened left to right) demonstrates how the X-rays are focused by reflection at shallow angles from two successive tapering cylinders; the four nested mirrors have the same focus. The turntable (green) can move any of the detectors (cubes) to this focus. The tubes below the X-ray mirrors are optical star sensors; the solar panels (top) provide power.*

negative. When an X-ray photon passed through the window into the gas-filled IPC, it created a burst of electrons which headed for the nearest point of the central grid. On its arrival, the pulse of electric charge produced an electric current there, and also induced pulses of current in the upper and lower grids. These grids were each formed of a single wire, strung backwards and forwards across the chamber. The induced current from an electron cascade ran both ways along – say – the upper wire, and was picked up at both ends. If the photon had entered the chamber exactly in the centre, the pulses would arrive at both ends of the wire simultaneously; otherwise, the current would reach one end before the other. From the time delay between the pulses of current reaching the two ends of the wire, the electronic circuits could register how far along the wire from the centre the pulse had originated. Knowing the layout of the wire in the grid, it was possible to work out where in the IPC the photon had entered.

The IPC was a very sensitive detector: when the turntable moved it into position behind the Einstein telescope it could see X-ray sources a thousand times fainter than any detected by previous X-ray satellites. But it could not discriminate the finest details in the image. The crisp image at the telescope's focus was slightly blurred by the IPC's method of operation. To complement the IPC, the Einstein Observatory carried a second detector. This High Resolution Imager (HRI) was five times less sensitive than the IPC, covered a smaller region of the sky (25 arcminutes) and could give no information on the wavelengths of the incoming radiation. But, as its name indicates, it could resolve very fine details of X-ray sources. The HRI was designed to show smaller details than the telescope mirror could actually focus, so that any 'blur' in the HRI images was entirely due to the limited resolution of the Einstein telescope itself. Giacconi's team built the X-ray mirrors to give a resolution of about 2 arcseconds.

**Fig. 11.9** *The Einstein Observatory's programme of observations was controlled from the Harvard-Smithsonian Center for Astrophysics, at Cambridge, Massachusetts. Astronomers used a computer to manipulate the data – for example, to produce coloured images, either in a single colour, or colour coded for intensity. Here, the computer has been programmed to separate the X-rays received when the Crab pulsar is 'on' (left of screen) from those picked up when it is 'off' (right) – the resulting images appear in Fig. 6.18.*

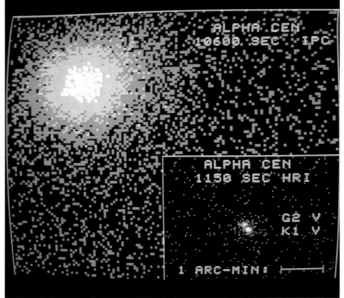

**Figs. 11.10–11.12** *The Einstein Observatory is seen with solar panels wrapped around in launch configuration (**Fig. 11.10**). Its length was necessary to accommodate the 3.44 metre focal length of the mirrors. Each of the mirror tubes (**Fig. 11.11**) was made of quartz, and took a year to construct before being shaped with a diamond point, polished and coated inside with nickel to reflect the X-rays. Observations of the nearest star system, Alpha Centauri (**Fig. 11.12**), reveal the comparative resolution of the two detectors. The more sensitive IPC blurs the image to a size of 1¹/₂ arcminutes; the HRI (inset) resolves it into two stars (labelled by their optical spectral types).*

The HRI images were some thirty times sharper than the IPC pictures, and showed details as fine as can most ground-based optical telescopes (Fig. 11.12).

No ordinary X-ray detector could resolve details as small as this. The HRI was, in fact, a development of the microchannel plate used as an image intensifier in ultraviolet astronomy (Fig. 9.5). It consisted of two microchannel plates, one above the other. The tiny tubes making up each plate were individually little more than a hundredth of a millimetre in diameter, and were about a millimetre long. An X-ray photon striking the metal coating on the front of the first plate knocked out electrons which were amplified by the two successive plates to produce a shower of ten million electrons from the back of the second plate – at a position corresponding exactly to the position of the original photon in the X-ray image. The HRI detected the position of this cascade of electrons with two very fine grids of wires, one running at right angles to

the other. Electronic detectors monitored the individual wires in the grids, and when an electron cascade ran into the grids the electronic assembly could tell exactly where it occurred by identifying the particular wires which picked up the pulse of current.

With the Einstein Observatory's HRI, X-ray astronomy leapt forward to the level of the much older fields of optical and radio astronomy. In optical astronomy, over three centuries elapsed between Galileo's first crude 'optick tube' telescope and the sensitive, precision instruments of the twentieth century. Radio astronomers covered the same ground in about forty years, from Jansky's first detection of cosmic radio waves in the 1930s to the Cambridge and Westerbork synthesis telescopes at around 1970, which first rivalled the optical telescopes in resolution and sensitivity. But in X-ray astronomy, progress was astonishingly rapid. The comparable improvement took only sixteen years, from the detection of Scorpius X-1 in 1962 to the launch of the Einstein Observatory in 1978.

The Einstein Observatory's instruments could detect hundreds of ordinary stars by the X-rays from their coronas. Even though it could not make out details in the stars' coronas, the Observatory surprised astronomers by showing that some very faint dwarf stars – like the nearest star, Proxima Centauri – emit X-rays as powerfully as the Sun's corona. As a result, astronomers are now having to rewrite their theories of how star coronas (including the Sun's) remain at multi-million degree temperatures. The Einstein Observatory has also picked out the very young and energetic massive stars, which have recently been formed in nebulae like the Orion Nebula (Fig. 4.15) and the Carina nebula (Fig. 4.32). At the other end of a star's life, it has perceived, for the first time, details in the hot gases of supernova remnants (Figs. 11.13–11.14), the debris of past star explosions. Its discoveries include a couple of dozen supernova remnants in our nearest neighbouring galaxy, the Large Magellanic Cloud.

Looking to further galaxies, the Einstein Observatory saw a view very different from the familiar optical photographs. We are used to seeing the Andromeda Galaxy as a spiral shaped, uniformly glowing expanse of stars too faint to be resolved individually. The Einstein Observatory saw the Andromeda Galaxy as a loose cluster of about a hundred individual bright spots (Fig. 10.13). These are the immensely powerful X-ray sources caused by gas streams falling from a star onto a compact companion star, like Scorpius X-1 in our own Galaxy.

Farther out in space we come across active galaxies – Seyfert galaxies, radio galaxies and quasars – and in these the Einstein Observatory's telescope has picked up powerful X-ray emission from the very centres (Fig. 11.15). In some galaxies, the X-ray output fluctuates dramatically in a matter of a few hours, indicating that the gas clouds producing the radiation are very small. This has strengthened most astronomers' conviction that a quasar consists of gas clouds held tightly around a central supermassive black hole. The Observatory has found that quasars are such prolific X-ray generators that the 'background' of X-rays in the Universe, discovered on the same rocket flight as Scorpius X-1 in 1962, may simply be the radiation from a huge number of very distant ones.

Finally, the Einstein Observatory has resolved details of vast clouds of hot intergalactic gas in distant clusters of galaxies (Fig. 11.16). In some clusters the X-rays come from hot gas surrounding individual galaxies; in others the gas forms a single immense 'pool' in which all the galaxies are immersed. Clusters of galaxies are the largest 'building blocks' of which the Universe is made. While an optical photograph shows merely a confusing scatter of individual

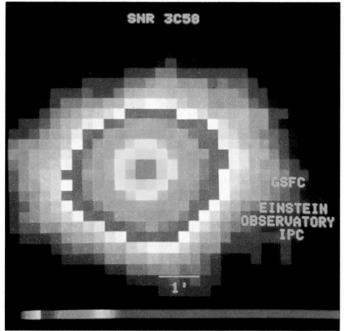

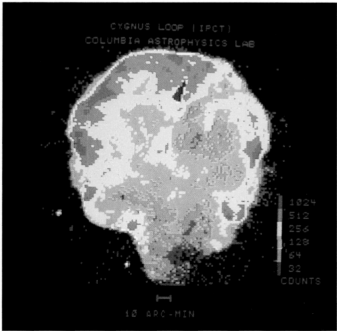

**Figs. 11.13–11.14** *First detailed views of supernova remnants came from the Einstein Observatory. 3C 58 (**Fig. 11.13**) is a relatively young remnant, the remains of a supernova seen in AD 1181. (Its radio emission is shown in Figs. 7.4 and 7.5.) The colours indicate intensity, from blue for pale regions to red for the brightest regions; the squares result from the method of analysis. 3C 58's bright centre suggests it is similar to the Crab Nebula (Fig. 6.8). The Cygnus Loop (**Fig. 11.14**), shown to a smaller scale, is a much older remnant, with an age of 20 000 years. It is so large that several IPC views were needed to cover it (a region missed out appears as the dark triangle near the top); colour coding is for intensity, as shown in the scale on the right.*

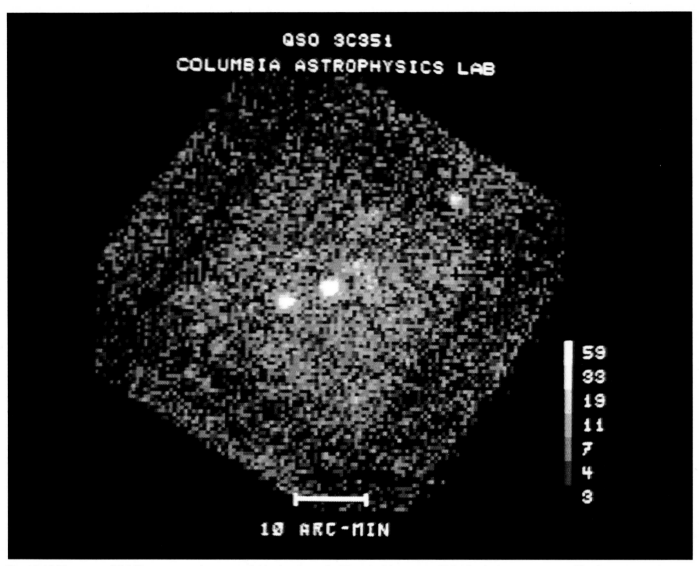

Fig. 11.15 *The quasar 3C 351 appears at the centre of this view from the Einstein Observatory IPC (the dark square is caused by the IPC's window supports). At least a dozen other sources appear here, mainly other distant galaxies and quasars.*

galaxies, the X-ray view of the enveloping gas pool reveals at a glance the essential unity of the cluster.

The Einstein Observatory failed in April 1981, after a two-and-a-half year life in which it revolutionised X-ray astronomy. Its immediate successor is Europe's Exosat (Fig. 11.17), which carries similar detectors. The German Rosat, named after Wilhelm Röntgen, the discoverer of X-rays, is due for launch in 1987. It will carry a larger telescope than Einstein and will be able to detect sources three times fainter. But the Americans have bolder plans for X-ray astronomy in the 1990s. Their Advanced X-ray Astrophysics Facility (AXAF) is costed at $500 million (1981 prices), five times more than the German Rosat. AXAF will see X-ray sources ten times fainter than the limit of the Einstein Observatory, and will pick out detail five times finer – with a resolution of around half an arcsecond. It should operate for at least ten years, with astronauts visiting the satellite in orbit to service it when necessary.

The grazing-incidence telescopes of the Einstein Observatory and its successors can only cope with X-ray wavelengths down to about 0.1 nanometres. The shortest wavelength X-rays and the gamma rays need new kinds of telescopes – and new kinds of detectors. The ultrashort waves can penetrate right through the gas of a proportional counter, or the front face of an HRI-type detector, without hitting any of the atoms – and so they are not detected at all.

The most versatile detector for these wavelengths works in exactly the same way as the luminous paint used in watch dials, which converts energetic radiation into light. The *scintillation detector* is basically just a large crystal of sodium iodide – a substance very like ordinary rock salt. It is surrounded by photomultiplier tubes, standard detectors of faint light (Chapter 3). When a gamma ray penetrates the sodium iodide crystal, it collides with the atoms and loses its energy in the form of ordinary light. The flash of light is picked up by the photomultiplier tubes. A shorter wavelength – higher energy – gamma ray generates a brighter flash of light. By measuring the brightness of the flash, astronomers can thus deduce the wavelength of the original gamma ray; and if enough gamma rays are

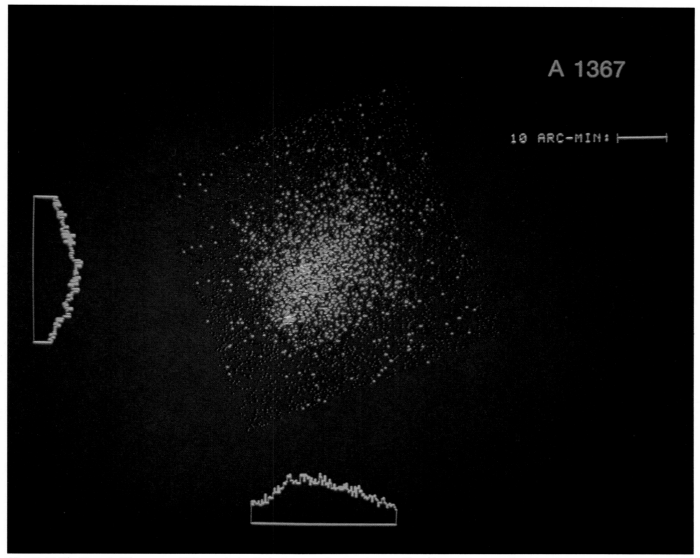

**Fig. 11.16** *The Einstein Observatory IPC reveals details (as image and as intensity profiles) of X-ray emission from hot gas within the Abell 1367 cluster of galaxies; the bright point at lower left is gas concentrated around a radio galaxy, 3C 264.*

collected from a source they can determine its spectrum. Early gamma ray satellite data in the 1960s suggested that the gamma rays were coming mainly from the band of the Milky Way. They could not, however, 'photograph' the gamma ray sky. The breakthrough came with the American SAS-2, which operated for only seven months after its launch in 1972, and with the highly successful European satellite COS-B (1975–82). These satellites carried the first real gamma ray telescopes.

The main problem with such a telescope is that gamma rays cannot be focused by reflection – basically because they are far smaller than the atoms making up any mirror. Astronomers can get by without any actual focused image, however, if, instead, they know the precise direction from which each gamma ray photon has come. They can then plot the position of each ray's origin, and so build up a picture of the gamma ray sky. The telescopes on SAS-2 and COS-B both operated in this way.

The COS-B (Fig. 11.18) gamma ray telescope looked less like an astronomical instrument than one of the detectors designed to catch energetic subatomic fragments

**Fig. 11.17** *The European successor to the Einstein Observatory, Exosat, carries several detectors. The four square boxes contain non-focusing proportional counters, and a gas scintillation counter (not seen here) observes short wavelengths. The metal circles cover (for prelaunch tests) the ends of two X-ray telescopes. Each has two nested mirrors, and two different detectors, similar to Einstein's IPC and HRI.*

Figs. 11.18–11.19 *The gamma ray satellite COS-B mounted within the rocket 'shroud' before launch in 1975 (*Fig. 11.18*); the slowly spinning satellite is powered by solar cells (blue).* Fig. 11.19 *shows COS-B maps of the Milky Way, at successively shorter wavelengths: 0.000 008– 0.000 018 nanometres (top), 0.000 004–0.000 008 nanometres (middle) and 0.000 000 2–0.000 004 nanometres (bottom). In each, the Milky Way runs horizontally, and the Galaxy's centre is in the middle (compare with Fig. 8.4). The intensity is contoured, at lowest levels by grey, then white lines on black, and black lines on white for the brightest regions.*

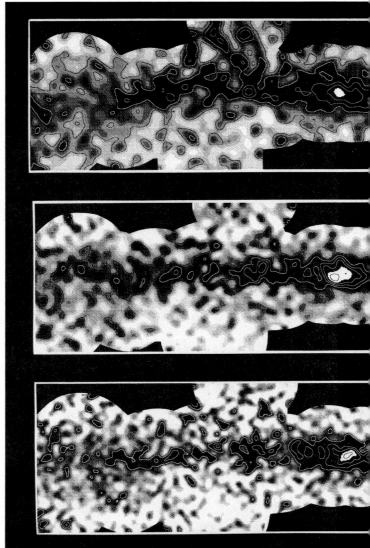

from the particle accelerators used to study the structure of matter. In a sense, that is what it was, except that the 'particle' here (the gamma ray) came not from an artificial source but from a natural 'particle accelerator' far off in space. The COS-B telescope consisted of a stack of flat *spark chambers*, interleaved with thin tungsten sheets. The spark chambers were similar in structure to the Imaging Proportional Counter on Einstein, except that they were sensitive not to radiation but to electrons and positrons (the antimatter equivalent of electrons). A gamma ray photon with a wavelength shorter than 0.001 nanometres can convert its energy into the formation of an electron and a positron, and in the process the gamma ray itself disappears. This is most likely to happen when the gamma ray is passing close to the nucleus of a heavy atom like tungsten. A gamma ray from space which entered the COS-B telescope penetrated down through the stack of spark chambers and tungsten sheets until it happened to pass too close to a tungsten nucleus. The gamma ray photon then changed into an electron and positron. These carried on travelling downwards on diverging paths, and as they passed through the stack, the spark chambers recorded their passage. At the bottom of the telescope, a

scintillation detector stopped the electron and positron, and measured their energies.

Thus the energy of the original gamma ray was known, because it was equal to the energy that the electron and the positron deposited in the scintillator. Also, the tracks of the electron and the positron were revealed by the successive spark chambers. The original gamma ray photon must have followed a path almost exactly half way between the two. Hence, it was possible to work out whereabouts in the sky that a particular gamma ray originated.

The COS-B telescope had extremely poor resolution by the standards of the telescopes of any other branch of astronomy – about 2°, or four times the apparent size of the Moon – but its pioneer efforts produced the first maps of the gamma ray sky. To gamma ray detectors, the sky is dominated night and day by the brilliant band of the Milky Way (Fig. 11.19); the Sun is totally invisible at gamma ray wavelengths except during a solar flare. The gamma rays from the Milky Way are generated when cosmic rays – high-speed, electrically charged particles from supernova explosions – plough into gas atoms in space. Where the gas is densest, the gamma ray sky should

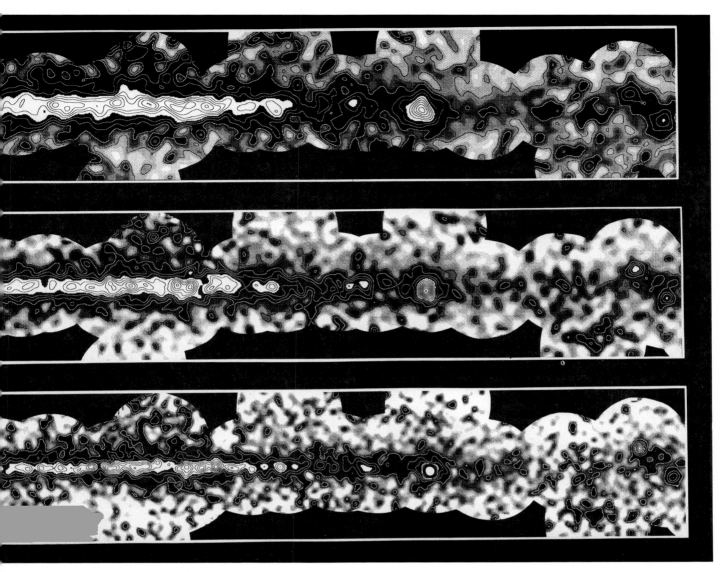

be brightest, and, indeed, COS-B has found bright patches corresponding to dense star-forming nebulae like the Orion Nebula (Fig. 4.6).

However, two of the most brilliant 'gamma ray stars' are not nebulous patches but regularly flashing points: the Crab and Vela pulsars. The Vela Pulsar is, in fact, the brightest star in the gamma ray sky. The COS-B satellite has found another two dozen gamma ray stars lying within the Milky Way band, but most have yet to be identified as radio, X-ray or optical stars.

Gamma ray astronomers are faced with one frustration that no other astronomers have to cope with. Their radiation comes in individual packets of exceptionally high energy; but these packets arrive very rarely. The total amount of energy over a long period of time might amount to about the same as that received by an optical telescope, but this energy arrives in a very different way. When COS-B looked at a bright gamma ray source like the Crab Pulsar, it picked up one or two photons every hour. When it looked at a really faint gamma ray source, the satellite was lucky to pick up one photon a day.

This 'hit rate' becomes even slower as we move to shorter wavelength photons, with their higher energies and

still rarer appearance from space. To detect them at all, we need a huge gamma ray detector, and some astronomers have used the biggest scintillation detector we have to hand – the Earth's atmosphere. When a gamma ray is absorbed in the air, it produces a flash of light, and the flash from the highest energy gamma rays can be picked up by an ordinary optical telescope equipped with an electronic light detector. The telescope needs a large mirror to see the faint flashes, but this mirror does not need to be accurately made. Using such telescopes, flashes that were produced by gamma rays with wavelengths of only 0.000 000 001 nanometres – the shortest radiation so far detected from space – have been found.

No gas cloud can be hot enough to generate such short wavelength radiation in the way that stars produce light and cool dust clouds shine in the infrared. Gamma rays come indirectly from the extremely fast subatomic particles generated in the most violent spots in the Universe: supernova explosions, the powerful magnetic fields around pulsars, the gravitational wells around huge black holes in the centres of quasars and radio galaxies – and perhaps from other, unknown regions of cosmic activity which gamma ray detectors may eventually reveal to us.

# 12 Active Galaxies

**12.1** *3C 273, 0.3–2.5 nm, Einstein Observatory IPC*

Not all galaxies are the relatively placid ensembles of stars and interstellar matter that we met in Chapter 10. A few per cent have violent activity occurring in their central cores.

Astronomers have discovered the activity of different types of galaxy in various ways, and so a variety of names have come into use to describe the types. *Radio galaxies* look like ordinary giant elliptical galaxies, but radio telescopes show that they are producing a tremendous amount of long wavelength emission – from a thousand to a million times the radio power of the Milky Way Galaxy. This radiation generally does not come from between the stars of the galaxy – as in the case of the weaker emission from normal galaxies – but from regions outside the galaxy, where optical telescopes see only empty space. There is usually a pair of these radio-emitting lobes, and in some cases they span millions of light years of space (Fig. 1.1).

When first discovered the emission from a radio galaxy's lobes was a great puzzle. Later observations have revealed in many cases a small radio source at the galaxy's core, and – leading out from it to one of the lobes – a narow 'jet'. It seems that the fundamental release of energy occurs in the radio galaxy's core, and it heads out to the lobes as a jet of fast electrons.

Radio galaxies are thus, despite first appearances, examples of galaxies with active cores. They are now grouped as 'active galaxies' along with other types

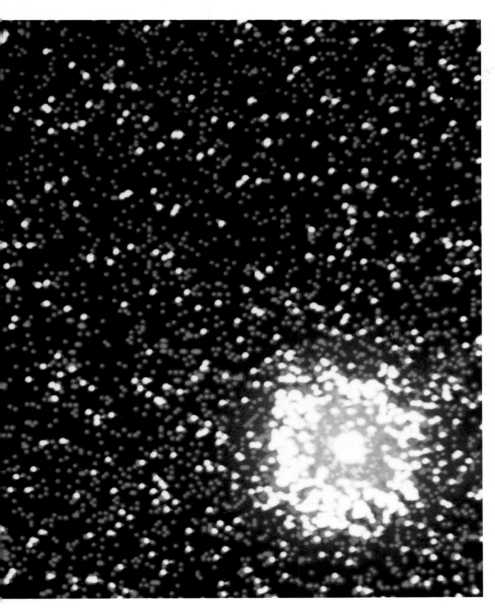

quasar, 3C 273 (see also Figs. 12.33–12.38). 3C 273 itself is the bright source at the lower right (the 'halo' is an artefact caused by its X-ray brilliance) and the scatter of dots over the picture is the general background of X-rays in the Universe. The concentration of dots at the top left has turned out to be another, much more distant quasar, unknown before this picture was taken.

Quasars are not only very luminous; they are extremely small, packing the power of up to a hundred galaxies into a volume no larger than the Solar System. Most astronomers explain the release of such a concentration of power as the effect of a disc of hot gas which is circling a very massive black hole right at the galaxy's heart. Friction within the gas disc makes the gas lose orbital energy and it gradually spirals down into the black hole. The friction heats up the gas disc. Various processes can transform this heat energy into radiation of all wavelengths, and into the energy carried away by electron jets.

The gas to 'fuel' the disc can come from any of several sources. In different galaxies and quasars, astronomers have attributed it to gas ejected by stars in the surrounding galaxy, cool gas condensing from the very hot and tenuous gas filling clusters of galaxies, and gas stripped from another galaxy passing close to the active galaxy.

Optical astronomy can reveal a lot about the gases around the centre of a quasar, particularly through the analysis of spectra, but other wavelengths can add much more information. Radio telescopes show how some of the power is channelled outwards as electron jets, and the extremely fine resolution offered by the technique of Very Long Baseline Interferometry can reveal details in quasar cores which are a thousand times smaller than can be seen at any other wavelength. Ultraviolet radiation comes from hotter gas, closer to the black hole, than the gas which produces the visible light; and the shortest wavelengths, X-rays and gamma rays, come from the smallest, innermost regions of all. Although we cannot obtain pictures showing directly fine details of these regions, future X-ray studies of quasars' fluctuations in brightness, their spectra and their polarisation should tell us much more about some of the most energetic matter in the Universe – gas at the inner edge of the hot disc, about to disappear for ever down the gravitational well of the black hole.

whose central activity is more obvious because their cores shine brilliantly at optical wavelengths. The first of these to be found were *Seyfert galaxies* and *N-galaxies*. But the most extreme type is the *quasar*. In quasars, the tiny central core can outshine the rest of the galaxy ten or a hundred times over. The centres of all other types of active galaxy are 'quasar cores', scaled down in activity.

Because the quasars are so luminous, they are the most distant objects we can

detect. They emit not only light, and sometimes radio waves, but also intense X-radiation. The Einstein Observatory was programmed to observe hundreds of quasars, and it detected X-rays from all of them – even those which are faint optically. Indeed, later studies of Einstein pictures taken for other reasons have shown that they often include the X-ray images of distant, and previously unknown, quasars. **Fig. 12.1** is the Einstein Observatory's wide-angle view of the first identified

12.2 *Optical, true colour, 3.9 m Anglo-Australian Telescope*

## Centaurus A

The nearest active galaxy is surprisingly close to us in space. At 16 million light years, Centaurus A is only seven times further away than the Andromeda Galaxy; it is about the same distance as the Whirlpool galaxy M51; and less than one-third the distance of the better-known radio galaxy M87 in Virgo. It is the third strongest radio source in the sky (after Cassiopeia A and Cygnus A), and a source of X-rays and gamma rays. But Centaurus A lies a long way South in the sky, and is always below the horizon (or rises to a very low altitude) as viewed from the traditional optical and radio observatories of Europe and North America. Only in the past few years, with the opening of major new observatories in Australia and Chile and satellites which can observe the whole sky, has Centaurus A been given the attention it deserves.

The true-light photograph (**Fig. 12.2**) shows the central concentration of stars in Centaurus A, crossed by an unusual band of dust which obscures the stars behind it. This ball of stars is 6 arcminutes in apparent size, about 30 000 light years in actual extent, and contains almost a million million stars. The combined light from these stars makes Centaurus A one of the brightest galaxies as seen from Earth; at magnitude 7 it is easily seen in binoculars.

British astronomer John Herschel noticed the galaxy's unusual appearance when observing from South Africa in the 1830s: 'a most wonderful object . . . cut asunder . . . by a broad obscure band'. It was listed in the New General Catalogue of 1888 as NGC 5128. The galaxy was relatively little studied until 1949, when pioneering radio astronomers in Australia identified NGC 5128 with a radio source discovered the previous year and named Centaurus A – the strongest radio source in the constellation Centaurus. (It was among the first three radio sources to be identified, the others being Virgo A as the galaxy M87 and Taurus A as the Crab Nebula.)

New optical techniques have revealed that the stars and dust seen in ordinary photographs (like Fig. 12.2) form only the central regions of a very much larger galaxy. David Malin has brought out the faintest extremities of Centaurus A in **Fig. 12.3**, where the inset, showing the familiar central regions, is to the same scale! The original photograph was taken with the 1.2 metre UK Schmidt Telescope in Australia, and the inset is a normal print from the negative. The lightly exposed image of the galaxy's faint extensions is recorded on the negative, but resides in the surface grains of the emulsion. On ordinary printing it is lost in a background caused by chemical 'fog' scattered throughout the emulsion's thickness. Malin's technique of

'photographic amplification', however, prints only the image in the surface grains, producing Fig. 12.3 from the same negative.

As seen here, the main body of the galaxy is 150 000 light years by 130 000 light years in extent, with an apparent size (33 by 27 arcminutes) as large as the Moon. In three dimensions, Centaurus A is a rugby football shape (a prolate spheroid). From either end, faint streamers of stars reach out from the galaxy's body to double its total extent to a third of a million light years.

Optical photographs cannot show regions of the galaxy which may be larger and fainter. Further 'photographic amplification' brings up extremely faint swathes of light which cover the entire picture; some of the brighter ones are seen across the bottom of Fig. 12.3 They are not related to Centaurus A, but are clouds of dust in our own Galaxy, raised a few hundred light years from the Milky Way's disc by supernova explosions, and illuminated by the glow of stars in our Galaxy.

**12.3** *Optical, photographically amplified (inset printed normally), 1.2 m UK Schmidt Telescope*

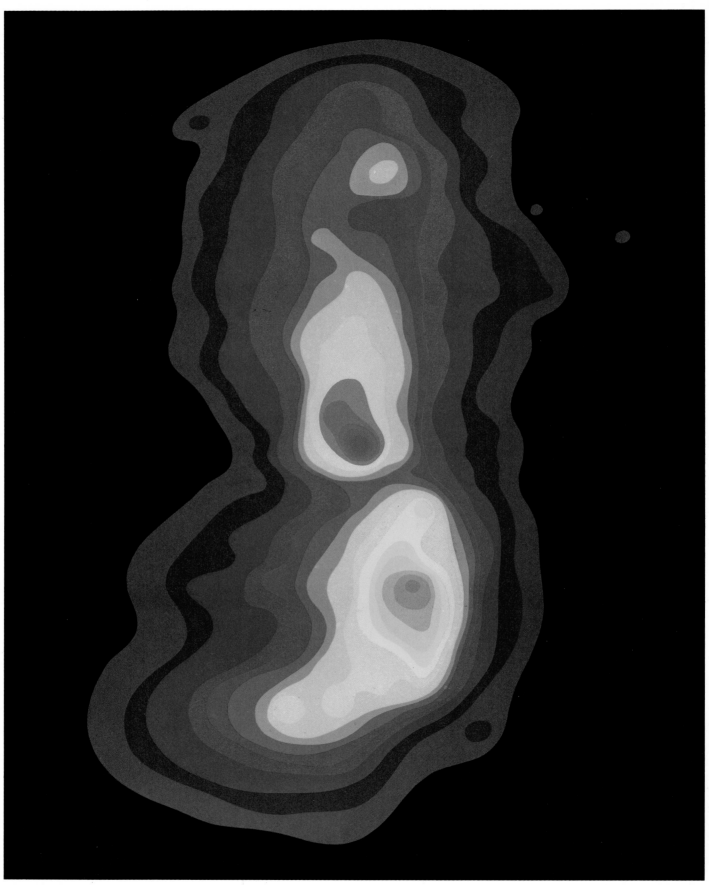

**12.4** *Radio, 21 cm continuum, 64 m Parkes Telescope*

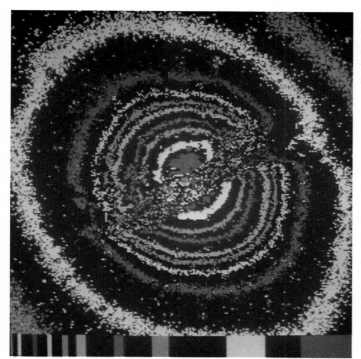

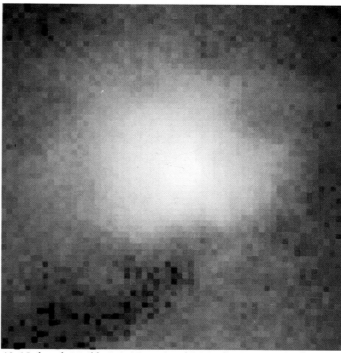

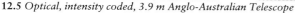

12.5 *Optical, intensity coded, 3.9 m Anglo-Australian Telescope*

12.6 *Infrared, 1.2 (blue), 1.6 (green) and 2.2 (red) μm, 3.9 m Anglo-Australian Telescope*

Radio telescopes show that the radio structure of Centaurus A is remarkably large. It comprises two huge 'lobes' stretching out from the galaxy itself, and seen clearly in the contour map (**Fig. 12.4**) where the colour coding runs from purple for the faintest regions through blue, green, yellow, orange and red to pink for the most intense region at the centre, coinciding with the galaxy itself. The radio lobes stretch over 9° of the sky – almost twenty Moon breadths, or one-tenth the way from the horizon to overhead. In real terms, the lobes cover 2½ million light years, a distance further than the separation of the Andromeda Galaxy and our Galaxy. Centaurus A is, in fact, one of the largest radio galaxies known. To compare with optical photographs, the galaxy as seen in Fig. 12.2 covers only the same area as the innermost pink contour; while the faintest regions in Fig. 12.3 cover only the central one-tenth of the radio span.

The lobes are produced by flows of fast electrons heading North and South from the central galaxy. These electrons generate two elongated clouds of magnetic field, which in turn trap the electrons. As the electrons whirl back and forth through the clouds, they produce synchrotron radiation in the form of radio waves.

The electrons now in the outermost part of the radio-emitting lobes were probably shot out from the centre about 100 million years ago. Since then, the flow of electrons has changed in direction, swinging 40° anticlockwise so that the most recent, innermost portions of the clouds lie along a line roughly 'five o'clock–eleven o'clock'.

Their strength has also varied. The flow of electrons to the South apparently ceased altogether for a while, leaving a gap in the radio emission between the galaxy and the bright spot at 'five o'clock'. During this period, the flow to the North was stronger, producing a bright region adjoining the galaxy at 'eleven o'clock'. Recent powerful activity within the galaxy causes the intense central emission (shown in finer detail on the next spread).

Although Centaurus A has unusually large lobes for a radio galaxy, its radio output is quite typical – about a thousand times stronger than the output of a spiral like the Andromeda Galaxy, but representing only one-thousandth of the power generated by the galaxy's stars in the form of light. Radio galaxies are almost invariably outbursts in giant elliptical galaxies; but is Centaurus A – with its peculiar dust lane – an elliptical galaxy?

In **Fig. 12.5**, an optical photograph from the Anglo-Australian Telescope has been colour coded to show the brightness of the inner part of the galaxy: the picture has the same East–West (left-to-right) extent as Fig. 12.2. The contours coloured purple, white, red, yellow, green, blue and then purple, white and red again, surround regions of increasing brightness. The dust lane shows as the jumble of contours across the centre. But apart from this, the successive contours do indeed form very smooth, undistorted ellipses.

The infrared photograph (**Fig. 12.6**), made with the same telescope covers just the central 1 arcminute of NGC 5128: the innermost one-twentieth of Fig. 12.2 or

Fig. 12.5. Colour coding for wavelength reveals subtle differences in the star and dust regions, but the important point is that all these wavelengths – 1.2, 1.6 and 2.2 microns – are unobscured by dust (except by a particularly dense cloud at the lower left), and so reveal the concentration of stars towards the galaxy's centre. The brightness increases in just the way expected for an elliptical galaxy (proportional to the fourth root of the radius) right into a central star cluster some 300 light years across. So the stars of Centaurus A form an undisputed elliptical galaxy from the centre (Fig. 12.6), through the main body of the galaxy (Fig. 12.5) and right out to a total extent of 150 000 light years (Fig. 12.3). Despite its dust lane – or perhaps with the dust lane as a clue – Centaurus A can help us understand why some giant elliptical galaxies become radio galaxies.

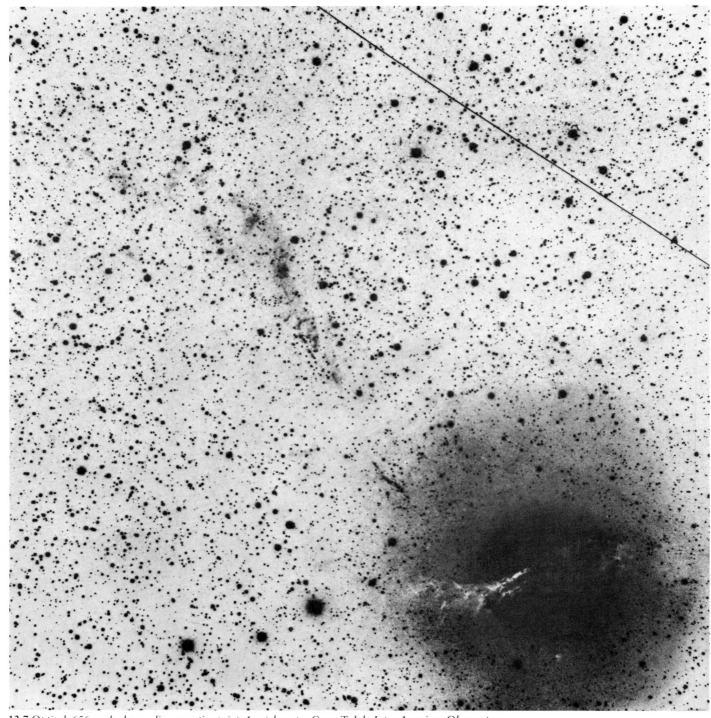

**12.7** *Optical, 656 nm hydrogen line, negative print, 4 m telescope, Cerro Tololo Inter-American Observatory*

The latest burst of activity in Centaurus A has left its mark at all wavelengths – optical, X-ray and gamma, as well as radio. The optical photograph (**Fig. 12.7**), taken at the red wavelength of hydrogen light, reveals unusual filaments of glowing gas stretching out some 140 000 light years to the top left of the galaxy. The region coincides with the radio 'ridge' running out ½° in this direction (Fig. 12.4). The flow of electrons may have compressed tenuous gas already existing out here, or it

may have carried gases with it from the galaxy's centre. Some of this compressed gas is turning into stars, and these filaments of stars and gas constitute the brightest regions of the elongated faint extensions seen on the 'photographically amplified' optical picture (Fig. 12.3). Another short bright filament is visible close to the galaxy, in Fig. 12.7. This has been caused by Centaurus A's latest burst of activity, occurring within the past 100 000 years.

This recent activity shows up most

clearly on the detailed radio picture (**Fig. 12.8**), which reveals the structure within the bright central peak of Fig. 12.4, on a scale sixty times larger. Here the colours represent radio intensity, with the brightest regions red, and less intense regions, yellow, green and blue. This picture is at almost the same scale as Fig. 12.2, but the radio telescope does not detect any of the galaxy's stars. The nucleus appears as the larger of the two red spots at the centre. Most of the emission comes from the lobes at the top left and lower right, lying about 20 000 and 10 000 light years out respectively. The outer edge of the top-left lobe coincides with the close-in optical filament seen in Fig. 12.7, and it is probably responsible in some way for this latest burst of star formation. Fig. 12.8 also reveals the flow of electrons heading towards this inner radio lobe, as a long, thin 'jet'. It appears mainly blue here, although an intense 'knot' in the jet forms the small, 'red' peak next to the nucleus.

The X-ray view (**Fig. 12.9**) reveals the jet of electrons rather better. It is roughly twice the scale of the radio picture (Fig. 12.8), with the nucleus of Centaurus A once again at the centre. The 15 000 light-year-long jet fades out at the top left of the frame, just where the radio lobe begins. The jet's radiation at both X-ray and radio wavelengths probably comes from the synchrotron process, as the electrons lose some of their energy in magnetic fields carried along with them. Thus the jet shows the same bright knots in the two pictures.

Most of the galaxy's X-radiation, however, comes from the tiny 'quasar' nucleus itself. Rapid changes in its X-ray intensity show that it is actually much smaller than it appears here. In 1973, the OSO-7 satellite found Centaurus A brightening in X-rays in just a few days, implying that the core is only a few 'light days' across – about one-hundredth of a light year. This compact core produces gamma rays too, and these require an even greater concentration of energy – believed by many astronomers to be due to gas at the inner edge of a disc surrounding a black hole at the heart of Centaurus A. The nucleus emits twice as much power in X-rays as the galaxy's total radio output; and a hundred times more power still in gamma rays. Despite its prominence in the radio sky, Centaurus A is not so much a 'radio galaxy' as a 'gamma ray galaxy'!

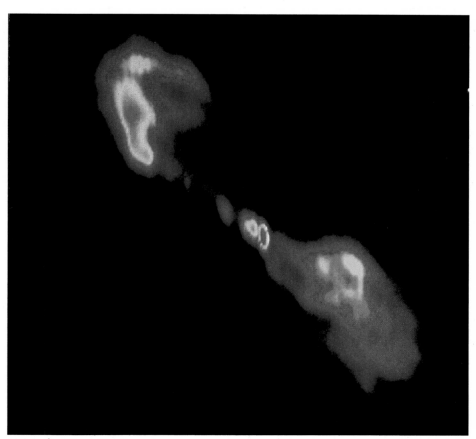

**12.8** *Radio, 20 cm, Very Large Array*

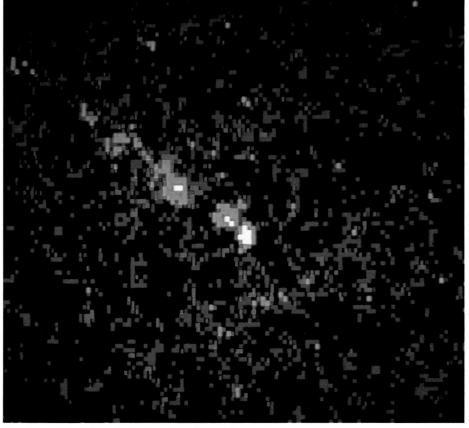

**12.9** *X-ray, 0.3–2.5 nm, Einstein Observatory IPC*

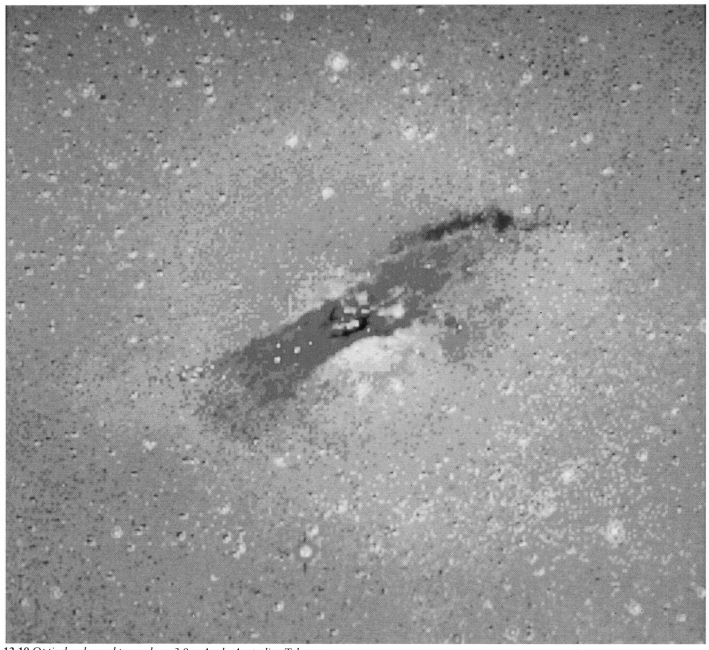

**12.10** *Optical, enhanced true colour, 3.9 m Anglo-Australian Telescope*

The most obvious feature of Centaurus A, the great dust lane across its centre, still remains its most perplexing feature. Elliptical galaxies generally consist only of very old stars. They contain none of the residual gas and dust which form new stars in the discs of spiral galaxies. But Centaurus A does have dust, and gas – and young stars which have formed from them.

The dust shows up prominently as deep 'canyons' in the bas-relief optical presentation, **Fig. 12.11**. Here a photograph has been scanned and its brightness levels stored in a computer. A copy of this 'map' has then been shifted very slightly towards the top left, and subtracted from the original. The composite is shown in photographic form, with grey representing zero, white for positive intensity and black for negative (where images on the shifted map are subtracted from fainter regions in the original). The technique has eliminated most of the galaxy's large-scale variation in brightness, while emphasising the small-scale brightness differences. These show up, apparently in three dimensions, as a bas-relief illuminated from the lower right, where dark regions of the galaxy – the dust lanes – appear depressed, and the foreground stars show up as spikes.

Optical photography at different wavelengths reveals other young components in Centaurus A. **Fig. 12.10** combines observations made through red, yellow and blue filters, printed with maximum contrast in their respective colours, to enhance slight colour variations. The main body of the elliptical galaxy comes out bluish-white. The dust lane absorbs yellow and blue light more strongly than red, so the enhancement brings it out in bright red. The bright blue regions shine in the bluish-white light from young, hot stars: they are especially obvious along the upper right-hand edge of the dust lane. These stars were probably born only 30 to 100 million years ago, and so they are less than one-hundredth as old as the rest of the galaxy.

The obscuring dust can only be detected

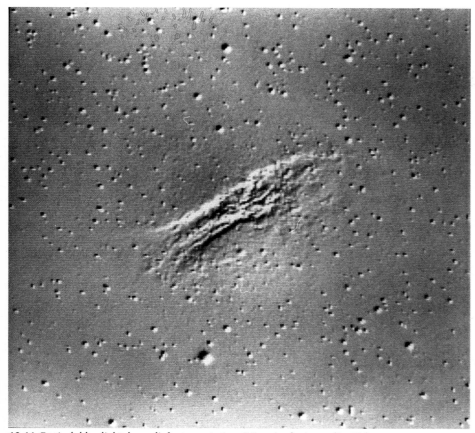

12.11 *Optical, blue light, bas-relief representation, 3.9 m Anglo-Australian Telescope*

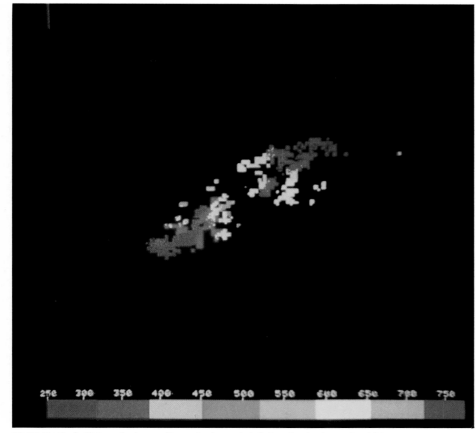

12.12 *Optical, 656 nm hydrogen line, velocity coded, 3.9 m Anglo-Australian Telescope*

in front of the galaxy, but the bright blue stars can be seen anywhere. The region of young stars at the top right curves sharply around at the right-hand end, and comes back inwards. Here it is dimmed by the dust in front, appearing purple in Fig. 12.10. At the lower left, there is another faint purple curve of young stars on the far side of the galaxy (almost a mirror image of the right-hand one). So the dust 'lane' is in fact a disc, tipped up 15° from edge-on as we see it, surrounding the centre of Centaurus A. It is 35 000 light years across, and its outer edge is lit by young stars.

The ultraviolet radiation from young stars within the disc lights up nebulae of hot gas around them, and the nebulae glow at the wavelengths of common elements like hydrogen. A filter which passes only the red light of hydrogen picks out the nebulae (**Fig. 12.12**). This view covers roughly the same area as Fig. 12.10, and the nebulae fill an oval region coinciding with the dust 'lane'.

The hydrogen-light observations in Fig. 12.12 were taken with an imaging spectrograph, which measures the precise wavelength of the hydrogen line at each point in the image. The wavelength will differ very slightly from the theoretical value if the gas is moving towards or away from us, because of the Doppler effect. The spectrograph thus reveals the speed of each nebula, and the colour coding for velocity here shows gas travelling away at the highest speed in red, and at the lowest speed in purple, with intermediate speeds shown on the scale at the bottom (marked in kilometres per second).

Centaurus A is receding from us at 550 kilometres a second (pale green in the code) with the expansion of the Universe; so *relative to the galaxy's centre*, the left-hand side of the disc is approaching us, and the right-hand side is receding, with speeds which increase outwards from the centre to about 250 kilometres per second. This pattern of speeds is remarkably similar to the run of speeds found in the disc of an ordinary spiral galaxy (for example, the Andromeda Galaxy, Fig. 10.4).

So what is the disc? It surrounds the galaxy's core, so it is not a spiral galaxy seen in front of Centaurus A, and the well-ordered rotation of the disc means that it is probably not the remains of a spiral galaxy which has collided with a giant elliptical. The disc is probably just a part of the galaxy's matter which has yet to turn into stars – although this begs the question of why other elliptical galaxies lack such a disc. Gases from the innermost region of the disc are probably responsible for the galaxy's 'quasar' activity, as they spiral into a massive black hole at the centre of Centaurus A.

# Cygnus A

The second brightest radio source is a galaxy so distant that it appears extremely faint to optical telescopes – at magnitude 15, less than a thousandth as bright as the dimmest stars visible to the naked eye. In this short exposure optical photograph (**Fig. 12.13**), the central regions of the galaxy appear as the small dumb-bell shape at the centre: the whole picture is only 2½ arcminutes across, less than one-tenth the apparent diameter of the Moon. Its discoverer Walter Baade was convinced that Cygnus A was a pair of galaxies in collision, and he placed a bet – a bottle of whisky – with his colleague Rudolph Minkowski that the spectrum would show light from hot gas produced in the collision. The spectrum turned out as he predicted, and Baade won his whisky – but in the long run the interpretation proved to be wrong. Cygnus A is a single galaxy, with an active quasar core which heats up gas in the centre. The peculiar double appearance in Fig. 12.13 is caused by a band of dust running across the galaxy, like the dust lane in Centaurus A (Fig. 12.2).

The Cygnus A galaxy resembles a larger and more massive version of Centaurus A, appearing small and indistinct because it lies fifty times further away, at 740 million light years. The central region of stars and dust seen in Fig. 12.13 is about 20 000 light years across, but long-exposure photographs reveal fainter outer regions of this giant galaxy extending to a total size of 300 000 light years (three times the diameter of the Milky Way); they would cover half of Fig. 12.13. Cygnus A is the main member of a small group of galaxies. (One companion is the fuzzy blob 2 centimetres to the right of Cygnus A.) Because Cygnus A's gravity must be holding the companions in orbit, the motions of the other galaxies give an indication of its gravitational pull, and hence its mass. The latter works out to be a staggering 100 million million Suns – making Cygnus A a thousand times more massive than our Galaxy, and one of the largest and most massive galaxies known.

The X-ray view (**Fig. 12.14**) reveals very hot gas pervading and surrounding the galaxy. The picture covers a region about three times larger than Fig. 12.13, with the dots showing the intensity of X-ray emission. The 100 million degree gas is concentrated into the same region as the galaxy's stars. Its distribution is seen more clearly in the graphs, which are scans of the X-ray intensity across the picture (the bottom graph) and up and down (the left-hand graph). The hot gas is evidently more closely packed towards the galaxy's centre, like the stars, and there is a faint extended 'atmosphere' around Cygnus A (as well as the general background of dots).

The most important aspect of Cygnus A

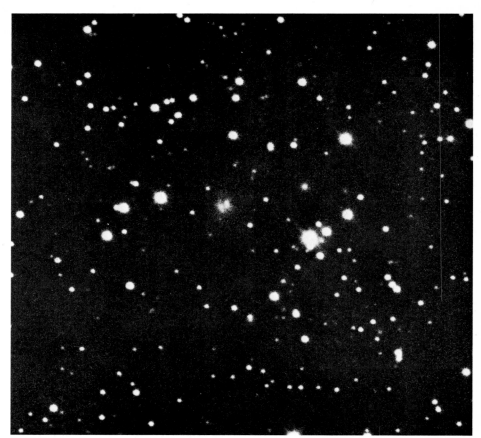

12.13 *Optical, 3 m telescope, Lick Observatory*

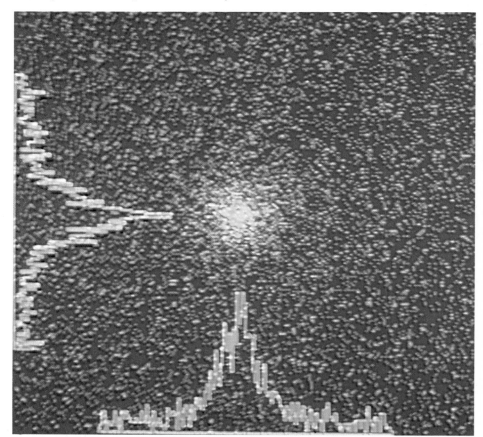

12.14 *X-ray, 0.4–8 nm, Einstein Observatory HRI*

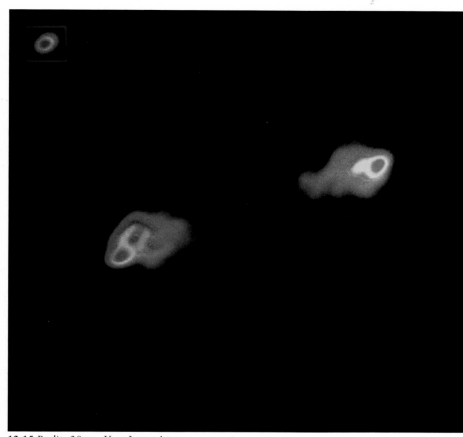

**12.15** *Radio, 20 cm, Very Large Array*

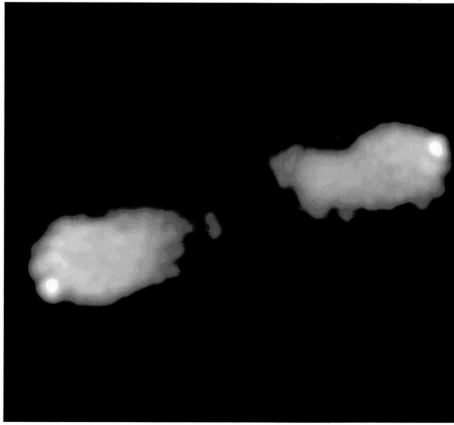

**12.16** *Radio, 73 cm, MERLIN*

is its exceptionally powerful radio emission. Whereas typical radio galaxies (like Centaurus A and M87) are a thousand times more powerful in radio waves than the Milky Way, Cygnus A is a thousand times more powerful still – equal in its radio output to a million Milky Ways. This intense emission comes not from the centre of the galaxy, but from two lobes, one on either side, as seen in **Fig. 12.15**. Here the red regions are brighter in radio waves, and yellow, green and blue denote successively fainter regions. (The spot at the top left shows the amount of smearing that the observing technique introduces into the map.) This radio view covers the same region of sky as the optical photograph (Fig. 12.13). The galaxy itself is not visible. Instead we see the lobes – clouds of magnetic field and electrons – stretching out 200 000 light years on either side to terminate in bright (red-coded) *hot spots*, where energy is being fed into the lobes.

The lobes' energy comes from the galaxy's active core – probably a disc of gases spiralling into a very massive black hole. The energy is ejected from the core as two oppositely directed beams or jets of fast electrons. These are too faint to be obseved directly but they carry energy outwards until they hit the surrounding gas in the galaxy's outer atmosphere. The orderly motions of the electrons are then disrupted; they generate magnetic fields, and as they move through these fields they radiate their energy as radio waves. The collision on each side is thus marked by a hot spot in the radio map.

The impact of the electron beams pushes away the gas at the collision point, and so the hot spots are moving outwards, at about 10 000 kilometres per second. The hot spots have hollowed out burrows which are relatively empty except for the tenuous magnetic field and electrons left behind. The electrons in the lobes lose energy as they radiate, with the high-energy electrons – which produce short radio waves – losing their energy first. As a result, a short wavelength radio picture (like Fig. 12.15) shows only the recently formed, outer parts of the radio lobes.

**Fig. 12.16** is a more detailed view, revealing just how small and bright the hot spots really are; and because it is made at a longer wavelength it also shows the lobes stretching further back along the burrows. The brightest regions are coded white here, and successively fainter parts of the source yellow, orange and red. The complex structures in the lobes have been created as the beam has swung around and changed in intensity. Comparison of different wavelength maps can date various regions of emission. The oldest regions near the galaxy date back some five million years, indicating that the activity in Cygnus A has been occurring for only a short period on the astronomical time scale.

**12.17** *Optical, red light, negative print, 1.2 m UK Schmidt Telescope*

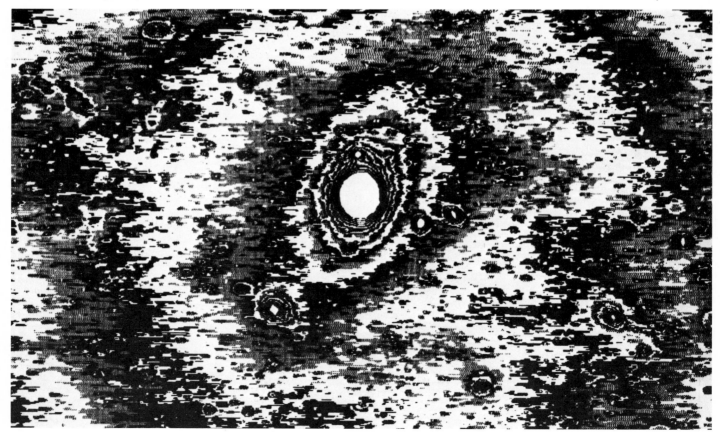

**12.18** *Optical, blue light, intensity coded, 1.2 m Palomar Schmidt Telescope*

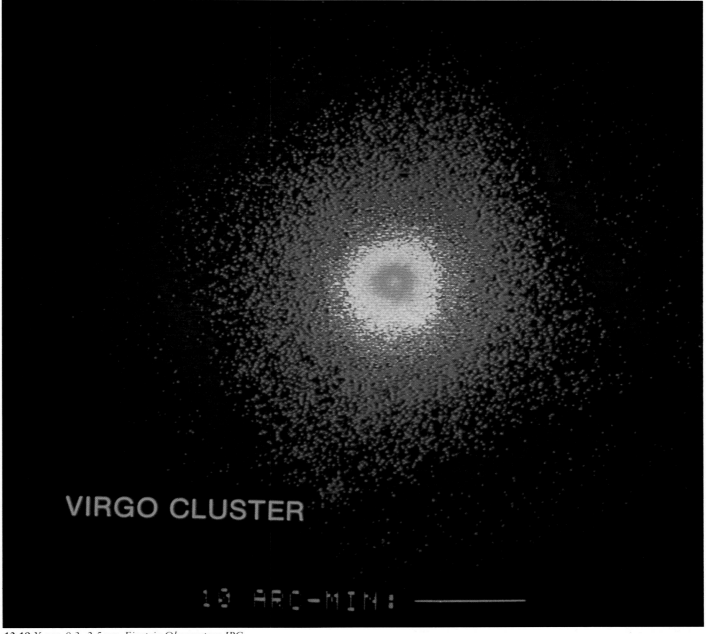

VIRGO CLUSTER

10 ARC-MIN:

**12.19** *X-ray, 0.3–2.5 nm, Einstein Observatory IPC*

# M87

The Virgo Cluster is the nearest major cluster of galaxies, some 50 million light years away. It consists of a thousand galaxies concentrated into a region 10 million light years across. It is also the centre of a loose swarm of galaxy groups, the Local Supercluster, which includes our Local Group of galaxies (Fig. 1.10).

The Virgo Cluster covers 12° of sky, and **Fig. 12.17** shows just the central 5° by 3°. The six brightest galaxies are visible in even a small telescope: from left to right they are M58, M90, M89, M87 (centre), M86 and M84. M87 appears fairly unremarkable here, but at radio and X-ray wavelengths it outshines the other galaxies a hundred

times over. It is, in fact, the most remarkable galaxy in the cluster, lying stationary at its centre while the other galaxies swarm around at speeds of up to 1500 kilometres per second.

**Fig. 12.18** covers the central part of the Virgo Cluster, at three times the scale of Fig. 12.17. The brightness levels have been converted into shades of grey, so that the bands of white, grey and black are like contour lines enclosing regions of equal brightness, building up towards the most intense region at the centre (the innermost, brightest part has been left white). The faint outermost regions of M87 reach over halfway to M86 and M84, and give M87 a total size of a million light years!

In the X-ray view (**Fig. 12.19**), the fainter

regions are blue, and the more intense central regions coded in red, green, purple and white. (The large, blue square is caused by the detector's window supports.) The purple and white region corresponds to Fig. 12.18's central white area. The Virgo Cluster as a whole contains hot X-ray emitting gas at 100 million degrees; but the gas here – at 30 million degrees – is associated with M87 itself, and extends to half a million light years. To retain this gas, M87 must have a mass of 30 million million Suns – ten times the mass of the galaxy as deduced by the light from its stars. Most of M87's mass must be invisible, in its faint outer parts – possibly as extremely faint stars, black holes or fundamental particles.

12.20 *Optical, 4 m telescope, Cerro Tololo Inter-American Observatory. Inset: optical, 3 m telescope, Lick Observatory*

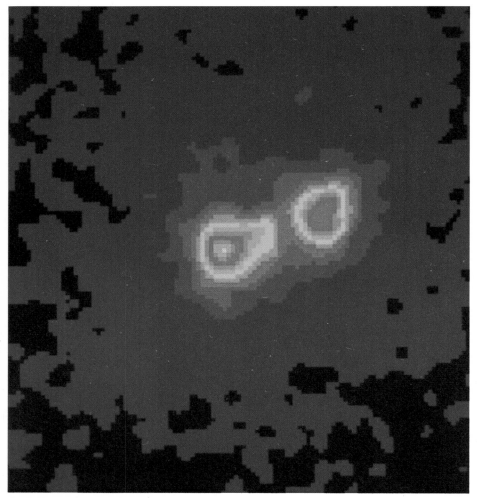

12.21 *X-ray, 0.4–8 nm, Einstein Observatory HRI*

On a conventional long exposure photograph (**Fig. 12.20**), M87 looks like a typical large elliptical galaxy, some 40 000 light years across. (This photograph covers roughly the central white area of Fig. 12.18.) The small, fuzzy objects are some of M87's 400 globular clusters, each cluster consisting of almost a million stars. The inset to Fig. 12.20 (to the same scale) is a short exposure, revealing only the brightest, innermost regions of M87. It shows that the galaxy contains an unusual 'jet' projecting from its nucleus for 5000 light years. The jet is very small compared to the total extent of the galaxy, but its discovery (by American astronomer Heber Curtis in 1916) was so unexpected that it has made M87 famous. The jet's light is synchrotron radiation emitted by fast electrons moving in a magnetic field, the brightest 'knot' being only 100 light years across but shining as brightly as 40 million Suns.

The colour-coded optical photograph (**Fig. 12.22**) shows that the jet extends right into the galaxy's small, bright nucleus. Here the brightest regions are coded white, and successively fainter regions green, red and blue. The distribution of intensity in the centre indicates how the stars become progressively more packed towards the galaxy's core. In 1978, British and American astronomers studied the central condensation of stars – coded white in Fig. 12.22 – with the 5 metre (200 inch) Palomar telescope, and concluded that the stars here are so tightly packed and so fast moving that they must be held in by the gravitational pull of a very massive black hole – as heavy as 5000 million Suns.

The X-ray view of M87's centre (**Fig. 12.21**) covers roughly the same region as Fig. 12.22 (It is a detailed view of the central X-ray emission coded purple and white in Fig. 12.19.) The colour coding runs from blue for the faintest regions, through yellow and red to white for the most intense spot at the galaxy's core. The red region to the right marks X-radiation from the electrons in the jet. The rest of the radiation is due to hot gas pervading the galaxy, and increasing in density towards the centre. The central gas should 'fuel' an accretion disc of gas surrounding the black hole, and the disc should appear as a tiny brilliant point in the X-rays – like the central X-ray source in NGC 1275 (Fig. 12.30). But Fig. 12.21 shows no sign of a point source, only a gradual increase in X-ray brightness to the level coded white. Although most astronomers accept the theory that black holes power 'quasar core' activity in galaxies, it is ironical that the galaxy where optical observations first provided independent evidence for a massive central black hole is also the galaxy where X-ray observations seem to refute the idea!

**12.22** *Optical, intensity coded, CCD, 5 m Hale Telescope*

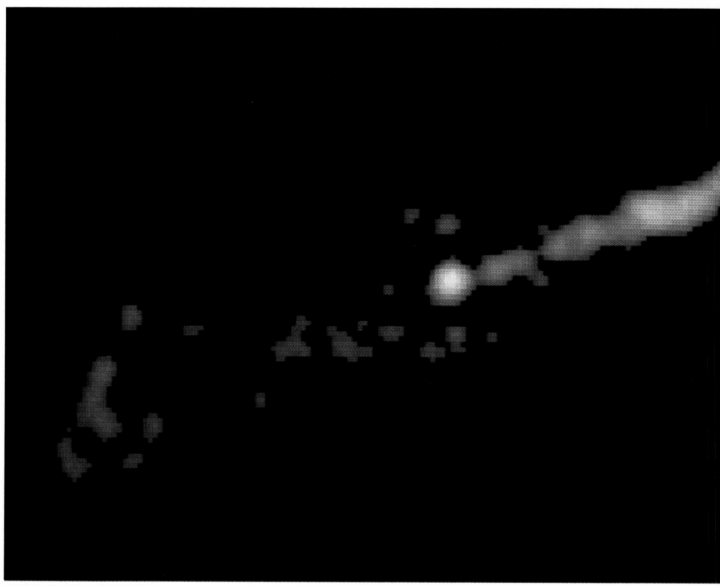

12.23 *Radio, 73 cm, MERLIN*

M87 is a prominent radio source, sometimes called Virgo A. Its radio output is about a thousand times that of our Galaxy, but only one-thousandth that of a powerful radio galaxy like Cygnus A (Fig. 12.15). However, M87 has the great advantage that the electron beam carrying its energy from the quasar core can be detected and studied at all wavelengths from X-ray to radio.

In normal optical photographs, details of this beam – the 'jet' – are blurred by the Earth's atmosphere, and the inner parts are difficult to discern against the galaxy's brightness. British physicists Steve Gull and John Fielden have used a computer technique (maximum entropy processing) on their image of M87 (Fig. 12.22) to remove the galaxy's light and sharpen up the jet (**Fig. 12.24**). The colour coding runs from white for the brightest regions, through blue, red, yellow and green, to blue

again for the dimmest parts. (The galaxy's core is artificially distorted into a rectangle.) The jet clearly runs straight for the first 3000 light years (12 arcseconds) from the nucleus to the brightest knot, but then wiggles slightly for its last 2000 light years, through two further knots.

A detailed radio picture (**Fig. 12.23**) from Jodrell Bank's MERLIN array shows that the beam continues further. The strongest radio regions are coded white, with less intense regions yellow, orange and red. The galaxy's core is the small, bright source in the centre, and the brightest regions in the jet to the right (half way along) coincide with the optical knots seen in Fig. 12.24. The jet's radio emission is produced by lower energy electrons than are needed to produce light, and these continue out to 8000 light years, revealing an increasingly wavy jet. A beam of fast particles should in fact become unstable

after a certain distance, and begin to wiggle and break up – like the water jet from a fireman's hose – and the knots seen in both optical and radio maps are probably spots where the edges of the jet are unstable and energy is released.

The jet, however, accounts for only a few per cent of the galaxy's radio power. The rest comes from electrons ejected in the past, which have now spread out into the larger, relatively dim region seen in **Fig. 12.25**. This region is coded blue, with successively brighter regions yellow, red and white. The galaxy's nucleus is the central bright spot (made oval by the observing technique), and the jet stretches to the right. The jet clearly bends right round at the end, through an angle of 160°. This is probably caused by a 'wind' of extragalactic gas blowing from the top right. The diffuse radio regions to the left of the nucleus contain electrons deposited

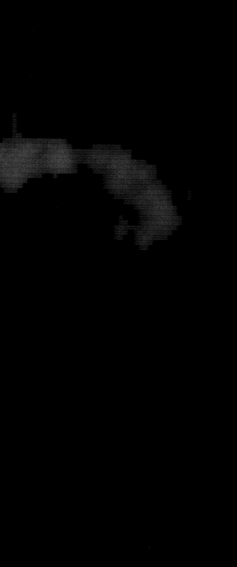

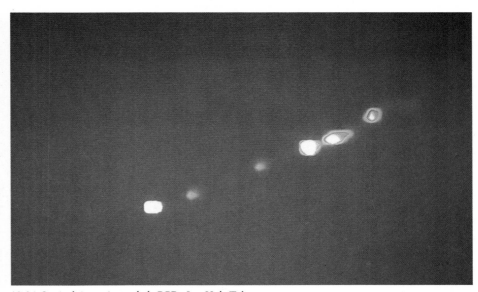

**12.24** *Optical, intensity-coded, CCD, 5 m Hale Telescope*

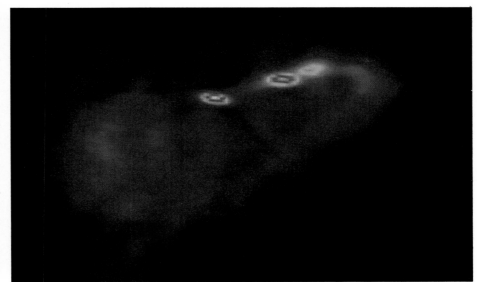

**12.25** *Radio, 6 cm, Very Large Array*

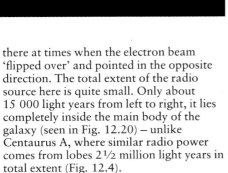

there at times when the electron beam 'flipped over' and pointed in the opposite direction. The total extent of the radio source here is quite small. Only about 15 000 light years from left to right, it lies completely inside the main body of the galaxy (seen in Fig. 12.20) – unlike Centaurus A, where similar radio power comes from lobes 2½ million light years in total extent (Fig. 12.4).

Radio telescopes connected over thousands of kilometres can probe much finer detail, and **Fig. 12.26** reveals the galaxy's core, on a scale a hundred times larger than Fig. 12.23. The colour coding shows the brightest region, at M87's centre, in shades of orange and brown (with a central black speck), while lower brightness regions are yellow, green and blue. Fig. 12.26 reveals the first few light years of the jet as it heads out from the still-enigmatic power-house at M87's heart.

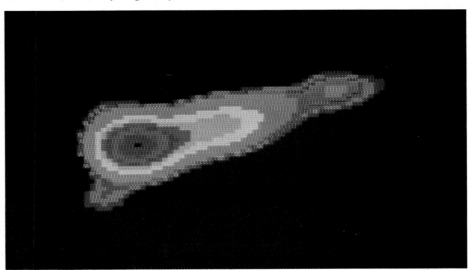

**12.26** *Radio, 18 cm, VLBI (Haystack, Green Bank, Fort Davis, Owens Valley, Very Large Array, Maryland Point, North Liberty, Effelsberg)*

12.27 *Optical, 3.8 m telescope, Kitt Peak National Observatory*

## NGC 1275

The Perseus Cluster of galaxies lies 230 million light years away, with its brightest members appearing as a ragged line of galaxies spanning a quarter of a million light years – roughly left to right in **Fig. 12.27**. The cluster's centre, however, (as determined by its content of faint galaxies and hot gas) lies at the left-hand end of this line, near to the brightest galaxy is the cluster, NGC 1275.

NGC 1275 is a giant elliptical galaxy, of the kind found at the centre of many clusters of galaxies, and, like M87 (Fig. 12.20) for example, it has an active 'quasar core'. NGC 1275 is unusually active. It is a radio source a thousand times more powerful than the Milky Way and a strong X-ray source. Even optical photographs betray its peculiarity. Fig. 12.27 shows ragged wisps of gas and dust, unusual for an elliptical galaxy. Short-exposure photographs reveal the core as a tiny intense spot of light. If NGC 1275 were so far off that the galaxy were invisible, the core would still be seen – and classified as a quasar (like 3C 273, Fig. 12.33).

The X-ray picture (**Fig. 12.28**) shows both the shape of the Perseus Cluster and the uniqueness of NGC 1275 within it. Green represents low-brightness regions and white the brightest region (the dark square is due to supports in the detector). The picture covers 1° of sky, roughly the same as the width of Fig. 12.27, but centred on NGC 1275. The X-rays come from hot gas (at 80 million degrees) filling the cluster, and the green region's oval shape indicates that the cluster is slightly flattened.

The intense white region is emission from hot gas in and around NGC 1275 itself. The galaxy lies just to the right of the cluster's centre (in the middle of the green oval). NGC 1275 is ten times more powerful in X-rays than M87 (Fig. 12.21), and at these wavelengths the Perseus Cluster outshines the Virgo Cluster fifty times over.

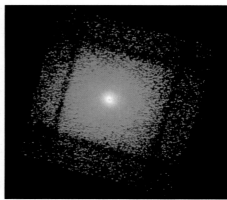

**12.28** *X-ray, 0.3–2.5 nm, Einstein Observatory IPC*

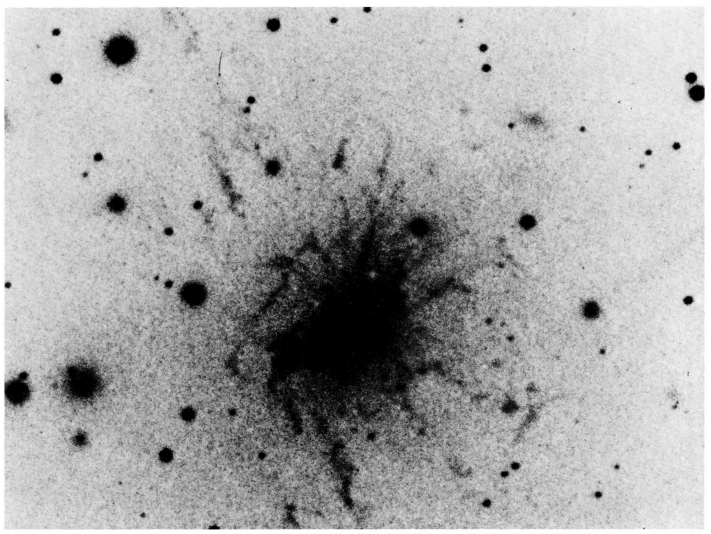

**12.29** *Optical, 656 nm hydrogen line, negative print, 2.1 m telescope, Kitt Peak National Observatory*

Strange filaments of glowing gas, which extend from NGC 1275 to a distance of a quarter of a million light years can be seen (at the top left) in **Fig. 12.29**, taken through a filter which passes only the red light from hydrogen gas. (The photograph is at six times the scale of Fig. 12.27, and rotated 40° anticlockwise.) These streamers are, however, not debris of an explosion; they are probably heading *into* the galaxy.

The Perseus Cluster contains about 50 million million solar masses of gas at a temperature of 80 million K, but the gas closest to NGC 1275 has cooled down to 10 million K. Denser clumps have cooled to 10 000 K, and shine at optical wavelengths as the gas filaments. This gas flows inwards, increasing the mass of NGC 1275 by 300 solar masses every year. Much of it forms into new stars.

The detailed X-ray view (**Fig. 12.30**) covers the same area as Fig. 12.29. The distribution of the 10 million degree gas seen here is similar to the gas concentrated around the central galaxy of the cluster Abell 2199. But NGC 1275 also has a

small, brilliant core – best seen as the sharp spike in the superimposed graph showing the intensity of X-rays across the picture. This X-ray point coincides with the bright quasar core visible optically, and it is generally believed to be an accretion disc of gas surrounding a massive black hole. In NGC 1275 the 'fuel' for the disc is probably inflowing gas from the filaments.

NGC 1275's quasar core also emits radio waves, and fine details can be mapped by combining the results from widely separated telescopes in the technique of Very Long Baseline Interferometry (VLBI). **Fig. 12.32** shows the innermost few light years of NGC 1275, mapped on five different occasions from (left) 1972 to 1976 (right) with radio telescopes ranging from California and Ontario to Germany. The most intense regions are red, and successively less intense parts yellow, green and blue. The galaxy's core is the top blob in each picture, varying in brightness as its activity changes from year to year. From the core, a 'jet' stretches downwards for almost 10 light years. The jet is increasing

slightly in length, extending outwards at about half the speed of light. It probably started in a burst of activity observed by radio astronomers in 1958.

**Fig. 12.31** is another VLBI map, made with a similar array of radio telescopes a few years later. The colour coding here runs from white for the brightest region, through red and yellow to blue for the faintest. This shorter wavelength map shows finer details. More importantly, the quasar core itself emits more powerfully at this wavelength, so it appears as the strongest part of the source. This small core (with the white centre) is elongated roughly horizontally, at right angles to the jet.

The accretion disc surrounding a massive black hole should be about the size of this core (just under a light year across) and (if seen edge-on) should appear as an oval perpendicular to the line of the jet. Hence, astronomers may here – for the first time – have resolved the disc of gas which theorists have long invoked to explain quasar activity.

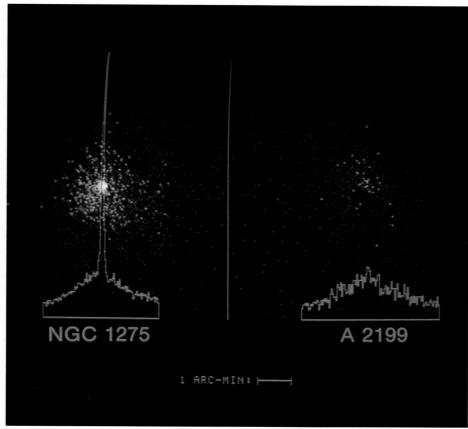

12.30 X-ray, 0.4–8 nm, Einstein Observatory HRI

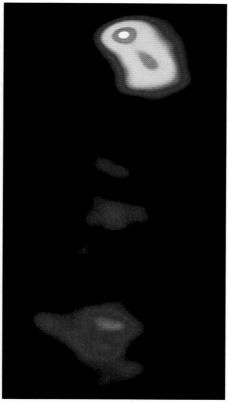

12.31 Radio, 1.3 cm, VLBI (Effelsberg, Haystack, Green Bank, Algonquin, Very Large Array, Owens Valley)

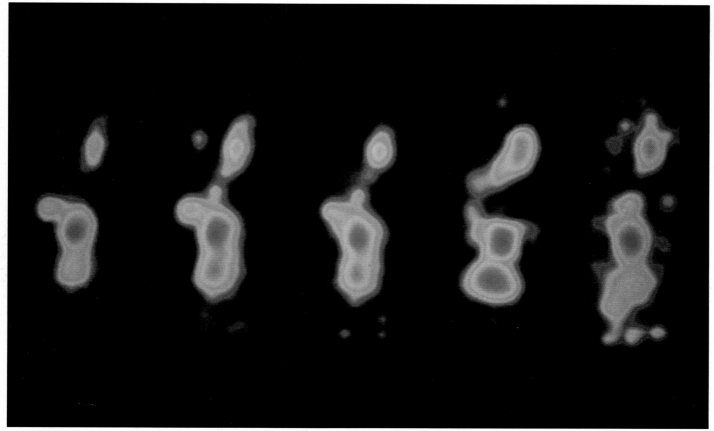

12.32 Radio (yearly maps 1972–6), 2.8 cm, VLBI (Effelsberg, Algonquin, Green Bank, Fort Davis, Owens Valley)

# 3C 273

The star-like object at the centre of **Fig. 12.33** is the nearest quasar, apparently the brightest at all wavelengths from radio to gamma ray, and the best-studied. It was first detected as the twentieth strongest radio source, and appears as number 273 in the third Cambridge catalogue. 3C 273's position was found accurately in 1962, by timing the exact moment that the Moon passed in front and blocked off the radio emission. The technique also showed that it consists of two closely spaced radio sources. On optical photographs, one radio source coincides with the bright 'star' in Fig. 12.33 and the other with the end of the faint 'jet' which extends 22 arcseconds to the lower right. The 'star' is relatively bright (at magnitude 13) and its image is overexposed here and blurred out to an artificially large size (as are other images of true stars).

Astronomers thought at first that 3C 273 was indeed a star in our Galaxy, but its spectrum turned out to be very different. Dutch-American astronomer Maarten Schmidt realised in 1963 that the lines seen crossing the spectrum of 3C 273 are due to hydrogen – the commonest element – but all with their wavelengths increased substantially. The strongest spectral line from hydrogen occurs at ultraviolet wavelengths, and this line (Lyman-alpha) was studied in detail by the International Ultraviolet Explorer satellite in 1978. **Fig. 12.34** shows a small part of 3C 273's ultraviolet spectrum running from the top left to the lower right. (The jet is too faint to show up in the spectrum.) The superimposed red dotted line marks the centre of the spectrum, with wavelengths marked in angstrom units from 1200 (120 nanometres) to 1400 (140 nanometres), and intensities coded from dark blue for the background, through pale blue to white for the brightest regions. The intense spectral line next to the figure 1200 is hydrogen's Lyman-alpha line at its normal wavelength (122 nanometres), and this comes from hydrogen surrounding the Earth in a huge, hot cloud called the geocorona. The Lyman-alpha from 3C 273 appears in the lower right-hand corner near the figure 1400, with its wavelength increased by 19 nanometres. The proportional increase is 16 per cent, and the same value is found for all of 3C 273's spectral lines.

The simplest explanation for the shift in the spectral lines is that 3C 273 is so far away from us that the expansion of the Universe is carrying it away at a speed of 50 000 kilometres a second, and the Doppler effect thus stretches all the wavelengths. 3C 273 must then lie at a staggering 2100 million light years away, and to appear as a relatively bright 'star' in our skies, it must be more luminous than the brightest galaxies known, as brilliant as

12.33 *Optical, 3.8 m telescope, Kitt Peak National Observatory*

12.34 *Ultraviolet, spectrum, 120–140 nm, International Ultraviolet Explorer satellite*

400 Milky Ways. In addition, the light comes from a very small region. 3C 273's star-like appearance on photographs, and indirect arguments based on changes in its brightness indicate that it is less than 1/100 of a light year across.

Since 3C 273 was clearly not a star, despite its appearance, it was dubbed a 'quasi-stellar object', later abbreviated to 'quasar'. Over a thousand other, more distant, quasars, are now known, and further investigations have shown that quasars are outbursts in the centres of distant galaxies, similar to the smaller outbursts found in galaxies like NGC 1275 (Fig. 12.29).

Quasars are so brilliant that it is usually difficult to detect the surrounding galaxy. In **Fig. 12.35**, a small disc has been placed over the image of 3C 273 (at the position of the cross) to block off its light, and the surrounding region has been recorded by a very sensitive light detector. Levels of brightness are colour coded, from black for dark sky, through blue, red and yellow to white for the brightest regions. (To give an idea of the detector's sensitivity, the bright yellow–white projection at the bottom right is the jet, barely visible in Fig. 12.33, while the bluish regions are only two-millionths as bright as the quasar itself.) The quasar, occulted by the black disc, is clearly surrounded by an oval-shaped galaxy. Although only one-twentieth as bright as its quasar core, this ranks with the largest and brightest elliptical galaxies known: it outshines the Milky Way twenty times, and is three times as large, with a diameter of one-third of a million light years – similar to Cygnus A (Fig. 12.13).

3C 273 is also a strong X-ray emitter, producing as much power in X-rays as it does in light, and over a million times more powerful at this wavelength than the Milky Way. Early X-ray observations showed that its output can vary in half a day, indicating that this huge amount of radiation comes from a region no larger than the Solar System. The Einstein Observatory looked at 3C 273 in roughly the same detail as optical photographs like Fig. 12.33. In **Fig. 12.36**, the X-ray brightness is colour coded, with dark blue, red and brown corresponding to the darkest regions, and successively brighter regions in buff, purple, grey, yellow, green and white. In this long-exposure picture, 3C 273 is smeared out to an artificially large size, and the background has false 'spokes' caused by the supports of the telescope mirrors. But the yellow and green blob at 'four o'clock' is real, and corresponds to the inner half of the jet as seen at optical wavelengths. This beam of electrons emits only 1/300 as much light and X-rays as the quasar's core. But even so, the jet alone emits as much light as our Milky Way Galaxy, and is 10 000 times more powerful in X-rays.

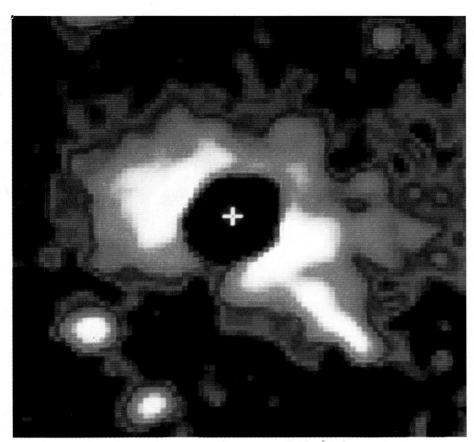

12.35 *Optical, intensity coded, CCD, 1.8 m telescope, Lowell Observatory*

12.36 *X-ray, 0.4–8 nm, Einstein Observatory HRI*

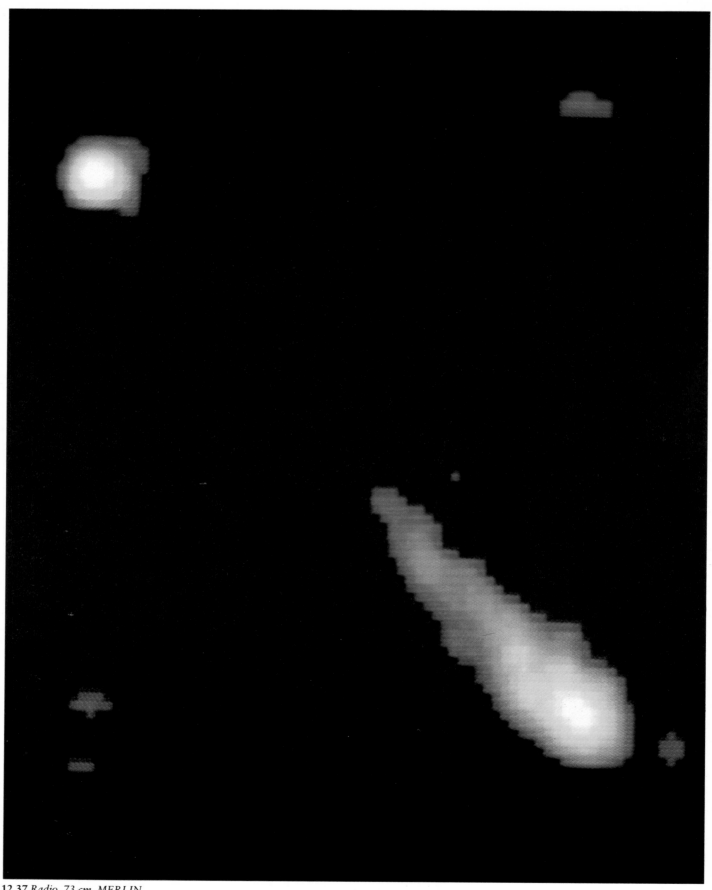

**12.37** *Radio, 73 cm, MERLIN*

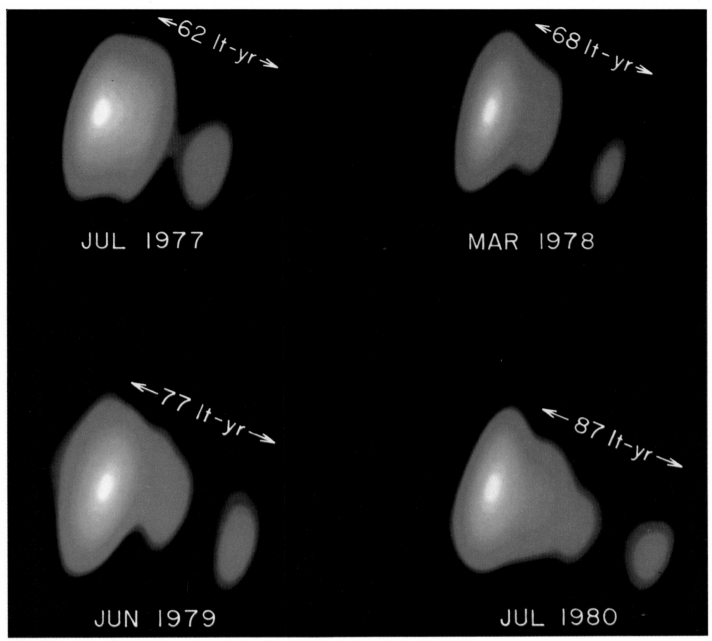

12.38 *Radio, 2.8 cm, VLBI (Effelsberg, Haystack, Green Bank, Fort Davis, Owens Valley, Hat Creek)*

The quasar 3C 273 and its jet emit roughly equal amounts of radio waves, each almost a million times the Milky Way's radio output. In the radio picture from the MERLIN radio telescope array (**Fig. 12.37**), the lowest brightness regions are red, and brighter parts orange, yellow and white. (The red blobs at the lower left and upper right are false images.) The jet seen at radio wavelengths coincides in position with the optical jet (Figs. 12.33 and 12.35). But the radio jet is a hundred times more powerful at the outer than the inner end. Comparison of the X-ray (Fig. 12.36), optical (Fig. 12.35) and radio (Fig. 12.37) views of the jet shows that as this electron beam leave the quasar's surrounding galaxy, it first releases its

energy at the short wavelengths of X-rays and light, then at the longer wavelengths of light and radio waves, and finally – at its tip – just at the long wavelength radio waves. The radio picture from MERLIN also reveals 'wiggles' in the jet, like those in the MERLIN view of M87's jet (Fig. 12.23).

By combining observations from widely spaced radio telescopes, astronomers can detect the beginnings of the jet in the core of 3C 273. **Fig. 12.38** shows four such maps produced in consecutive years, (with colour coding as for Fig. 12.37). This 'mini-jet' is clearly expanding, as shown by the lengths marked on the successive pictures. It apparently grew by 25 light years in only three years – so it was expanding more than eight times faster

than the speed of light!

According to Einstein's theory of relativity, motion faster than light is impossible. We can, however, explain the apparent 'superluminal' velocity by a trick of geometry, if the beam of electrons is moving just slower than the speed of light, but is heading almost straight towards us. As a result, both the 'mini-jet' and 3C 273's main jet must appear very foreshortened. The jet visible to the optical, X-ray and radio astronomers appears to be 230 000 light years long; allowing for the foreshortening, its true length is 1 200 000 light years – over two hundred times the length of the other famous jet, in M87, and as large as the radio lobes of giant radio galaxies like Centaurus A (Fig. 12.4).

## Double quasar

The famous 'double quasar' at the centre of **Fig. 12.39** is, in fact, two images of just one distant quasar, its light focused by a massive galaxy. Albert Einstein's general theory of relativity, published in 1916, predicted that gravity can bend light, and in such a way that a massive body's gravity can act rather like a lens to focus the light into several different images. Quasars are very distant, very brilliant objects, and the light from some quasars must – purely by chance – pass close to intervening galaxies before it reaches the Earth. The double quasar was the first example of a gravitational lens, discovered shortly before the centenary of Einstein's birth in 1979.

Radio astronomers at Jodrell Bank had detected radio waves from this double quasar some years earlier. In 1979, they used large optical telescopes in the United States to split up the light of these 'two objects' into spectra, and were astonished to discover not only that both objects were quasars, but also that their spectra were identical. Spectral lines of the same elements appeared with the same intensity and with the same width along the spectrum; and both spectra were shifted to longer wavelengths by precisely the same amount. The light must be coming from the same object.

The shift of its spectral lines, due to the expansion of the Universe, indicates that the quasar lies some 10 000 million light years away. It is a fairly average quasar, one-tenth as luminous as 3C 273 (Fig. 12.33) and fifty times brighter than our Milky Way Galaxy. Its brilliant emission undoubtedly comes from a small region in the centre of a galaxy too faint to discern at this distance. If the double appearance is due to a galaxy along our line of sight, however, the intervening galaxy should be visible as a faint fuzzy object between the two images of the quasar. **Fig. 12.40** shows the geometry. Earth is at right (with our view shown on the screen), and the quasar (white spot) at left, with the radio lobes (blue) seen in Fig. 12.43. The intervening galaxy (yellow) has little effect on the main quasar image (top of screen): but it bends radiation passing nearby to form a second quasar image just below the galaxy's position.

British astronomer Peter Young first discovered the galaxy's image in November 1979; Alan Stockton took the picture shown in Fig. 12.39 a few days later. The upper-left edge of the *lower* quasar image looks slightly fuzzy, giving the appearance of a galaxy overlapping the bright quasar image. To see this galaxy more clearly, Stockton scanned the photograph, and computer processed it. In **Fig. 12.41**, the left-hand image shows the central part of Fig. 12.39 colour coded for brightness: the background sky is dark blue, and brighter

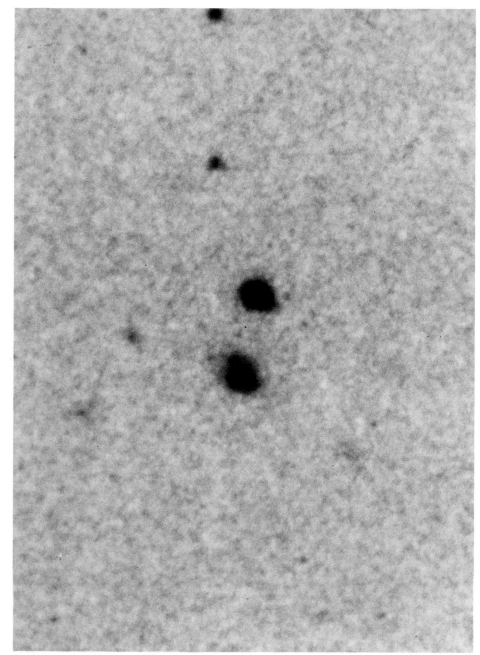

12.39 *Optical, red light, negative print, image-intensifier, 2.2 m telescope, Mauna Kea*

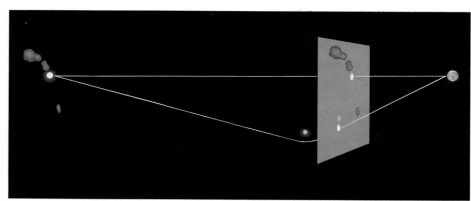

12.40 *Diagram, gravitational focusing (angular scale exaggerated).*

parts of the images pale blue, yellow, and red, with the most intense regions black. The quasar images are elongated because the telescope mirrors were slightly out of line. But the two quasar images should be identical in shape, and Stockton has subtracted a duplicate of the upper image from the lower, to reveal the galaxy: the lower, fuzzy patch in the right-hand image in Fig. 12.41.

The galaxy lies about 4000 million light years from us (just over one-third of the way to the quasar), and it is a giant elliptical galaxy of the same type as M87 (Fig. 12.20), some ten times brighter than the Milky Way and over a hundred times more massive. The black and red image in Fig. 12.41 (right) shows its central 20 000 light years, but the fainter (pale blue) regions reveal that it extends out to a total diameter of at least 100 000 light years. This galaxy is the central, brightest member of a cluster of at least 100 galaxies. The gravitational lens bending the quasar's light is mainly due to this one galaxy, but is strengthened by the gravity of the cluster as a whole.

The double quasar was first detected by its radio emission, and a highly detailed radio picture from the Very Large Array (**Fig. 12.43** overleaf) reveals a rather complex structure. Here the background sky is coded green and black, with regions of successively more intense radio emission red, yellow, dark blue, light blue, purple and white. The two small, white-centred objects at the top and bottom of the picture's centre are radio images corresponding to the two optical images of Fig. 12.39. The image of the quasar's core has been split in two by the lens. The quasar's elongated radio structure shows to either side of the top image. Like many quasars and radio galaxies, it has two lobes of radio emission. They stretch out 100 000 light years to either side, similar in size to those of Cygnus A (Fig. 12.15).

Just above the lower image of the quasar, there is a small faint (yellow and red) radio source, which coincides with the galaxy seen in optical pictures (Fig. 12.41). This small source and the neighbouring quasar image are shown in more detail in **Fig. 12.42**, where the colour coding runs from purple and pale blue for the background sky, through the range of colours seen in the narrow scale at the top, to black for the most intense region at the centre of the quasar image. The small source is only 1/15 as intense as the quasar image, and astronomers are not yet certain what it is. It could be radio emission from the core of the giant elliptical galaxy which causes the lens effect. On the other hand, it could be yet another image of the quasar. A small amount of radiation should come through the centre of the lens – the galaxy here – and produce a third, weaker image of the distant quasar.

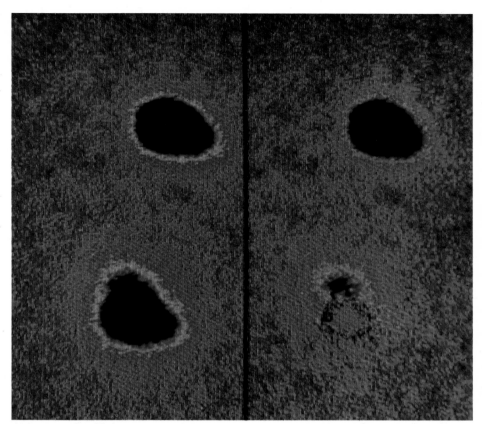

**12.41** *Optical, red light, intensity coded (left) and intensity coded with lower quasar image subtracted (right), 2.2 m telescope, Mauna Kea*

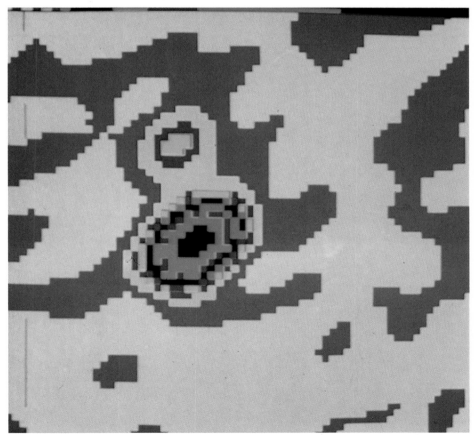

**12.42** *Radio, 6 cm (detail), Very Large Array*

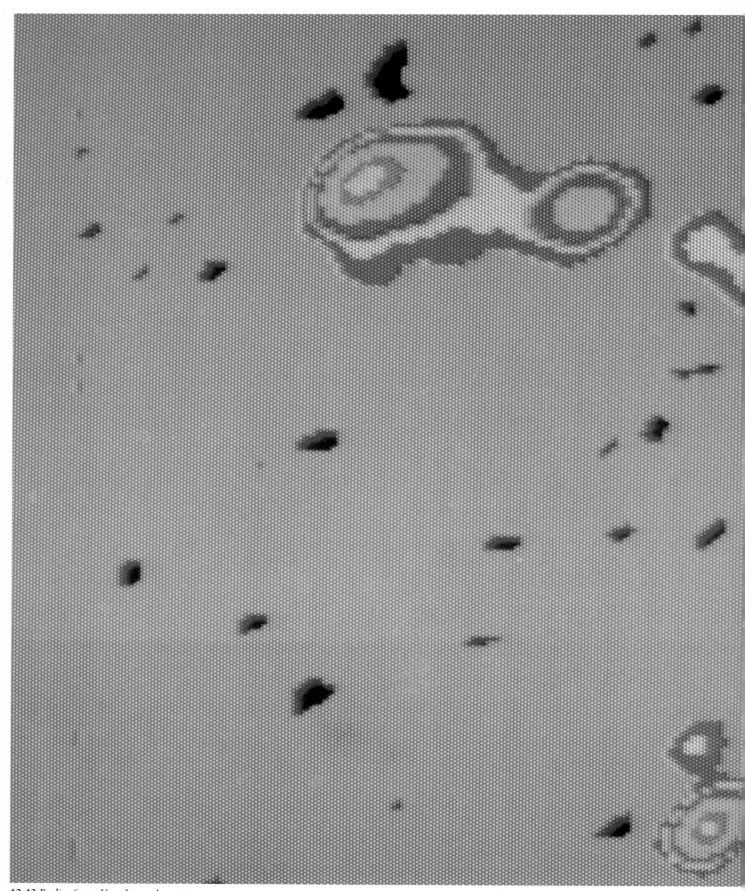

**12.43** *Radio, 6 cm, Very Large Array*

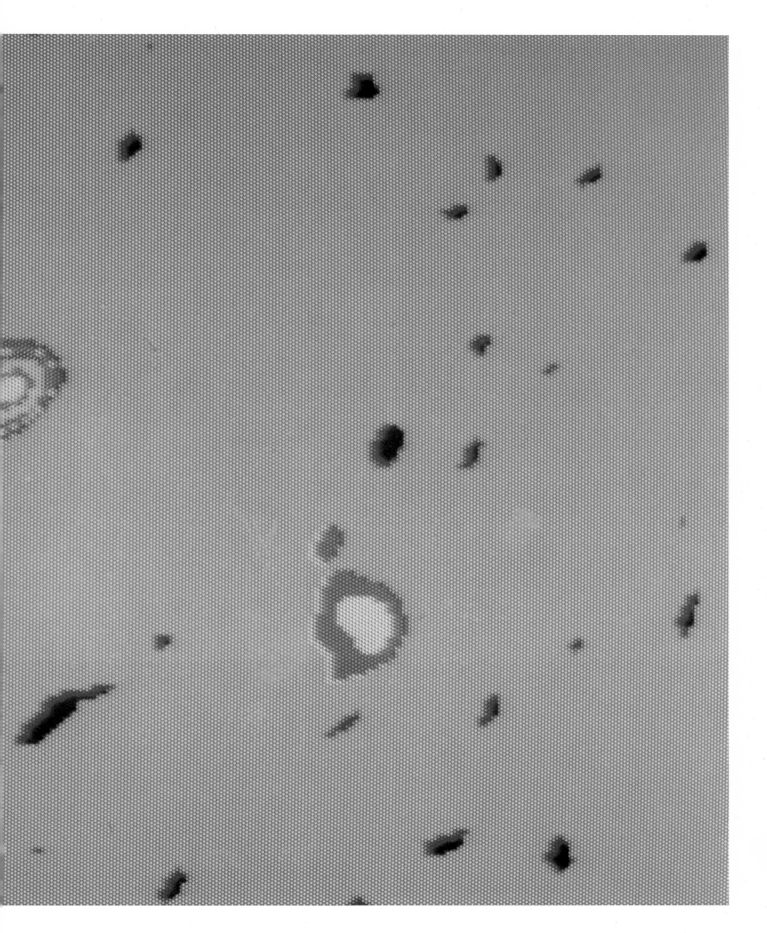

# Picture Credits

1.1 M. Berkinshaw, R.A. Laing & J.A. Peacock, *Mon. Not. Roy. Astron. Soc.* **197**, 253 (1981); image-processing by Steve Gull & John Fielden.
1.2 David Parker.
1.3 W. Hsin-Min Ku, Columbia Astrophysics Laboratory.
1.4 Original plate taken in Egypt, 1910, supplied by D. Klinglesmith III, NASA/Goddard Space Flight Center.
1.5, 1.6, 1.7 Image-processing by D. Klinglesmith III & J. Rahe, using the laboratory for Astronomy & Solar Physics' Interactive Data Analysis Facility at NASA/Goddard Space Flight Center.
1.8 Five College Radio Astronomy Observatory; image-processing by John C. Good.
1.9 David Parker.
1.10 David Parker.

2.1 NASA.
2.2 Mount Wilson Observatory photograph, © California Institute of Technology.
2.3 E. Fuerst & W. Hirth, Max-Planck-Institut für Radioastronomie, Bonn. Colour radio map produced at the computing centre of Rheinisches Landesmuseum, Bonn, by R. Beck.
2.4, 2.5 NASA.
2.6 Mount Wilson Observatory photograph, © California Institute of Technology.
2.7 R. Stachnik, P. Nisenson & R. Noyes, Harvard/Smithsonian Center for Astrophysics, 'Speckle Image Reconstruction of Solar Features' (submitted to *Astrophys. J. Lett.* (1983)).
2.8 NASA.
2.9 Courtesy of G. Carey Fuller.
2.10 NASA.
2.11 Courtesy of G. Carey Fuller.
2.12 L. Golub & G.S. Vaiana, Center for Astrophysics.
2.13 J. Dürst, Swiss Federal Observatory.
2.14 National Center for Atmospheric Research/High Altitude Observatory, sponsored by the National Science Foundation.
2.15 L. Golub & G.S. Vaiana, Center for Astrophysics.
2.16, 2.17, 2.18, 2.19, 2.20 NASA.
2.21 Michael Kobrick, NASA/Jet Propulsion Laboratory.
2.22 NASA.
2.23 David Allen, © Anglo-Australian Telescope Board, 1983.
2.24 NASA.
2.25 A. Metzger, D. Gilman, K. Hurley, J. Luthey, H. Schnopper, F. Seward & J. Sullivan, 'Detection of X-rays from Jupiter', *J. Geophys. Res.* (in press).
2.26 Fred Espenak, NASA/Goddard Space Flight Center.
2.27 S. Kenderdine, Mullard Radio Astronomy Observatory, *Icarus* (in press); image-processing by Steve Gull & John Fielden.
2.28 J.A. Roberts, University of California, Berkeley, & Division of Radiophysics, CSIRO, Sydney, G.L. Berge, California Institute of Technology, & C. Bignell, US National Radio Astronomy Observatory. The National Radio Astronomy Observatory is operated by Associated Universities Inc., under contract with the National Science Foundation.
2.29 I. de Pater, Lunar & Planetary Laboratory, University of Arizona, using the Westerbork Synthesis Radio Telescope, The Netherlands.
2.30, 2.31, 2.32, 2.33 NASA.
2.34 G.L. Grasdalen, R.D. Gehrz, J.A. Hackwell & A. Herzog, University of Wyoming.
2.35 I. de Pater, Lunar & Planetary Laboratory, University of Arizona, & J. R. Dickel, University of Illinois at Urbana-Champaign, obtained with the Very Large Array of the National Astronomy Observatory.
2.36 David Allen, © Anglo-Australian Telescope Board, 1983.

3.1 David Parker.
3.2 Dr E.I. Robson.
3.3 David Parker.
3.4 Yerkes Observatory photo, University of Chicago.
3.5 David Parker.
3.6 B.W. Hadley ABIPP, Royal Observatory, Edinburgh.
3.7, 3.8, 3.9 © Anglo-Australian Telescope Board, 1977, 1980 & 1981 respectively.
3.10 Photo by David Malin of the Anglo-Australian Telescope, from original negative by the UK Schmidt Telescope Unit, © Royal Observatory Edinburgh.
3.11, 3.12 Jean Lorre, NASA/Jet Propulsion Laboratory.
3.13 Freeman D. Miller, University of Michigan
3.14 David Parker.
3.15, 3.16 Reproduced with permission of the Royal Greenwich Observatory.
3.17 John Walsh.
3.18, 3.19 Reproduced with permission of the Royal Greenwich Observatory.
3.20 MMT Observatory
3.21, 3.22 NASA.
4.1 J.A. Hackwell, G.L. Grasdalen & R.D. Gehrz, University of Wyoming.
4.2 Harvard College Observatory.
4.3 David Malin, © Anglo-Australian Telescope Board, 1981.
4.4, 4.5 R.J. Maddalena, M. Morris, J. Moscowitz & P. Thaddeus, Columbia University; colouring & additional work on Fig. 4.5 by David Parker.
4.6 J.B.G.M. Bloemen, Leiden University, using data from ESA's Cos-B satellite.
4.7 W. Reich, based on observations with the Stockert 25 m telescope of Bonn University.
4.8 Photolabs, Royal Observatory, Edinburgh, Original negative by UK Schmidt Telescope Unit.
4.9 European Southern Observatory
4.10 K.F. Hartley, RGO/Starlink; reproduced with permission of the Royal Greenwich Observatory.
4.11 Fred Espenak.
4.12 Lick Observatory photograph.
4.13 R. Bohlin & T. Stecher, NASA Goddard Space Flight Center.
4.14 David Malin, © Anglo-Australian Telescope Board, 1981.
4.15 X-ray: W. Hsin-Min Ku, Columbia Astrophysics Laboratory. Optical: Lick Observatory photograph.
4.16 W. Hsin-Min Ku, Columbia Astrophysics Laboratory.
4.17 The National Radio Astronomy Observatory, operated by Associated Universities Inc., under contract with the National Science Foundation.
4.18 Guido Garay, Harvard University: VLA image produced at the Center for Astrophysics.
4.19 F.P. Schloerb, Five College Radio Astronomy Observatory; image-processing by John C. Good.
4.20 Courtesy of M. Werner, NASA/Ames Research Center.
4.21 D. Allen (AAO), A.R. Hyland (Mt Stromlo) & J. Bailey (AAO), © Anglo-Australian Telescope Board, 1983.
4.22 R.D. Gehrz, J.A. Hackwell & G.L. Grasdalen, University of Wyoming.
4.23 John F. Arens, Gerald Lamb & Michael Peck, NASA/Goddard Space Flight Center, with assistance of W. Hoffman (Steward Observatory) & G. Fazio (Smithsonian Astrophysical Observatory).
4.24 Tricolour photo by Ronald E. Royer.
4.25, 4.26 Optical data by H. Dickel & T. Gull, using 0.9 m telescope of The Kitt Peak National Observatory; radio data by R. Harten, Westerbork Synthesis Radio Telescope.
4.27 J.A. Hackwell, R.D. Gehrz & G.L. Grasdalen, University of Wyoming.
4.28 P. Scott, Mullard Radio Astronomy Observatory,

*Mon. Not. Roy. Astron. Soc.*, **194**, 23P (1981); image-processing by S. Gull & J. Fielden.
4.29 Mark Reid: VLBI image produced at NRAO.
4.30 Photolabs, Royal Observatory Edinburgh; original negative by U.K. Schmidt Telescope Unit.
4.31 © Anglo-Australian Telescope Board, 1977.
4.32 X-ray Astronomy Group, Leicester University, & Harvard/Smithsonian Center for Astrophysics.
4.33 J.A. Hackwell, R.D. Gehrz & G.L. Grasdalen, University of Wyoming.
4.34, 4.35 K. Davidson, Dept. of Astronomy, University of Minnesota, using 1.5 m telescope of the Cerro Tololo Inter-American Observatory.
4.36 Palomar Sky Survey, © California Institute of Technology.
4.37 R.A. Stern, M.C. Zolcinski, S.K. Antiochos, J.H. Underwood, *Astrophys. J.* **249**, 647.
4.38 Freeman D. Miller, University of Michigan.

5.1 David Parker.
5.2 J.A. Hackwell, G.L. Grasdalen & R.D. Gehrz, University of Wyoming.
5.3 L.R. Shaw, University of Wyoming
5.4 Michael Peck.
5.5 John F. Arens, Gerald Lamb & Michael Peck, NASA/Goddard Space Flight Center, with assistance of W. Hoffman (Steward Observatory) & G. Fazio (Smithsonian Astrophysical Observatory).
5.6, 5.7 Dr E.I. Robson
5.8 NASA.
5.9 Courtesy of Fokker.
5.10 NASA.

6.1 E. Kallas & W. Reich, based on observations with the Effelsberg 100 m telescope of the Max-Planck-Institut für Radioastronomie & processed with the Astronomical Image Processing System (BABSY) at Bonn University.
6.2 A. Condal & G. Walker, University of British Columbia.
6.3 Fred Espenak, NASA/Goddard Space Flight Center.
6.4 S. Harris & P. Scott, Mullard Radio Astronomy Observatory, *Mon. Not. Roy. Astron. Soc.* **175**, 371 (1976); image-processing by Steve Gull & John Fielden.
6.5 John F. Arens, Gerald Lamb & Michael Peck, NASA/Goddard Space Flight Center, with assistance of W. Hoffman (Steward Observatory) & G. Fazio (Smithsonian Astrophysical Observatory).
6.6 Palomar Observatory photo, © California Institute of Technology.
6.7 U.R. Buczilowski & R. Tuffs (Cambridge), Max-Planck-Institut für Radioastronomie, Bonn; colour radio map produced at computing centre of Rheinisches Landesmuseum, Bonn, by R. Beck.
6.8 Harvard/Smithsonian Center for Astrophysics.
6.9 U.S. Naval Observatory photograph.
6.10. Smithsonian Astrophysical Observatory.
6.11, 6.12 Palomar Observatory photographs; © California Institute of Technology.
6.13 T. Gull & R. Fesen, NASA/Goddard Space Flight Center
6.14 G. Pooley & E. Swinbank, Mullard Radio Astronomy Observatory, *Mon. Not. Roy. Astron. Soc.* **186**, 775 (1979); image-processing by Steve Gull & John Fielden.
6.15, 6.16 Reproduced with permission of the Royal Greenwich Observatory.
6.17 Lick Observatory photographs.
6.18 Harvard/Smithsonian Center for Astrophysics.
6.19 J.D. Scargle, using Palomar Observatory photographs.
6.20 Palomar Observatory photograph by S. van den Bergh, © California Institute of Technology.
6.21 F. Seward, P. Gorenstein & W. Tucker, *Astrophys. J.* **266**, 287.

**6.22** S. Gull & G. Pooley, Mullard Radio Astronomy Observatory ; D. Green & S. Gull, *IAU Symposium* 101; image-processing by Steve Gull & John Fielden.

**6.23** Palomar Observatory photo, © California Institute of Technology.

**6.24** T. Bell, S. Gull & S. Kenderdine, Mullard Radio Astronomy Observatory, *Nature*, 257, 463; image-processing by S. Gull & J. Fielden.

**6.25** False-colour composite prepared at David Dunlap Observatory, Ontario, by Karl Kamper, from plates obtained by S. van den Bergh with 5 m telescope, Palomar Observatory.

**6.26** X-ray Astronomy Group, Leicester University, & Harvard/Smithsonian Center for Astrophysics.

**6.27** J.R. Dickel, S.S. Murray, J. Morris & D.C. Wells, first published in *Astrophys. J.* 257, 145 (1982). Optical data by S. van den Bergh & Dodd, 5 m Hale Telescope, *Astrophys. J.* 162, 485 (1970); radio data by J.R. Dickel & E.W. Greisen, NRAO interferometer, *Astron. Astrophys.* 75, 44 (1979); X-ray data by S.S. Murray *et al.*, Harvard/ Smithsonian Center for Astrophysics, *Astrophys. J. Lett.* 234, L 69 (1979).

**6.28** Photolabs, Royal Observatory, Edinburgh; original negative by UK Schmidt Telescope Unit.

**6.29** D.K. Milne, CSIRO, *Austr. J. Phys.* 21, 203; colour coding by David Parker.

**6.30** Harvard/Smithsonian Center for Astrophysics.

**6.31** Photolabs, Royal Observatory, Edinburgh; original negative by UK Schmidt Telescope Unit.

**6.32** B.A. Peterson, P.G. Murdin, P.T. Wallace, R.N. Manchester, Anglo-Australian Observatory/ Starlink.

**6.33** Harvard/Smithsonian Center for Astrophysics.

**6.34** Palomar Observatory photograph by S. van den Bergh, © California Institute of Technology.

**6.35** B. Geldzahler, Max-Planck-Institut für Radioastronomie, Bonn; colour radio map produced at computing centre of Rheinisches Landesmuseum, Bonn, by R. Beck.

**6.36** The National Radio Astronomy Observatory, operated by Associated Universities Inc., under contract with the National Science Foundation.

**6.37** David Parker.

**6.38** X-ray Astronomy Group, Leicester University, & Harvard/Smithsonian Center for Astrophysics.

**6.39** Palomar Sky Survey, © California Institute of Technology.

**6.40** Einstein X-ray telescope image of the supernova remnant G109.1.–1.0 (courtesy P.C. Gregory).

**6.41** Radio image of the supernova remnant G109.1.– 1.0, obtained with the Very Large Array at a wavelength of 21 cm (courtesy P.C. Gregory).

**7.1** David Parker

**7.2** Jerry Mason.

**7.3** Michael Marten.

**7.4, 7.5** Kurt W. Weiler, National Science Foundation.

**7.6** G. Hutschenreiter, Max-Planck-Institut für Radioastronomie, Bonn

**7.7** R. Beck & R. Gräve, Max-Planck-Institut für Radioastronomie, Bonn; colour radio map produced at computing centre of Rheinisches Landesmuseum, Bonn, by R. Beck.

**7.8** R.J. Dettmar, Astronomical Institute of the University of Bonn, & R. Beck, Max-Planck-Institut für Radioastronomie, Bonn.

**7.9** R. Beck, Max-Planck-Institut für Radioastronomie, Bonn.

**7.10, 7.11** Kurt W. Weiler, National Science Foundation.

**7.12** A. Downes & J. Fielden, Mullard Radio Astronomy Observatory, using the Very Large Array, NRAO; *Mon. Not. Roy. Astron. Soc.* (in press).

**7.13** David Parker.

**7.14** R. Saunders, G. Pooley, J. Baldwin & P. Warner, Mullard Radio Astronomy Observatory, *Mon.*

*Not. Roy. Astron. Soc.* 197, 287 (1981); image-processing by S. Gull & J. Fielden.

**7.15** Jerry Mason.

**7.16** © 1982 by Douglas W. Johnson, Battelle Observatory.

**7.17** Dr E.I. Robson

**8.1** Courtesy of Lund Observatory.

**8.2** David Parker.

**8.3** G. Haslam *et al.*, Max-Planck-Institut für Radioastronomie (West Germany), using observations from Effelsberg (West Germany), Jodrell Bank (UK), & Parkes (Australia); colour radio map produced at computing centre of Rheinisches Landesmuseum, Bonn, by R. Beck.

**8.4** Courtesy of European Space Agency.

**8.5** Photolabs, Royal Observatory, Edinburgh.

**8.6** Map of X-ray sky produced at NASA/Goddard Space Flight Center by DeAnn Iwan based on data obtained with A2 instrument on HEAO-1 satellite. Principal investigators: E. Boldt (Goddard) & G. Garmire (Penn State).

**8.7** J.B.G.M. Bloemen, Leiden Observatory.

**8.8** Carl Heiles, University of California, Berkeley.

**8.9** Carl Heiles, University of California, Berkeley, & Edward B. Jenkins, Princeton University.

**8.10** US Naval Observatory.

**8.11** X-ray Astronomy Group, Leicester University, & Harvard/Smithsonian Center for Astrophysics.

**8.12** The National Radio Astronomy Observatory, operated by Associated Universities Inc., under contract with the National Science Foundation.

**8.13** P.R. Jorden (RGO) & M. Scarrott (Durham), reproduced with permission of the Royal Greenwich Observatory.

**8.14** A. Bentley, A. Herzog, J.A. Hackwell, G.L. Grasdalen & R.D. Gehrz, University of Wyoming.

**8.15** David Allen (AAO) & John W.V. Storey (University of New South Wales), ©Anglo-Australian Telescope Board, 1982.

**8.16** European Southern Observatory.

**8.17** © Mark Paternostro, 1981.

**8.18** Photolabs, Royal Observatory, Edinburgh; original negative by UK Schmidt Telescope Unit.

**8.19** Five hour exposure with the SERC Schmidt Telescope through Dr J. Meaburn's mosaic interference filter. Courtesy Dr J. Meaburn, Manchester University.

**8.20** George R. Carruthers, U.S. Naval Research Laboratory.

**8.21** © Association of Universities for Research in Astronomy Inc., The Kitt Peak National Observatory.

**8.22** W. Hsin-Min Ku, Columbia Astrophysics Laboratory.

**8.23** NASA

**8.24** Courtesy of J.V. Feitzinger, Ruhr University Bochum. Description of R 136a as a supermassive object first published in *Astron. Astrophys.* 84, 50 (1980) by Feitzinger, Schlosser, Schmidt-Kaler & Winkler.

**9.1** David Parker.

**9.2** NASA.

**9.3, 9.4** R. Bohlin & T. Stecher, NASA/Goddard Space Flight Center.

**9.5** David Parker.

**9.6** European Space Agency.

**9.7, 9.8** Fred Espenak, NASA/Goddard Space Flight Center.

**9.9** Fred Espenak.

**9.10, 9.11** NASA.

**10.1** © Anglo-Australian Telescope Board, 1980.

**10.2** Lick Observatory photograph.

**10.3** Smithsonian Astrophysical Observatory.

**10.4, 10.5** E. Brinks: 1983, Ph. D. thesis, Leiden Observatory.

**10.6** S. Unwin, Mullard Radio Astronomy Observatory, *Mon. Not. Roy. Astron. Soc.* 190, 551; image-processing by S. Gull & J. Fielden.

**10.7** R. Beck, E.M. Berkhuijsen & R. Wielebinski, Max-Planck-Institut für Radioastronomie, Bonn; colour radio map produced at computing centre of Rheinisches Landesmuseum, Bonn, by R. Beck.

**10.8, 10.9, 10.10.** R. Bohlin & T. Stecher, NASA/ Goddard Space Flight Center.

**10.11** L. Van Speybroeck, *et al.*, *Astrophys. J. Lett.* 234, L45.

**10.12** Palomar Observatory photo, © California Institute of Technology.

**10.13, 10.14** L. Van Speybroeck, *et al.*, *Astrophys. J. Lett.* 234, L45

**10.15** Jean Lorre, NASA/Jet Propulsion Laboratory.

**10.16** R. Bohlin & T. Stecher, NASA Goddard Space Flight Center.

**10.17** D.M. Elmegreen, *Astrophys. J. Supp. Series*, 47, 229 (1981). Photograph taken at Palomar Observatory.

**10.18** K. Newton, Mullard Radio Astronomy Observatory, *Mon. Not. Roy. Astron. Soc.*,190, 689 (1980); image-processing by Steve Gull & John Fielden.

**10.19** R. Beck, E.M. Berkhuijsen & R. Wielebinski, Max-Planck-Institut für Radioastronomie, Bonn; colour radio map produced at computing centre of Rheinisches Landesmuseum, Bonn, by R. Beck.

**10.20** W. Hsin-Min Ku, Columbia Astrophysics Laboratory.

**10.21** U.S. Naval Observatory,

**10.22** D.M. Elmegreen, *Astrophys. J. Supp. Series*, 47, 229 (1981). Photograph taken at Palomar Observatory.

**10.23** R. Bohlin & T. Stecher, NASA Goddard Space Flight Center.

**10.24** J. M. van der Hulst, P.C. Crane, R. Kennicutt & R.J. Allen, using the Very Large Array, NRAO; image-processing by Gröningen Image Processing System (GIPSY).

**10.25** Palomar Schmidt photo by M. Burkhead, image-processed by D. Klinglesmith III, *et al.*, using Laboratory for Astronomy & Solar Physics' Interactive Data Analysis Facility, NASA/ Goddard Space Flight Center.

**10.26** Courtesy of Los Alamos National Laboratory, derived from Palomar Schmidt photographs by M. Burkhead.

**10.27** R. Bohlin & T. Stecher, NASA Goddard Space Flight Center.

**10.28** © Association of Universities for Research in Astronomy Inc., The Kitt Peak National Observatory.

**10.29** R. Gräve, Max-Planck-Institut für Radioastronomie, Bonn; colour radio map produced at the computing centre of Rheinisches Landesmuseum, Bonn, by R. Beck.

**10.30** D.M. Elmegreen, *Astrophys. J. Supp. Series*, 47, 229 (1981). Photograph taken at Palomar Observatory.

**10.31** Colour representation made by R.J. Allen, R. Ekers, J.P. Terlouw & J.M. van der Hulst of the Kapteyn Astronomical Institute, The Netherlands, T.R. Cram & A.H. Rots of the US National Radio Astronomy Observatory, & the computing & photographic services of Gröningen University. The Westerbork Radio Observatory is operated by the Netherlands Foundation for Radio Astronomy with the financial support of the Netherlands Organization for the Advancement of Pure Research (ZWO).

**10.32** Palomar Observatory photo, © California Institute of Technology.

**10.33** Phil Appleton, Dept. of Astronomy, University of Manchester, & staff of Manchester Starlink node.

# Index

Bold figure numbers indicate that the picture is available from the Science Photo Library, 2 Blenheim Crescent, London W11 1NN.

Other major sources of pictures in this book are:
Photolabs, Royal Observatory, Blackford Hill, Edinburgh EH9 3HJ
National Radio Astronomy Observatory, Edgemont Road, Charlottesville, Virginia 22901.
Anglo-Australian Observatory, PO Box 296, Epping, New South Wales 2121, Australia.
Kitt Peak National Observatory, 950 North Cherry Avenue, PO Box 26732, Tucson, Arizona 85726.
Lick Observatory OP, University of California, Santa Cruz, California 95064.
Hansen Planetarium (Mt Palomar & Mt Wilson photographs), 15 South State, Salt Lake City, Utah 84111.